ALSO BY HENRY PETROSKI

Success Through Failure: The Paradox of Design

Pushing the Limits: New Adventures in Engineering

Small Things Considered: Why There Is No Perfect Design

Paperboy: Confessions of a Future Engineer

The Book on the Bookshelf

Remaking the World: Adventures in Engineering

Invention by Design: How Engineers Get from Thought to Thing

Engineers of Dreams: Great Bridge Builders and the Spanning of America

Design Paradigms: Case Histories of Error and Judgment in Engineering

The Evolution of Useful Things

The Pencil: A History of Design and Circumstance

Beyond Engineering: Essays and Other Attempts to Figure Without Equations

To Engineer Is Human: The Role of Failure in Successful Design

The Toothpick

The Toothpick

TECHNOLOGY AND CULTURE

Henry Petroski

ALFRED A. KNOPF · NEW YORK · 2007

THIS IS A BORZOI BOOK
PUBLISHED BY ALFRED A. KNOPF

Copyright © 2007 by Henry Petroski
All rights reserved. Published in the United States by Alfred A. Knopf,
a division of Random House, Inc., New York, and in Canada by
Random House of Canada Limited, Toronto.
www.aaknopf.com

Knopf, Borzoi Books, and the colophon are
registered trademarks of Random House, Inc.

Library of Congress Cataloging-in-Publication Data
Petroski, Henry.
The toothpick: technology and culture/by Henry Petroski.
p   cm.
"A Borzoi book."
Includes bibliographical references and index.
ISBN-13: 978-0-307-26636-1
1. Toothpicks. I. Title.
GT2952.P48 2007
674'.88—dc22   2007005712

Manufactured in the United States of America
First Edition

to Catherine,
granddaughter of the "toothpick twirler"

SOME STUFF RUSTLED UP
FROM THE WILD, WILD WEB

The toothpick was first used in the United States at the Union Oyster House. Enterprising Charles Forster of Maine first imported the picks from South America. To promote his new business he hired Harvard boys to dine at the Union Oyster House and ask for toothpicks.

—History of the Union Oyster House
http://www.unionoysterhouse.com/Pages/history.html

The toothpick was allegedly invented by some students at Harvard.

—Biography.ms
http://toothpick.biography.ms

According to Forster Manufacturing, the commercial toothpick [was] invented by Charles Forster in his basement in Boston, Massachusetts.

—Science and Technology in the Year 1869
http://www.answers.com/topic/1869

Toothpicks were invented in Bangor by Charles Forster.

—Do You Know Maine?
http://www.maine.gov

Remarkable as it may seem in this synthetic age, the modern toothpick is made out of unreconstituted virgin white birch, just as its predecessors have been since after the Civil War, when Charles Forster invented the automatic toothpick-making machine.

—*"The Straight Dope"*
http://www.straightdope.com/classics/a1176a.html

In 1872, Silas Noble and J. P. Cooley patented the first toothpick-manufacturing machine.

—*History of Dentistry and Dental Care: Toothpicks*
http://inventors.about.com/library/inventors/
bldental.htm

Founded in 1869, Forster Manufacturing grew its toothpick, clothespin and wood products business into one of New England's largest, earning Strong, Maine the title "Toothpick Capital of the World." The plant processed birch logs to make toothpicks; wood waste from the process generated steam and power for the plant, and excess power was sold to the New England grid.

—*Renewable Energy*
http://www.cleavco.com/energy-forster.htm

1887 . . . that's when someone named Charles Forster started the first wooden toothpick factory in this country.

—*Lincoln Daily News*
http://archives.lincolndailynews.com/2001/
Mar/30/Featuresnew/perspectives.shtm

CONTENTS

PREFACE

I have never been a regular user of toothpicks, though there has always been a box or two of the little wooden things about the house. Occasionally they have come in handy for applying a dab of glue or oil to a small part, cleaning dust out of a tight crevice, plugging up an empty nail hole or two, serving as shims, testing the doneness of a batch of brownies, and the like. Now and then, the indispensable disposables have even been used to coax an especially stubborn sesame seed out from between the teeth. But mostly the tiny sticks remain in a faded and tattered pasteboard box in a dark corner of the kitchen cupboard.

For whatever reason, the usually forgotten toothpick came to my mind one day when I was searching for an engagingly simple device that would serve to illustrate some basic principles of engineering and design and that at the same time would help reveal the inevitable interrelationships between technology and culture. In this common wooden object I thought I had identified the simplest tool and thereby a perfectly accessible example for my purposes.

The story of the toothpick and its significance was to be but a brief chapter in an earlier book of mine. However, when I sought out the history of the thing and the nature of its manufacture, I found a paucity of readily accessible material. What information I did find was largely elliptical or, when not that, contradictory. Rather than allowing myself to be sidetracked while writing that previous book, I scrapped the chapter and put the unused toothpick back in its box, to be retrieved at a later time. I resumed my pursuit of the toothpick the next summer in Maine, where I retreat to write and where the mechanized wooden toothpick industry once flourished.

Uncharacteristically, that summer I spent much of my time away from my study, traveling to small towns among the Maine woods, where toothpick mills once operated. They are closed now, victims of changing habits, corporate buyouts, and offshore manufacturing competition, but the plants are remembered by oldtimers and their spirit is preserved in the assorted obscure memoirs and ephemera that can be found in small local libraries and historical societies. These independent institutions, with tiny staffs and limited hours of operation, proved to be invaluable resources, especially since large research libraries full of books and scholarly journals have little to offer on this humble but elegant device. The standard Internet search engines, so often now the research tool of first resort even in our largest libraries, produced a tangled web of sticky contradictions. However, the archival databases to which the Internet provides access ultimately proved to yield the most valuable resources, including complete and searchable runs of historic newspapers and other periodicals, as well as easily retrievable patents. The Internet also brought a surprisingly rich trove of artifacts through eBay, the online auction house. Many of the most fruitful resources came ultimately from individuals with unique knowledge and information. The list of helpful archivists, clerks, collectors, curators, dealers, docents, historians, librarians, managers, and manufacturers is much too long to include in a preface. I have named them in the acknowledgments in the back of this book.

However, there is one person who devoted so much time, thought, and energy to this project, as she does to all of my projects, that I cannot wait to acknowledge her. Catherine Petroski, my wife and friend of forty years, has been my copilot, navigator, photographer, scanner, digitizer, eBay agent, and—as always—first reader. She has watched me immerse myself in this growing project and has seen me, in the interests of research, become a sometime chewer of toothpicks.

Arrowsic, Maine
Summer 2006

The Toothpick

Prologue

THE PLAIN WOODEN TOOTHPICK, it may be argued, is among the simplest of manufactured things. It consists of a single part, made of a single material, intended for a single purpose—from which it gets its simple name. It is also among the most convenient and ready of things. It can be used directly out of the box—there being no instructions to read, no parts to assemble, no priming or booting required, and no maintenance expected. When it has served its purpose, it is simply discarded.

Such simplicity of design and use might lead one to expect an equally simple and straightforward history, one easily researched and explicated by a student doing a term paper. In the twenty-first century, such a student would very likely navigate around the World Wide Web via Google or some other digital search engine and come up with enough snippets to stitch together a plausible story—as long as the sources were unquestioned, the gaps glossed over, and the contradictions ignored. Of the quotations rustled up from the Web and corralled at the beginning of this book, every one but the statement about generating power from wood waste is at best a half-truth.

In fact, the full and true story of something even so simple as the toothpick cannot easily be gleaned from the Internet alone. Unfortunately, more traditional sources of information and scholarship, such as manuscripts, articles, books, and other written materials in archives and analog libraries, also often provide sparse, erroneous, and contradictory information for topics considered too banal for and thus neglected by scholars seeking to pursue grander things and themes.

The very simplicity and banality of the thing made the toothpick and its manufacture an artifact of tacit knowledge and trade secrets. Even in the late twentieth century, Japanese visitors who showed up at a Maine toothpick factory were turned away, lest they see the tricks of the trade. An American scholar, who should hardly have been seen as a potential competitor, was similarly denied entrance to a Minnesota counterpart. He had to go to Sweden to see some toothpicks being made.

Secrecy coupled with a dearth of reliable, confirmable documentary material makes the task of uncovering the real story of a common object a challenge for ordinary scholarship relying on the usual scholarly sources. But there are other sources of information, not least of which is the artifact itself and the documented social and cultural context in which it has been made and used. Much of the story of the toothpick must be coaxed out of the thing itself and its milieu. With patience, slivers of it can be teased out of even a closed box of toothpicks the way a stubborn seed eventually can be dislodged from between the teeth. Insights into the use and misuse of things can be gleaned from both the froth and the detritus of society.

Whatever its history, the toothpick-manufacturing process has become so automated and efficient that no human hand touches the product until it is taken up to be used. An antiseptic toothpick costs but a fraction of a fraction of a cent, and it can be tossed away after a single use. Since it is made of untreated wood, the biodegradable toothpick presents no substantial danger to the environment. At first glance, it seems not easily implicated in global warming—until we remember that trees have been sacrificed for and energy consumed in its production.

But as neglected, small, insignificant, and inconsequential as the artifact might seem to be, the story of the toothpick holds great potential for revealing often hidden and frequently overlooked relationships among the people and things of the world. As Archimedes asserted that, if he were given a long enough lever and a place on which to stand, he could move the earth, so we can imagine that, given a toothpick and a sense of its place in history, we can nudge our understanding of technology and culture a bit farther.

Since the toothpick is a technological and cultural artifact, its use and significance are determined by its producers and consumers as they have over time been embedded in the life of business and the business of life. Individuals can develop a dependency on toothpicks, a preference for certain shapes and sizes, and a set of habits and rituals surrounding their use. Society, ever subject to the fads and fashions that it itself creates, imposes changing expectations on the availability of toothpicks and on the manner and acceptability of their use. Different classes of people, being attuned to different social rhythms and cultural clocks, develop different relationships with toothpicks. This is naturally part of the story.

Because common things so easily transcend limits of time and place, their story is not readily confined to a single period or to a single culture. The history of the toothpick is as old as mankind and as universal as eating. Its story knows no disciplinary bounds, and it is revealed in the records of anthropology as surely as in the annals of etiquette. It is an international story, with chapters set in prehistoric Africa, ancient Greece and Rome, medieval Portugal, and modern Brazil, China, Japan, Sweden, and the United States, to name but a few of its backdrops. The story of the toothpick is the story of Everyone and Everything at Everytime.

Things get their names and reputations from people, and it is people who also dictate how things are spoken of and used. As much as its name defines a single intended purpose, the toothpick has been adapted to countless other uses. Like any tool or device, the toothpick has been called into service when something else was not available or suitable to the task at hand. This, too, is part of the story of the thing, as is its propensity to spawn an infrastructure of boxes, cases, dispensers, holders, and other contraptions that can be as extraordinarily diverse and complex as the one-part machine that they support is simple.

To an engineer, the challenge of mass-producing something like a toothpick with sufficient efficiency that it can be sold at a profit holds a special fascination. The origins and rise of the mechanized toothpick industry in the latter half of the nineteenth century make for a fascinating chapter in the history of manufacturing, as do the human stories of inventors and innovators such as Benjamin Franklin Sturtevant and Charles Forster, along with their inventions, their patents and patent rights, and their struggles through failures on their way to achieving successes. It is in this chapter of the story that the history of the modest toothpick assumes heroic proportions and provides especially poignant lessons for the technological enterprises of today and tomorrow.

CHAPTER ONE

The Oldest Habit

NOTHING CAN BE more annoying than having a piece of food stuck between our teeth. As tiny as it might really be, in time it can seem to grow out of all proportion to its place in the mouth. As the pea under the princess's mattress prevented her from enjoying a night's sleep, so a tiny seed between molars can deny the diner much-anticipated postprandial peace and satisfaction. Like a grain of sand between two millstones, the foreign matter grates on us until it is worked free.

We have all devised our own preferred methods for dealing with the problem, but when we are not alone some of us may be constrained by social strictures to work within a closed mouth. Our tongue is often the instrument of choice, but the tongue's soft, blunt tip is usually ineffective. We have to flex and strain the muscles that harden and point it, and the process can be excruciatingly trying, tiring, and not really so private or inconspicuous as we might wish.

When wooden matches were commonly found near the kitchen stove, they were convenient to be split or whittled into toothpicks. One uninhibited character in a 1920s novel entered a shop "still helping the breadcrust out of his teeth . . . with his tongue," supplemented by a split matchstick, which was a sure giveaway of his plight.[1] However, even without opening our mouth to use a pointed tool, whenever we proceed to drag the tongue across and thrust it between our teeth at a repast's tenacious residue, we reveal our mission by the bulge moving around our lips and cheeks like a mole beneath the lawn.

Sucking at the stuck debris can sometimes be effective, but not always easily for stubborn little things. It takes more than eight pages in James Joyce's *A Portrait of the Artist as a Young Man*, involving "sucking at a crevice in his teeth," among other efforts, for the character Cranly to dislodge fig seeds.[2] We can also try squirting saliva between the teeth to flush out some unfriendly food debris. However, like vacuuming a rug or washing windows with a water hose, such actions can be noisy. The overzealous tooth sucker whose lips slip apart can sound like a wet kisser bussing the air, the too-eager spit squisher like someone squeezing a wet sponge.

The finger can be an effective lever to move what will not otherwise budge, but in many circles its use points to the defeat of other means. Besides, like the tongue, the finger is usually too blunt an instrument for the task at hand, and some people have been known to "grow a long finger nail especially for picking teeth."[3] Sometimes, even an ordinary fingernail can be enlisted successfully, but implementing it as a solution can seldom be done with grace.

The most common alternative to natural and self-contained means is, of course, the familiar wooden toothpick. Where social strictures do not censure its use, the toothpick can be a most effective tool to succeed where tongue and fingers fail. As all tools are extensions of our bodies and their extremities, so the toothpick is an extension of the finger. It allows us to reach into the back of our mouth more easily and effectively, a need that has existed coevally with the need for food itself. Indeed, it has been suggested that "the ability to sense and remove food particles between teeth" dates from a couple million years ago, and that toothpicking represents "the earliest currently known nonlithic tool use by hominids."[4] Hence, picking one's teeth is believed to be the oldest human habit.[5]

The evidence comes from fossilized teeth, which have been called "the most durable relics of early life," certainly outlasting any toothpicks that may have been used on them.[6] Unlike stone tools, implements made of grass, thorn, wood, or other vegetable materials would decay or erode over time, leaving no artifacts recognizable as toothpicks. However, for nearly a century, anthropologists around the world have noted curious striated grooves on fossilized teeth from a large number of diverse locations and covering a great span of time. As early as 1911, a French anthropologist described grooved teeth found at the La Quina Neanderthal site and proposed the hypothesis that it was the use of abrasive toothpicks that caused the grooves.[7] Subsequently, similarly grooved teeth have been found among the remains of Australian Aborigines,

North American Indians, Canary Islanders, and Upper Dynastic Egyptians, as well as other populations.[8] The oldest examples of such grooving have been found in Africa. A tooth from the Olduvai Gorge archeological site in Tanzania "bears a series of tiny parallel lines scraped by a sharp, thin object pushed into the narrow space between teeth."[9] An example from the Ethiopian site Omo has been estimated to be almost two million years old.[10]

It has been speculated that the use of toothpicks may have commenced with meat eating among early hominids, and that "the intent of primitive man was by no means the artificial cleaning of his teeth but simply the removal of an unpleasant subjective sensation."[11] Even today, after ages of evolution, our teeth are still "not well designed for eating meat," as those of us who are not vegetarians know all too well.[12] It is stringy pieces of meat that can be the most difficult to remove from their lodging place. Still, that the grooves in fossilized teeth were due to toothpick use has been debated among anthropologists, some of whom did not believe that simply picking food particles from between the teeth, even over a lifetime of eating, could cause such distinct grooves, some of which are as much as two millimeters wide.[13]

To produce such damage, it has been believed, there would have to be the prolonged working back and forth of a toothpick or toothpick-like device. One theory was that toothpicks were used not simply to remove bits of food but to serve the therapeutic and palliative function of easing the pain of periodontal disease and dental caries. Extended use might have worn away the decay and left a clean groove. Regular toothpick use may have caused the triangular opening bounded by neighboring teeth and gums to grow, thereby allowing more food to become impacted in the space, thus requiring further probing. Sensitive gums might have encouraged prolonged toothpick use, which might also have resulted in grooving.[14] It has even been proposed that the practice "may have been a largely unnecessary, non-functional pastime that was cultural rather than practical."[15] Whatever its cause or motivation, the grooving of teeth appears to have been ubiquitous.[16]

There remained skepticism among scientists, however, that even prolonged use of wood or bone implements could cause such pronounced grooves in teeth. But there are numerous examples in nature and technology of the erosive power of repetitive action. On the geological scale, the effects of flowing water have produced such dramatic wonders as the Grand Canyon, and the combined forces of water, wind, and sand have resulted in countless natural bridges. On the scale of industry, the abrading action of mooring and tow ropes used along quays and waterways has grooved stone and cast iron, as can be seen in countless harbors and on an icon of the Industrial Revolution, Iron Bridge, which still spans the Severn River in western England. But, for anthropologists at least, it remained a mystery as to how soft wood and bone could groove hard enamel and dentin, for "grooves have never actually been documented in the molars or premolars of modern industrialized populations, even among heavy toothpick users."[17] One theory has held that the grooves were due to the erosive action that resulted when "saliva or other fluids containing sand, soil, or other abrasive particles are sucked" into the mouth through the teeth.[18]

Among other theories was one that held that grooves were created during the process of preparing strands of sinew by drawing them through the teeth, as had been observed in South Australia.[19] A similar activity is believed to have been practiced among prehistoric California Indians, who evidently used their teeth "as tools for cultural activities such as stripping leather thongs, reeds, or other plant fibers."[20] The idea that toothpick use was culturally or socially driven, "without practical use," gained some adherents, for it explained erosive forces working on the teeth well beyond the time that would have been required to simply dislodge food particles.[21] But the true explanation remained a mystery, since the proffered hypotheses did not seem readily testable.

The old mystery came to her mind while Leslea Hlusko, a biological anthropologist and paleontologist, was on a picnic with a friend, whom she was watching "use a piece of grass as a toothpick."[22] She wondered if the use of "long, round, tough stalks like those from prairie grass" explained the grooved fossil teeth, and so she conducted experiments to see.[23] After running tests that consisted of working back and forth "grass stalks between human and baboon teeth for three hours and eight hours, respectively," using up at least forty different stalks in the former case, Hlusko was able to produce grooves similar to those found on fossils.[24] Though even a lifetime of using smooth modern floss or soft wooden toothpicks was not considered capable of leaving such marks, "unlike wood, grass contains large numbers of hard, abrasive silica particles."[25]

The ability of a gritty substance to produce an effect that a smooth one would not was demonstrated more recently in the county jail in Okanogan, Washington. A prisoner there claimed that he was claustrophobic and persuaded the sheriff to allow him to spend more time outdoors in the exercise yard, which was understandably enclosed by a chain-link fence. The prisoner spent a lot of time at the fence, looking out at the open spaces beyond. In fact, what he was doing was using dental floss coated with toothpaste to saw through the steel fence links. In time, he had severed enough links to squeeze through the fence and escape. A prison official doubted that only floss and toothpaste (and time) could cut through steel and so, like Hlusko, performed tests and demonstrated to his amazement that they could. It turned out that a few years earlier an inmate in a Texas prison had also used the innocent-seeming tools of oral hygiene to cut through the bars of his cell to get out and attack a rival prison gang member.[26] Even the simplest of things can be used, with patience, to accomplish the most difficult of tasks.

Perhaps with good reason, "next to the wheel, the toothpick has often been called man's most universal invention."[27] Artifacts found in the Aleutian Islands certainly suggest the priority of toothpicking above other grooming habits, for among the things found in a cave was a "comb, made by tying splinters like toothpicks across a flat piece of wood the size of a lead pencil."[28] It has been said that picking the teeth is "one of the few primitive pleasures left to man."[29]

There is also evidence that the toothpick was not known only to hominids. Chimpanzees have been observed to "use bits of straw in much the same way as does a farmer."[30] A chimp named Belle had become accustomed to serving as dentist to Bandit, her patient, using her fingers, sometimes supplemented by twigs, to groom his teeth. In one case, Belle stripped the leaves off a twig of red cedar to produce a pen-sized device that she proceeded to use as a toothpick.[31] Animals kept in captivity are known to imitate their captors, but their behavior sometimes suggests deeper roots. A man who lived in nineteenth-century New Orleans reportedly taught his pet monkey to pick its teeth "with a regularly-organized toothpick."[32]

A tamed orangutan kept in Holland was, after eating and drinking, "in the habit of wiping its mouth with the back of the hand, as men sometimes do, and it generally used a toothpick."[33] According to another report regarding an "ourang outang," after meals "it always wiped its mouth, and when presented with a toothpick, always used it in a proper manner."[34]

Though the tongue and fingers may well have been the first toothpicks, it appears that very early in their development hominids put into service such found tools as grass stalks, wood splinters, and chipped bones. But found things are seldom entirely suitable for the task at hand, and in time they or their use would have been modified and worked to the point of being more effective.[35] Even chimpanzees, like Belle, have been observed to alter their found tools to increase their efficacy. Humans carry modification to the next level—by fabricating their implements from scratch. Such activity falls into the categories of invention, design, engineering, and manufacturing, which have come a long way since prehistoric times.

CHAPTER TWO

Artifacts and Texts

ALTHOUGH THE EVIDENCE for toothpicking by our hominid ancestors is indirect—depending as it does on hypotheses about what caused the grooves in fossilized teeth—the evidence that toothpicks were used at least as long as five thousand years ago is direct, in the form of surviving artifacts. These pointed implements no doubt represent the products of a long evolutionary process that led from the biodegradable grass stalks and splintered sticks that no longer survive to the durable metal implements that rest in museums and private collections today.

The evolution of technology seldom involves only a single primitive thing developing along a straight line into a single sophisticated thing. Rather, the evolutionary trail can take many twists and turns. It can be circuitous, leading back upon itself, albeit with tangents forking off in all sorts of new directions. It can also be tree-like, with myriad offshoots from a single trunk.

One branch of the evolutionary tree might develop by form rather than by function. Privileged classes of people, especially the aristocracy, would not likely have been expected to hunt down their own fresh toothpicks of grass stalks or splintered twigs. That task naturally would have fallen upon servants, who would have shouldered the continual pressure of producing each new toothpick that would be at least as good as the last. This would drive their ingenuity to come up with ways to ensure that such would be the case. Offering toothpicks made of special materials, such as scarce or precious metals, would provide a predictable implement that not only held its point but also gave the privileged something distinguished from what was used by common people.

Other implements for the personal toilet would have evolved in similar ways, and these things would naturally come to be kept together in a kit of sorts. Among the other things such a kit might contain were an earscoop and tweezers. What is considered the first known toilet set, dating from about 3500 B.C., was found during the excavation of a king's tomb in the old Mesopotamian city of Ur, located in what is now Iraq. The set of gold implements includes a tweezers, an earspoon, and a "spatulate, stiletto-like instrument running to a point," which is taken to be a toothpick. These are attached—like a set of keys—to a silver ring, and the lot was designed to be inserted into a conical case made of gold and "richly decorated with ribboned filigree work."[1] Similar but less ancient toilet sets made of other metals, such as silver, copper, and bronze, have been found in Europe and the Far East. Toothpicks were also "certainly used in ancient China, Japan, and other Eastern countries."[2] A combination toothpick and earspoon found in northern France was made of bronze. It was "two inches long, with the middle part wrought in spiral form, so as to increase the solidity of the article, and also to enable the hand to keep it easily firm in all positions."[3] Such structural and functional considerations continue to drive the design of toothpicks made of all kinds of materials.

The ancient Greeks were sometimes called "toothpick chewers," which would be consistent with their use of "little wooden sticks" to rid their teeth of food particles.[4] The Greeks and Romans also had more durable toilet sets, which could be carried on chains or pins, so that the implements were always at hand. One popular Roman combination of toilet items united by "a common ring and

From ancient times, toothpicks were incorporated into toilet sets and jewelry, such as this silver filigree example from the region between Tibet and the Chinese province of Yunnan.

comprising a toothpick, an earscoop and a shovel-shaped tongue-scraper became a standard design well into the nineteenth century."[5] A half dozen or more specialized devices might have been collected into these personal kits, including nail cleaners and toothpicks of different designs to better reach different parts of the mouth, much as the professional-quality stainless-steel ones of a modern dentist do. Large numbers of such personal accessories have been found in Etruria, which is modern Tuscany.[6]

Some Roman noblemen are believed to have had slaves who specialized in cleaning the teeth, thus making them the "earliest dental hygienists."[7] But most people tended to their own mouth and thus took a direct interest in toothpicks. At Rome's zenith, "a lady of fashion customarily included a tooth-pick in her etui," the small ornamental case that she would wear on her person.[8] The scholar Pliny, his nephew Pliny the Younger, and the poet Martial each "discussed the advantages and disadvantages of different types of toothpicks."[9] In the *Natural History*, Pliny offered much and varied advice on the care of the teeth, including recipes and prescriptions: "If the head of a dog that has died mad be burnt, the ashes obtained may be advantageously used against toothache, mixing it with cyprine oil and then dropping the mixture into the ear, on the side of the pain. It is beneficial also to pick the sick tooth with the longest tooth, on the left side, of a dog; or with the frontal bones of a lizard, taken from the head of the animal at full moon, and which have not touched the earth."[10]

That Martial also mentions toothpicks (*dentiscalpia*) several times in his epigrams suggests that the cleaning implements were commonly used in Rome. In fact, their use appears to have been so familiar and habitual that Martial ridiculed "the old dandy who, stretched at length on the triclinium" (a kind of sectional couch that nearly encircled the dining table and upon which Romans reclined while eating), stuck a toothpick in his mouth even though he was toothless. According to Martial, the old dandy did so "to give himself the air of a man not too far stricken in years."[11]

Ancient Hebrews are said to have been "instructed to refrain from carrying any personal item heavier than their toothpick on the Sabbath," which has been taken as evidence that they were accustomed to carrying reusable toothpicks in their clothing.[12] Although no dental cleaning tools are explicitly mentioned in the Talmud, a "stick of wood used as a toothpick, and also occasionally for eating," is. This was "no artificially fashioned toothpick but the simplest type of natural product, a wooden chip that the orthodox believer was permitted to pick up from the ground even on Sabbath if he needed it for his teeth, and which he customarily held between his teeth when he took his walks."[13]

Hindus and other Indians are known to have carried a fig-tree splinter in their mouth, using it as both toothpick and toothbrush.[14] The Parsi, Zoroastrians descended from Persian refugees who settled in Bombay over a millennium ago, are said to have "elevated toothpick use to a religious rite of sorts," and the Dravidian Gonds in India have been "known to bury the dead with their toothpicks."[15] The Prophet Muhammad is believed never to have traveled without toothpicks,[16] which he preferred to be "carved from aromatic aloe wood which had been dipped into the holy water fountain in Mecca." A servant, designated "the master of the toothpick," was instructed "to carry this sacred object behind his ear."[17] The Prophet also requested that "his funeral preparations include placing a toothpick in his mouth."[18] According to the Koran, cleaning the mouth before praying is "a means of praising God," and "a prayer which is preceded by the use of a toothpick is worth seventy-five ordinary prayers."[19]

Just as works and styles of literature come in and out of fashion, so do articles of technology. It has been said that the toothpick itself has "had its declines and revivals." In Ben Jonson's play *Every Man Out of His Humor,* a citizen's wife observes of a court gallant, "How cleanly he wipes his spoon at every spoonful of any whitemeat he eats! and what a neat case of pick-tooths he carries about him still. O Sweet Fastidius! O Fine Courtier!"[20] And there is no dearth of references in Shakespeare's works to toothpick use. According to Parolles in the play *All's Well That Ends Well,* "Virginity, like an old courtier, wears her cap out of fashion: richly suited, but unsuitable: just like the brooch and the tooth-pick, which wear not now."[21]

The toothpick had indeed been worn on and off as an object of fashion. As much as the custom may have flourished in ancient times, "after the fall of Rome and until the Middle Ages, toothpick usage was largely neglected."[22] In medieval times it did become "a frequent subject of comment,"[23] and "keeping the toothpick in the mouth all day was a common habit." Indeed, the "practice was regarded as a desire for personal cleanliness,"[24] and "toothpicking became an integral part of the after dinner ritual for many European families," a custom that "persists in many working class families in America." More privileged Europeans carried toothpicks of gold or silver, often keeping them in fancy cases hung from a chain.[25]

It was during the Renaissance that the toothpick, alone or in a toilet set, exposed or in a decorated case, appears to have been worn and used most conspicuously and proudly. That is not to say that all users were equally ostentatious. The material of one's toothpick was a matter of personal choice. The Persian poet Omar Khayyam apparently possessed a gold toothpick, while the humanist Erasmus "had a quill from a chicken or a cockerel."[26]

James IV of Scotland once bought "two gold toothpicks with a chain, for the custom was to wear the picks suspended around the neck." And in 1541 his son James V had the royal goldsmith make him "a little silver case for toothpicks."[27] But the propriety of wearing such a case was evidently the subject of some disagreement in 1560, when Giovanni della Casa, archbishop of Benevento, wrote of the subject in his *Galateo*. Among his observations and opinions was the following:

> They also are undoubtedly mistaken in their notions of politeness who carry their tooth-pick cases hanging down their necks; for besides that it is an odd sight for a gentleman to produce anything of that kind from his bosom like some strolling pedlar, this inconvenience must also follow from such a practice,—that he who acts thus discovers that he is but too well furnished with every instrument of luxury and too anxious about everything that relates to the belly; and I can see no reason why the same persons might not as well display a silver spoon hanging about their necks.[28]

In 1570, Queen Elizabeth "received a gift of six gold toothpicks as well as 'tooth cloths' edged in silver and black." At a time when the toothbrush was not in common use, such cloths could be wrapped around a finger and used to rub the teeth clean. The anonymous portrait *Queen Elizabeth as an Old Woman* shows her wearing multiple chains around her neck, from one of which would likely have hung a gold toothpick or a case in which one was kept close at hand. In spite of what Shakespeare said through Parolles, the dangling toothpick, in or without a case, was definitely in evidence during the sixteenth century and well into the nineteenth. In fact, among commoners the teeth were picked publicly with impunity, evidently with little regard for the occasion, as demonstrated by a 1565 illustration showing a Venetian gondolier picking his teeth with a large sharpened stick, perhaps an exaggeration but not likely an aberration for his class. He appears to be waiting, like a taxi driver, for his passenger to finish serenading his inamorata.[29]

A large pointed stick was used as a toothpick by this sixteenth-century gondolier, who was evidently passing the time while waiting for his passenger to finish serenading his love.

It may be that Shakespeare's reference to the toothpick being out of style reflected the fact that the use of toothpicks had become so common that members of the fashion vanguard avoided them, just as trendsetters of all times abandon the latest things as soon as they cease to be exclusively theirs. Nevertheless, numerous portraits and paintings from all ages survive that show fashionable people with a bare toothpick or a fancy case containing toothpicks or a toilet set dangling from a chain around their neck or hung from their girdle. One collector of dental artifacts, Hans Sachs, has illustrated some of these, along with numerous examples of the toilet sets themselves, in his book on the toothpick and its history.[30]

Among "the last actions of Charles the first, when preparing for his execution, was to give away his gold toothpick as a present or memorial to some individual on the scaffold," and his generosity was not forgotten.[31] The 1649 act was recalled in a story relating to a toothpick case of special significance and was described in a Boston newspaper in 1762 in anticipation of the object being presented to the University of Cambridge: "Remember was the last word that K. Charles spoke to Bishop Juxon before his martyrdom. And likewise a tooth-pick case, curiously ornamented with silver, made of the piece of the oak which K. Charles II. cut from the tree while secreted there from the pursuit of his enemies; on the top is engraved a crown, and the words ROYAL OAK. His Majesty wore it for twenty years."[32] Contemporaneously with this newspaper item, advertisements regularly announced the importation from England of such items as gold toothpicks and "bone and jappan'd tooth-pick cases."[33]

With the introduction of the widespread use of pockets and pocketbooks, the toothpick went from something dangling from a chain to something concealed in a pocket, often in a case of some distinction. The practice continued among some classes well into the twentieth century. In the meantime, another kind of toothpick case had also become widely available. It was the work of silver- and goldsmiths and operated much like a Victorian pencil case. A cylindrical tube (the case) concealed and protected the toothpick point when not in use. When wanted, the pick could be extended by sliding a lug or collar toward the open end of the tube. Many such cases had a ring by which they could be attached to a chain. Such items came to be so common that the gold toothpick ceased to be a mark of great distinction, and it continued to be used mainly by the sentimental and those so privileged that they paid little heed to fad and fashion, doing things the way their ancestors did.

But in earlier times, it was not only the material of which the toothpick and its ancillary equipment were made that distinguished the common from the grand. The manner and custom of using the device differed widely among classes and cultures, sometimes serving as a signifier or shibboleth. The aristocratic display and use of a toothpick in plain view of other diners had been a practice of upper-class Romans, for "when guests were invited to dinner they were provided not only with spoons and knives but also with elaborately decorated toothpicks of metal, often of gold, which they took home with them," as if they were favors. "And it was considered quite proper to pick the teeth between each course of the meal!"[34]

Before Shakespeare, and even before him the Elizabethan dramatist Thomas Kyd, told the tale of Hamlet, the Danish chronicler Saxo Grammaticus set down the "true story." According to it, when the king of Britain received Hamlet at a banquet, the prince neither drank nor ate a thing; when asked why, he gave a cryptic but defensible answer and then proceeded to insult the king and queen personally, saying that the king had the eyes of a serf and that the queen was of "slavish origin." In support of the latter assertion, Hamlet gave three reasons, the first two having to do with how she wore her cloak and gown. The third reason had to do with her table habits, for "when, after dinner, she used her toothpick, she swallowed the extracted particles of food instead of spitting them out with royal dignity."[35]

The Renaissance has been called the "golden age of toothpicks," when they "were freely employed at court dinners by the best mannered individuals, and the food particles which they dislodged were spit out with gusto. At that time, such behavior was viewed as a compliment to the host."[36]

CHAPTER THREE

Sucksacks and Whiskers

O NE HISTORICAL SURVEY of toothpick use provides a comprehensive, but not exhaustive, list of materials and objects that
have served "for ceremonial or oral hygiene purposes."[1] The materials include those from the animal, vegetable, and mineral classifi-
cations. Among the metals used have been bronze, copper, gold,
silver, and iron. At least several dozen kinds of wood have been
employed at one time or another, and virtually any wood may have
done in a pinch.

The use of the rays of the umbelliferous plant *Ammi visnaga*,
which harden after flowering, has earned it the names "Spanish
toothpick" and "toothpick bishop-weed."[2] Stalks of grass and the
spines of cacti have also been used as toothpicks, especially in and
near Mexico.[3] Those from a New Mexican cactus were said to be
"as hard as bone and as tough as iron," and one would "last a man a
year."[4] One Mexican cactus is "ribbed and thickly set with teeth-
like spines, which furnish the natives with combs."[5] Indeed, the
individual teeth of combs of all kinds might be broken off for use as
toothpicks.[6]

Many parts of animals have served as toothpicks directly or as
material from which toothpicks could be carved. These have
included claws (of the bittern and hawk, and of the South American
bird-catcher spider), bills (even, it has been written, of young birds
that had yet to leave the nest), and bones of chickens and other
birds (including the "small bones taken from the drumsticks of
cocks and hens"); bones of fish, such as the cod; bones from the
thighs of rats, once valued in social clubs. They have been carved
from deer bone by American Indians; by a "Russian Savage" from
human bones, "for the British Ladies"; from bones of the hare and
the vulture, the latter to help toothaches; and, supposedly, from the
penis bones of such animals as the raccoon, the coyote, and the red

fox.[7] Also used were tusks (elephant and walrus), shell (tortoise), and quills (chicken, crow, vulture, goose, swan, porcupine, and that "walking bunch of tooth-picks," the hedgehog).[8] A creature called the great West India spider, which reportedly could cover a man's hand when its legs were stretched out, was said to have a "crooked tooth on each side" of its mouth. These teeth were "often set in gold or silver, to serve for tooth-picks."[9]

No matter of what it is made, the toothpick is characterized by its slender, pointed profile, and this has long led to a variety of fanciful images, ranging from the ridiculously small to the unimaginably large. In one of its early issues, *Scientific American* reported, without further comment, that "out West, they dry mosquitoes for tooth picks."[10] Beavers have "a double claw on one toe," which has been referred to as the "tooth-pick."[11] At the other extreme, the captain of a ship that had been to Greenland reported seeing "a very large fish, whose tail reached to the North Pole." The "most monstrous monster" commonly fed on whales the way humans do on shrimp. A 140-foot-long whale might stick between its teeth, causing the monster fish to use "its fore fin as a tooth pick to take it out."[12] On a similarly large scale, according to Rabelais, Gargantua "picked his teeth with an elephant's tusk" and also "used a young walnut tree."[13]

Perhaps the most persistent image of an animated toothpick relates to the birds that are said to clean parasites and leeches from the mouth of the crocodile. More generally, they feed on the food debris left on the tongue and between the teeth, especially of those crocs who might be found basking in the sun on a Nile River sandbank. Aristotle commented on such a bird, which he identified as a kind of plover or sandpiper. In "On Marvelous Things Heard," one of the minor works attributed to the philosopher, he recounted the bird's peculiar behavior: "In Egypt they say that sandpipers fly into the mouths of crocodiles, and pick their teeth, picking out the small pieces of flesh that adhere to them with their beaks; the crocodiles like this, and do them no harm."[14] Since the reptile cannot move its tongue to clean its own teeth, the bird serves an indispensable function. According to reports, having finished cleaning the crocodile's teeth, all the bird need do is signal its desire to leave, and "the reptile immediately opens its jaws, and permits the animated toothpick to fly away."[15] A traveler on the Congo River was a bit skeptical, however, and wrote of "plovers, with yellow wattles and spurs to their wings, who hop on the crocodiles' bodies, and if they do not, as some suppose, pick the teeth, they at any rate linger strangely and, as one would think, rashly round the jaws of the grim saurians."[16]

Whether real or imagined groomers of crocodiles, such birds have become creatures of lore, and they have been termed "nature's toothpick." Birds performing such dental hygiene have also been described as "about the size of a dove, or perhaps rather larger, of handsome plumage, and making a twittering noise when on the wing." The story of their grooming activity has continued to be told and retold, including by the writer Elspeth Huxley, who was raised in Africa. In her book *Out in the Midday Sun,* she describes "a crocodile lying on a sand-spit with its jaws wide open. The reptile paid absolutely no attention. Some locking mechanism enables crocodiles to lie for hours with open jaws, which cools them down and allows access to a small plover who cleans their teeth by plucking bits and pieces from their great big molars."[17] The bird has also been called variously the "trochilus or crocodile bird," a "winged toothpick," a "living toothpick," a "dentistical bird," a sucksack, and a zic-zac.[18]

Another living toothpick is the Pederson's shrimp, which inhabits the clear waters of the Bahamas. Also known as Pederson's cleaner shrimp, it is said to have a symbiotic relationship with some of the fish in the area, swaying back and forth when they approach it. A fish in need of a tooth cleaning stops a few inches from the shrimp, which then enters the fish's mouth. After picking around inside, it is allowed to leave. Reportedly, on occasion the "services of the shrimp are in such demand that fish line up awaiting their turn."[19]

Being and making toothpicks have long been considered remarkable occupations, and so fanciful stories about them abound. According to a newspaper report on the odd ways in which some Parisians made their living, the writer who uncovered the many unusual jobs had "learned to be surprised at nothing, and that if he were told that there are people who earn their bread by making tooth-picks out of old moons, he should accept the narrative with equanimity, and believe it with fanaticism."[20] The widespread popularity of toothpicks of all kinds by the end of the nineteenth century led to a rash of jokes involving even more imaginative ones, usually suggested by something that was slender and pointed. A humor column in the *Saturday Evening Post* asked why a cat swallows a mouse headfirst. The answer: "In order to save its tail for a toothpick."[21]

Among the reasons there exists such a variety of real and imagined toothpicks and toothpick materials is the physical and technical reality that nothing found or made ever works perfectly. So there is a constant search for a better way to produce a better toothpick or anything else. Opinions and assertions about what makes the best toothpick seem to have been around as long as the implements themselves. There is also the question of economics, supply and demand, and other extra-technical factors.

Sometimes, what is used as a toothpick can be a matter of making use of something that might have no apparent value for anything else. For centuries, Eskimos hunted walrus for their meat; their blubber, which provided oil for heat and light; their tusks and bones, from which tools and devices such as harpoon heads were made; and their hide, which covered boat frames. The Eskimos believed that "everything has a useful purpose."[22] By the late nineteenth century, Alaskan natives were engaged in the "peculiar but profitable industry" of making toothpicks out of walrus whiskers. The whiskers can grow to three or four inches, and their stiffness increases with age. The individual whiskers, which could number in the hundreds on a single animal, were pulled from the dead walrus and, after being thoroughly dried, were "arranged in neat packages and exported to China." There, they were "considered a necessary appurtenance of a Chinaman of the upper class."[23]

In one early-twentieth-century recounting of a sea-lion hunt, a ship captain reprimanded his crew for not shooting enough of the animals. He also complained that they took no whiskers, saying, "They're worth two bits a-piece from Chinamen, who gild them for toothpicks."[24] The British also valued walrus whiskers, even when ungilded. In 1925, a steamer carried a case of them to London, where they were to be "utilized for toothpicks at fashionable hotels."[25]

When properly dried, the whiskers can be quite durable. They were described as "lifetime" toothpicks by Wien Consolidated, the Alaskan airline that offered them to its passengers. Because the airline operated under the name Wien Consolidated only from 1969 to

At one time, airline passengers were provided with a toothpick made from a walrus whisker.

1973, any such toothpicks can be dated from that period.[26] The one I obtained through eBay may have come from a young walrus or a sea lion, for it measures only a bit more than an inch and a half across its graceful curve.[27] Dried walrus whiskers have also been sold to Alaska tourists as "Eskimo toothpicks."[28]

CHAPTER FOUR

Poor Goose!

N EXT TO METAL and wood, the most common raw material for
oral hygiene devices has been animal feathers and quills. The
Romans employed plumes, cutting the quill end to serve as a toothpick and using the feather end as a kind of toothbrush.[1] But not any
quill might do. Pliny cautioned "against the use of vulture feathers
as toothpicks because they cause a bad smell," rendering the breath
sour.[2] Rather, he recommended the use the "bone of a hare, sharp
as a needle . . . to prevent bad breath." He also promoted the use of
porcupine quills, which were believed to make teeth firm and
strong.[3]

When George Washington was concerned about the condition
of his teeth, his dentist urged him to use a "quill toothpicker."[4]
Indeed, at the end of the eighteenth and the beginning of the nineteenth centuries, quills were often the toothpick of choice, and they
appear to have been readily available for that specific purpose.
Thus, an anonymous contemporary diarist simply "walk'd about
the streets—purchased 25 quills—for tooth-picks."[5] Perhaps they
were bought from a young street vendor, who held them fanned out
in his hand, already formed into toothpicks. But many people
shaped their own, recycling their writing instruments or just idling
their time away, as Charles Dickens noticed in an American court-
room, where he found it difficult to pick out the accused. The prisoner was sitting among his lawyers, "making a toothpick out of an
old quill with his pen-knife."[6]

Evidently American congressmen were also accustomed to cutting quill pens into toothpicks. In 1827, a newspaper report on
Congress's annual "contingent expenses" noted that in addition to
spending almost seven hundred dollars on sealing wax ("nearly *two
pounds to a man*"), about the same amount was expended on over
twenty thousand quills and pens ("two thirds of which must have

A young boy selling quill toothpicks in France around the year 1800 fanned them out in his hand.

been used for *tooth-picks!*"). The report considered this to be "a very convenient way of *pocketing* the people's money."[7] But the misuse of a pen did not start in the U.S. Congress.

The feather quill was employed as a writing instrument as early as the fifth century and certainly from the seventh. Like the reed pens that it mimicked, the root end of the quill was formed into a split point. With the prepared quill it was possible to achieve exceptionally "clean and fine strokes, and to write so long and so conveniently" compared with the reed. Though the quill soon became the writing tool of choice, reeds continued to be used at least into the sixteenth century.[8]

The quill pen (from the Latin *penne*, meaning "feather," making "quill pen" somewhat redundant) was the writing instrument used with ink well into the nineteenth century. The best quills for the purpose came not from vultures but from a "particular breed of geese found in the Hudson bay territory." However, quills from swans were larger and more durable. Toward the end of the century, a hundred of them could command as much as twenty dollars on the market, whereas even that many of the best goose quills could be had for under five. They came to be imported from Russia in quantities of seventy or eighty bales, containing six million quills in the aggregate.[9]

The demand for quills dropped when the steel pen was introduced. This virtually permanently pointed alternative was invented by James Perry, "an English schoolmaster, who drudged at whittling his urchins' quills." By 1823, Perry was employing fifty men to make steel pens, which were sold retail under the name Perryian at a pricey sixpence apiece. But the instrument was made really popular by Josiah Mason, a carpet weaver who "went to Birmingham and manufactured pins, needles, shoe strings, and other infinitesimal essentials." One day, after seeing Perryian pens in a shop window, he bought three of them and proceeded to make "better and lighter ones" at about a fifth of the cost. He showed his pens to Perry, and the two eventually formed a successful partnership.[10] By about 1840, steel pens were in general use. Their rise was so fast and so nearly complete that one could read in 1862 that there were "no quills, now-a-days, except for tooth-picks."[11]

At midcentury, when the first of America's clean-living movements was in high gear, imported toothpicks were selling in New York for about a dollar a thousand.[12] A contemporary lecture titled "The Hair and Teeth" contained an endorsement and supplied a primer on the quill toothpick. According to the reporter, the lecturer urged,

> in the strongest terms, that no one should be unprovided with a good tooth-pick; and by a *good* tooth-pick he did not mean a costly one in a commercial sense, made of gold, silver, or any other metal. The best tooth-pick ever invented was made of a simple goose-quill. Get a quill, and with a knife you can construct an article for yourself; one that is pliable enough to remove particles of matter from the teeth without injuring the enamel, and one which you need not waste any valuable time hunting for, should you lose it. Make a dozen of them at a time, and then you need never be without one.[13]

Cutting a feather quill (or even a quill pen) into a toothpick was much easier than forming a pen nib, though both must have taken considerable practice to produce the desired effect. One end of the typical quill toothpick was formed with a single knife cut that left a long tapering point that could be as sharp as a pin but more flexible. The other end was fashioned by a cut that left a shorter, blunter, and stiffer end that resembled more a scoop than a spike. The entire pick was typically no more than two and a quarter or two and a half inches long, but quill toothpicks were also available in larger sizes, some being as long as three and a quarter inches with barrels of a correspondingly larger diameter.[14] (Forming a good pen nib from a feather may have started with a cut not unlike that used for making the blunter end of a toothpick, but the finishing touches required the more delicate operation of slitting the tip and shaping the nib.)[15]

However well or poorly cut, the quill would become the standard for many a toothpick user. A box of a dozen Hygienic Quill Tooth Picks of the "finest quality" would boast that they were "for the Perfect care of the Teeth."[16] When public awareness of good dental hygiene was being promoted widely, an item in the *Saturday Evening Post* was not untypical of the advice and instruction that was available to readers:

> *The tooth-pick* should be a quill, not because the metallic picks injure the enamel, but because the quill pick is so flexible it fits into all the irregularities between the teeth.
>
> Always after using the tooth-pick the mouth should be thoroughly rinsed. If warm water be not at hand, cold may be used, although the warm is much better. Closing the lips, with a motion familiar to all, everything may be thoroughly rinsed from the mouth.[17]

The hollow nature of the quill has on occasion led to the alternative notion that the particles picked from between the teeth should be collected within the pick, but then it would have to be rinsed out if it was to be reused, as many if not most were. The hollowness, and thereby the compressibility, of the thin-walled shaft of the quill also made it popular among those who had a habit of holding in the mouth something on which to chew. In one anecdote, set in a bureaucrat's office, "the death-like stillness is broken by the crushing between his teeth of the ornithological toothpick, which is [his] almost constant companion."[18] The pleasure of the habit was described elsewhere as when a "nice young man has his degree of bliss when he chews a tooth-pick—poor goose! (not the nice young man, but the fowl which gave the quill)."[19] In 1900, a synthetic toothpick, consisting of a quill-like tubular body and stuffed with a pleasing or beneficial substance, was invented for the purpose of providing "a substitute for cigarettes, chewing-tobacco, etc., for the use of which there is a strong inclination after eating."[20]

The making of quill toothpicks in quantity began as piecework at home. The feathers came largely from Russia and Sweden, and they were fashioned by hand into toothpicks in those countries, as well as in France and Germany. According to one account, "Peasant-folk, after their daily work is done, sit down with a sharp knife and add to their 'pin-money' by cutting these quills." The handwork kept the price of quills relatively high, and as late as 1885 a wooden toothpick manufacturer speculated that "some inventive American" would someday devise machinery capable of mass-producing them and so "make them as popular as the wooden picks" appeared then to be.[21]

In fact, the quill toothpick has been described as having become "one of the first mass-market toothpicks."[22] The production of quills had already gone beyond pure handwork. As the processing of just about everything made in the nineteenth century was eventually mechanized, so had been the quill toothpick, albeit largely with hand-operated machinery. When the steel pen displaced the quill, pen makers had to find other uses for their goose feathers. In France, just outside Paris, Monsieur Bardin "found himself the owner of two million geese and

In nineteenth-century France, different parts of the wing feathers of a goose were separated for use in making different articles de plume. This drawing shows how thinly the feathers were sliced to serve different ends, ranging from bonnet trimming to brush making. The barrel of the feather was used for making toothpicks.

without a ready market for goose quills."[23] The situation led to the establishment of the firm led by Monsieurs Bardin and Soyez, thus founding a new industry engaged in the manufacture of *articles de plume,* or "feather articles," which served for making bonnet trimming, fine brushes, artificial flowers, flocked wallpaper, and, of course, toothpicks. The enterprise was so successful that, by the late 1870s, there were "not enough geese raised in France to supply its needs, and hence large quantities of the feathers are imported into that country from Siberia and Russia."[24] The quills were formed into toothpicks in the Paris suburb of Joinville-le-Pont.[25]

As the French feather-articles industry became more sophisticated and specialized, only wing feathers were used, "classed by numbers, according to their position in the wing," with each having a distinct application. The exterior feathers, for example, because

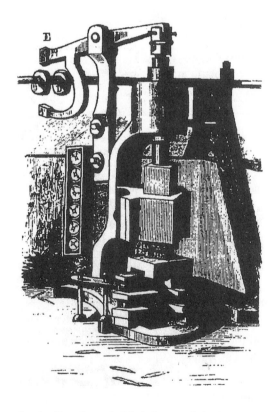

In the French toothpick industry of the late nineteenth century, a press was used to cut the points of a quill toothpick.

they had short barbs on one side of the shaft and long on the other, did not balance in the hand, thus leaving them useless as pens but suitable for making brushes.[26]

The separated barrel of a feather shaft was converted into a toothpick through a series of operations, which in the 1870s was partly mechanized. First, the opaque skin was removed by immersing a large quantity of barrels in a vat full of water, which was "violently agitated." The pointing of the quill toothpick was accomplished by means of a "stamp or press,"

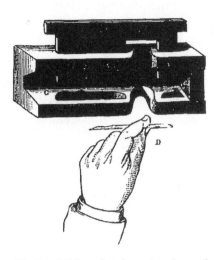

The hand-fed toothpick-cutting die used in the press of the previous illustration enabled both ends of the toothpick to be cut simultaneously.

which was most likely operated by foot power through a pedal, linkage, and lever system. The press employed dies by means of which both ends of the toothpick were cut into points simultaneously, but the machine still had to be fed by hand. After they were cut, the quills were put into a wire basket that was moved back and forth in water to wash out the light pith. The hollow toothpicks were then "dried in a centrifugal apparatus and by heat." A bunch of them were then gathered up by hand and inserted into something that looked somewhat like a blowtorch, with the picks suggestive of flames issuing from it. A copper ring was then slipped over the measured quantity to hold them in place until a red string could be tied about them to bind them tightly together like the handle of a fasces.[27]

Quill toothpicks destined for a restaurant were often imprinted with its name. Like quill pens, they could also have designs stamped into the shaft. The daily output was a quarter of a million picks.[28] When quill and other toothpicks began to be individually sealed in paper there was little need to imprint them directly, for the name of a restaurant or of the toothpick brand itself could be printed on the wrapper. One company that specialized in producing toothpicks "from the plumage of Young Geese" was the Hygeia Antiseptic Toothpick Co., based in New York. Hygeia quills were often individually wrapped, but the company also produced Mother Goose brand picks, which came in a transparent pocket container about the size of a small box of matches, and bore the advice that using a quill toothpick "will give you great comfort; recommended by dentists; prevents decay of teeth; protects your health."[29]

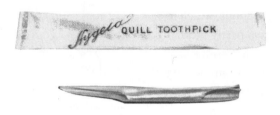

Quill toothpicks were typically cut differently at each end. Even when individually wrapped they could become dried out and split before ever being used.

In 1890, the largest quill toothpick factory in the world, which was then the one located near Paris, was reported to have had an annual output of twenty million picks, which would suggest that it was operating under capacity, no doubt due to the rise of the wooden toothpick.[30] As early as 1883 it was reported that the "goose-quill toothpick has gone out of fashion, the days of the years of its pilgrimage are ended, and the curbstone merchants who used to sell them have retired millionaires."[31] Despite the exaggerations, and in spite of the economic disadvantages, some users continued to favor quill over wooden toothpicks, even though quills were inferior in at least one functional way: after a couple of years they could dry out, get brittle, and crack, thus resulting in a limited shelf life and a dangerously sharp splinter. To address this limitation, quill toothpicks would eventually come to be made of plastic, which at the Soyez plant could be produced at a rate exceeding one million per day.[32]

As large as the French quill factory might have been, it was not necessarily always the largest, and early in the twentieth century that distinction belonged to a factory in Romania. That country's crown princess, who was King Edward VII's favorite niece, was known to be among the best horsewomen on the Continent, a reputation that was only fed by an incident in which her life was threatened by a runaway steed. She had been riding in the mountains when a severe thunderstorm arose and frightened her mount. The bolting animal did not yield to the reins, so the princess rode on in the hope of its tiring itself out. Before that could happen, however, she saw that the horse was carrying her toward a precipice. Just twenty feet before the horse went over, she threw herself from the saddle and landed safely on her hands and knees—certainly an exaggeration, since her momentum would likely have caused her to tumble over the precipice too. Nevertheless, in addition to engaging in reputed daring riding and pursuing her hobbies of collecting "scent bottles and vinagrettes," the princess founded, financed, and ran a quill toothpick factory that was reported to be the largest in the world and the source for "most of the toothpicks found in neat paper wrappers in New York hotels."[33]

Not all quill toothpicks had the cachet of being made near Paris or by a princess, and not all were of equally high quality. Nor were they always honestly presented to be what they were. One con man advertised that he would provide a hundred "useful table articles" to anyone who sent him two dollars. Those who fell for his scheme received in the mail a small box containing a hundred "cheap quill toothpicks."[34]

All things can be used for more (or less) than their intended purpose, and the quill was no exception. In 1787, during debates and editorial exchanges over the drafting of the Constitution, the possession of a quill pen served as a literary vehicle for discussing private property and personal freedom. In a pseudonymous political exchange between one Senex and one Scribble-Scrabble involving the declaration of rights, Senex put forth the case "that he can have no right to hold the pen with which he was scribbling, if Scribble-Scrabble had a right to take it from him." Scribble-Scrabble, arguing that "so great is the advantage which truth has, in most cases, over error, that the very examples and illustrations made use of by those who endeavour to support an erroneous hypothesis, will when properly stated, point out the mistake." And so he seized upon the very metaphor of possession of the quill to assert that "in a state of nature, the people have a right to make use of quills for pens to write with, or for toothpicks, pin-cases or needle-cases . . . as they please." And he continued:

> These may be called so many rights; or so many uses of the same right. Suppose then, there had been an article in the declaration of rights that runs thus—*The people have a right to keep, and make use of quills, for pens to write with;* would this article contain a negative, that takes from the people the other rights of keeping and using quills for toothpicks, pin-cases, needle-cases, &c.? or, if any should think these to be only uses of the same right, would these uses be taken from the people by such an article? Surely they would not.

Scribble-Scrabble went on to discuss the article relating to the right to keep and bear arms, arguing that because that right is granted for the common defense, it does not preclude retaining the right to hunt. But what is more germane in a book on toothpicks is the footnote to the passage, which reads:

> This instance of the quill, &c. may be looked upon by some, as trifling: If any such there be, I wish them to consider whether great things may not, in many respects, be compared to small; and whether small matters do not often illustrate things of the greatest importance.[35]

What is of greatest importance is relative, of course. On an outing in Colorado, a Victorian sportsman who had stopped at a house for a noontime dinner describes what he did afterward: "Being in want of a toothpick I hunt for the woodpile and find a chicken feather which answers my purpose." Returning to the table, he ponders the circumstances of the household while he trims the feather to suit his needs, evidently giving little thought that he might be producing a rather unsanitary implement.[36]

Quill toothpicks have been pressed into service in a variety of ways other than toothpicking and philosophizing. On a fishing trip, a pair of campers found their only fishing rod bent into a useless configuration. In an attempt to continue their good luck, they "straightened it out, split a quill tooth pick and bound it firmly about the rod with a bit of the line and started down stream toward the lake."[37] A similar repair was independently invented for fixing the split end of a bamboo pole, and the mended tip apparently held up through more than ten years of use.[38] We can be most resourceful in times of need.

When we do come up with new uses for old things, we often borrow the older thing's name, which can lead to confusion. The phenomenon was experienced in the late nineteenth century by a young Texas doctor who had been called to help a woman who was about to give birth but was "unable to make the supreme effort for final expulsion." An old woman, rocking nearby, said to the tentative and ineffective physician, "Doc, I wouldn't bother any longer with that woman. I believe I'd quill her and have done with it." Not knowing exactly what she was referring to, but suspecting that she meant to put the younger woman out of her misery, the doctor refused to do anything, even after repeated pleadings. Exasperated, he finally said, "Madam, I'll be d——d if I will do it. If you want to quill her you can do so, but I won't." At that point, "the crone took from the wall a turkey wing, and drawing a feather from it proceeded to fashion something like a long quill toothpick, and filling this with snuff from her own private stock leaned over the patient, and as the next pain came blew the snuff into the woman's nostrils. Quick as a flash the woman responded with a giant sneeze, and the child was born with the sneeze." The physician had been introduced to an effective new procedure, and a new use for a quill toothpick.[39]

CHAPTER FIVE

Out of the Woods

F ORTUITOUSLY, THE NATURAL bacteria-killing organisms in wood and its ability to soften make it the perfect material for exploring the oral landscape."[1] There can be little doubt that the earliest wooden toothpicks began as small twigs conveniently picked up off the ground. In the absence of a twig, the primitive toothpicker might have had to resort to breaking off what could not be found already in a convenient size. Larger branches would not have been directly useful, of course, but the frayed ends of those broken off or hanging from bushes and trees—perhaps due to a lightning strike or too heavy and aggressive a climber—might have presented an easy source of splinters. With the development of cutting implements, splinter-like toothpicks could be crafted at will, as long as there was a source or supply of wood at hand.

We can expect that the basic twig or stick became increasingly refined in how it was splintered and pointed. Twigs as torn from the bush or tree would prove generally inadequate and wear down relatively quickly in the moist environment of the mouth, so a new one would constantly have to have been found and shaped. A search for a new toothpick would not necessarily yield one even as good as the last, leading to the search for new woods or other materials from which to fashion a more long-lasting, effective, and predictable toothpick.

Not all woods are suitable for making good toothpicks, and even of those that are, the advice has been to use only the best. As Ben Jonson wrote in his play *The Devil Is an Ass*,

> *What diseases and putrefactions in the gummes are bred*
> *By those [toothpicks] made of adultrate and false wood?*[2]

Even among the "true" or preferred woods, some kinds are naturally better than others for carving out toothpicks, whose desirable qualities can provide a wish list for the ideal wood. It should be easily worked, of course, and be capable of providing a smooth finished surface, one free of mini-splinters. In toothpick proportions, it should have sufficient stiffness to provide the appropriate leverage but not be so stiff as to be unyielding in the mouth. It should be free of unpleasant tastes and odors, and it should possess an even color. In the millennia since the prehistoric toothpicks, a staggering variety of woods has been tried, some producing toothpicks closer to the ideal than others. According to one survey of documented sources, the following woods have been used: aloe, arak, birch, camphor, cherry, chestnut, cypress, dracuncula, fig, juniper, lentisk, linden, mahlab, mandarin orange, mango, pine, poplar, reeds, walnut, and willow, as well as roots of the lily, rosemary, and sugarcane.[3] But even this list is incomplete. Other sources name additional plants and trees whose wood has been used to make toothpicks: the root of marshmallow and lucerne (alfalfa), bamboo, balsa, balsam, white-wood (a generic term for trees that yield pale lumber, including the silver fir, basswood, and tulip tree), maple, gum, aspen, Nordic pine, and Spanish willow.[4] According to one dentist, "the use of wood to pick or clean teeth has always been selective towards the more fragrant of woods," such as those "with the aroma of garlic, or myrtle."[5]

Lentisk, the "beautiful evergreen" mastic gum tree of southern Europe that is also known simply as mastic, mastick, or mastix, had long been said to make the best toothpicks.[6] The Romans even used the word *lentiscus* to refer to a toothpick.[7] This wood was able to "take a highly ornamental polish, and tooth-picks made of it were worn in the hat by men, and in the hair by ladies."[8] The mastic gum tree is of the cashew family and is closely related to the pistachio. To the Romans, it was also known as the "toothpick tree." The word "mastic" derives from the Greek for "chew" or "masticate," and besides being good for making toothpicks, its wood provides, among other things, tooth-cleaning sticks, its sap tooth powders, and its seed-corn (an almond-like nut) "an aromatic chewing substance." Like its name, the tree originated in Greece, where it was native to the island of Chios. Originally, the Romans imported mastic toothpicks from there, but the supply could not keep up with demand, which drove up prices. So the tree was transplanted to Italy, along the coast north of Naples, which groves Ovid mentions in his *Metamorphoses*.[9]

The ancient Greeks are also known to have used wooden toothpicks, for the rhetorician Alciphron, who wrote as a commoner of domestic life and manners in pre-Christian Athens, mentioned "the little wooden sticks with which a person removed the bits of food remaining in his teeth." When the Romans exchanged gifts during the Saturnalia festival, they were sometimes accompanied by a poem, such as an epigram by Martial, who said that toothpicks were "best when made of mastix wood."[10] Indeed, in the epigram "Dentiscalpium," the poet insists that toothpicks of lentisk "are to be preferred, but that, in their absence, quill toothpicks may be used."[11] Seventeen centuries later, Samuel Johnson repeated the advice, also referring to the wood as lentisk.[12]

Ancient references to wooden toothpicks can refer both to splinters that were the precursors to the finished kind we know today and to what has been called a "chew stick." According to one theory, "the development of this device is thought to have been a natural sequence arising from the use of tasty twigs for toothpicking, and represents a transitional step toward the development of the bristle toothbrush."[13] It is easy to imagine that "twigs or splinters of wood, unraveling at their end from rubbing and the softening action of saliva, could have evolved into a chewing stick and a primitive brush for removing of deposits from the broad flat surfaces of the teeth and the beginnings of an ablution that could be recognized as oral hygiene."[14]

According to one description, a typical chew stick is "about the size of a carpenters pencil with one end beaten into a soft fibrous condition for application to the teeth in a brushing or scrubbing manner. The other end is sometimes pointed for interdental toothpicking." The material for such an implement, whose end could also be prepared simply by mastication, typically has come from a variety of plants, many called "toothbrush trees." Chew sticks were evidently used in Egypt as long as five thousand years ago, and they continued to be employed well into the twentieth century among groups ranging from tribes in Africa to rural southerners in the United States. As late as 1915, the practice of using dogwood twigs was documented in remote areas of North Carolina.[15] One elderly man from the vicinity of Shreveport, Louisiana, who employed "frayed white elm sticks" throughout his life, was found to have healthy gums and teeth free of plaque.[16] The practice is believed to have been effective because the "bicarbonate of soda and other aromatic juices found in many of the woods used to fabricate the 'chew stick' (licorice, lucern, dogwood, peach) acted as an astringent causing an increased flow of saliva, a natural mouthwash."[17]

Alternatively referred to by some dental historians as a "dental fiber pencil," this primitive kind of toothbrush has been used at one time or another in virtually all cultures. In Arab lands, it has been known as a *siwâk* or *miswâk*, which is a word connoting a brush, and its making has been described as follows: "From the end of a properly cut little stick of wood, flexible and knot-free, a piece of the bark is removed. After the stick has been soaked in water for twenty-four hours, the peeled portion is pounded with a stone or hammer until the plant fibres unravel to form a kind of small paint brush. The unbeaten part of the stick serves as a handle. Such a toothstick bears no real resemblance to the modern toothbrush, since the raveled fibres forming the brush end are a natural *continuation* of the handle."[18] According to one botanist, the "best and most sweet-scented" *siwâks* were made from the branches and roots of the arak, known as the "tooth-brush tree."[19] Sugarcane and the root of the lily were also used.[20]

The proper use of the *siwâk* was codified: "It must be held in the right hand, between the little finger, which should point downward, and the index, middle and ring fingers; the thumb should point toward the raveled end of the brush."[21] A similar "tufted toothpick," known as a *fusayoji*, is believed to have been introduced into Japan by Buddhist priests who came from India via China. The Indians tended to use linden, but the Japanese preferred willow.[22] The Japanese used the *fusayoji* in much the same way that the Arabs used the *siwâk*.[23]

A nineteenth-century *ukiyoe*—a Japanese colored woodblock print—of a Meiji-period woman using a *fusayoji* shows her fingers to be arranged in precisely the prescribed manner. In Africa, an analogous implement (called *msuaki* in some parts of the continent) has been used. Besides their burden, caravan carriers have been known to take with them only a gourd to hold water and a *msuaki*.[24]

Muhammad was an ardent user of the *siwâk* and developed

In this Japanese woodblock print, a woman of the Meiji era holds a chew stick known as a fusayoji *with her fingers in a precisely prescribed pattern.*

"rules and rituals for the correct and effective use" of it. According to one of his biographers:

> Even the approach of death did not keep the Prophet from demanding the siwak because it is the most elegant thing that one can use and the most fitting to be found beautiful, for it makes the teeth white, clarifies the understanding, makes the breath fragrant, extinguishes the gall, dries up the phlegm, strengthens the gums around the teeth, makes the glance clear, sharpens the power of the vision, open the bowels and whets the appetite.[25]

Sometimes a single thing is called upon to accomplish a multitude of tasks.

With the development of the modern toothbrush, chew sticks no longer played the role that they did for millennia. But the common wooden toothpick, whether hand-carved or machine-made, has yet to be displaced. Still, "wood need only be mentioned to be condemned" as toothpick material, because of the danger—even when the wood approached the ideal—of the point splitting and breaking and getting jammed between teeth or in the gums.[26] Other critics have been less absolute, warning only against using toothpicks made of "a piece of hard wood, like a sliver of beech, oak or walnut," recommending instead "a thin goose quill, soft basswood picks, or broomcorn."[27] In spite of such opposition historically, and even though "wooden toothpicks have a tendency to split and splinter, and frequently are too thick," today the use of wood for toothpicks overwhelms the use of all other materials.[28]

CHAPTER SIX

Made in Portugal

I N THE EARLY nineteenth century, there was considerable debate
in the still-young United States over the question of "foreign
manufactures." The issue could be reduced to a single question:
"Shall we manufacture for ourselves, or shall Britain manufacture
for us?" Among the objections to developing a domestic manufacturing base was the contention that factories "demoralize and
deprave those employed in them," with opponents claiming that in
Britain's factories were found "disgusting exhibitions of human
depravity and wretchedness."[1] The other side of the argument was
bolstered by the fact that the number of paupers per capita in English manufacturing districts was but a third that number where no
manufacturing took place.

Proponents also put forth less-quantitative evidence by alluding
to well-known examples of success outside Britain. Thus the
rhetorical questions, "What manufactures debase Portugal? Is it the
manufacturing of tooth-picks at the university of Coimbra?"[2] No, it
certainly was not. Perhaps the oldest and proudest surviving tradition of handmade wooden toothpicks that are known and admired
throughout the world exists in Portugal, in the district of Coimbra,
through which flows the Mondego River. Here, not far from the
town of Penacova, is the small hamlet of Lorvão, where the Mos-
teiro de Lorvão is located. It is one of Portugal's oldest monasteries,
having been founded by the Benedictines before the Moorish inva-
sion. By the end of the twelfth century, it had been taken over by
the Order of St. Bernard, and these nuns supported themselves, at
least in part, by making "a special confection," which evidently was
sticky to the fingers and tacky to the teeth. In the sixteenth century,
the nuns got the idea of making toothpicks for picking up the delicacies and for cleaning the teeth after eating. Making the famous

toothpicks came to be practiced as a cottage industry by the laypeople of Lorvão and neighboring villages.[3]

Such "beautifully chiseled" toothpicks, some ranging from six to ten inches long and known as *palitos de Coimbra*,[4] were considered to be the "best toothpicks made anywhere in the world." At the end of the nineteenth century, Portuguese toothpicks continued to be "whittled by hand from orangewood splints by peasant girls, the only tool used being an ordinary jackknife." The toothpicks were often described as being "smooth as ivory," and they had the reputation for not breaking into splinters.[5]

The Portuguese orangewood toothpick was said to be preferred by the "aristocracy of Europe," but Americans would increasingly be satisfied with a "Maine article" that was believed to be "good enough" for them. Once the American machine-made product became established, it would have "little difficulty in holding the market" against the handmade competition. Even so, to further steer Americans away from the imports, domestic toothpick users would be reminded that Portuguese girls received the "munificent sum of 5 cents a day" for their toil. Although the advantages of American machinery might not be able to compete with such cheap labor, and even if it could mass-produce the orangewood product (which it could not) and "fancy" foreign picks could not be duplicated here at four times the price, the American machine-made toothpick would eventually prevail.[6]

Nevertheless, Portuguese toothpick making continued to flourish on its own scale, and in the middle of the twentieth century there were about nine thousand people occupied at the trade. (In contrast, those employed in all the mechanized toothpick mills in Maine would be counted in the hundreds, and perhaps never more than about twelve hundred.)[7] Argentina and Brazil were the principal consumers of the hand-crafted Portuguese picks, but at midcentury the implements were inexplicably left out of trade treaties between the countries. Though "good manners in Latin America" were jocularly presumed to have lessened the demand for toothpicks, and thus threatened the occupations of thousands of Portuguese, they continued to carve them out of orangewood.[8]

According to the curator of a Dutch dental museum who visited Lorvão in the late 1970s to see if toothpicks were still being made in the traditional way, the process had changed little if any, in spite of the vagaries of markets and treaties. He found that simple Portuguese toothpicks were also made from willow and poplar, woods that grow on the banks of the Mondego and Tagus rivers. In preparation for making toothpicks, men cut the trunks of felled trees into twenty-inch lengths, "which were then stripped of their bark and split into triangular pieces," looking not unlike firewood. The "pieces were tied into bundles and labeled with a name for delivery somewhere in the surrounding countryside."[9]

Once the men delivered the wood, the work of toothpick making was taken over by women and children. The wife of one of the woodsmen demonstrated the process, which was recorded by the curator:

> A strip about 2 cm. in width and of the thickness of a toothpick was split from the piece of wood and three parallel incisions were made into it, reaching nearly as far as the end. The woman then put a leather belt on her lap tightening it with a piece of string underneath her feet. With an ordinary potato-knife she pointed the ends of the, by now, four parallel strips; by turning the strips about 30° to the right and then 30° to the left, one side of each was rounded. The strips were then turned through 180° and the process was repeated. By one horizontal cut four pieces were split from the strip. Four toothpicks were born! The woman repeated the process on the shortened strip till it was finished. The toothpicks thus carved were all of the same length and fitted exactly in the boxes standing ready at hand.[10]

The Lorvão toothpick carvers were at one time the only ones who produced in this way the ordinary, everyday picks commonly referred to simply as *palitos*, which look remotely similar to but are longer, broader, and thicker—as well as being considerably more sharply pointed, more irregular, and more individual—than the flat tapered machine-made toothpicks that can be bought in American supermarkets today. The handmade ones are also smoother and generally more flexible and lighter in color. According to the dental historian José de Paiva Boléo, "The choice of the wood and the method of production guarantee the good quality of the toothpicks.

They are supple, give sufficient resistance, the wood has neither taste nor smell and the handiwork has a high grade of perfection."[11]

There is another kind of Portuguese toothpick, which is usually longer and rather more elaborately carved and decorated with tiny curls of wood shavings, which are produced by means of miniature wood planes.[12] They were made not in Lorvão but in nearby Penacova and in Poiares, and also on the Portuguese island of Madeira and in Laranja, Brazil.[13] The fancy toothpicks, which were made by young girls who in 1903 were paid as little as three cents a day to turn out sticks "sharp as needles and smooth as ivory," tended to be used for special occasions.[14] Though often referred to by the same simple Portuguese name, *palitos*, as their plainer counterparts, they were more properly called *palitos especiales*, and have been referred to less descriptively as "cocktail-picks."[15]

Whatever they were called, they were always remarkable. According to a guest at a mid-nineteenth-century dinner party hosted by the president of an antiquarian society, the toothpicks used were "made of orange wood by shepherds, and are of various qualities, according to the labor spent on them. Those before us were of the first chop—each being ornamented at the blunt end with scrolls like those of an Ionic column, the minikin involutes being delicate shavings left adhering to the body."[16] Such elaborately crafted toothpicks were considered "too expensive for ordinary commerce," but on occasion "small quantities" of them were imported "for use at notable banquets."[17] Outside Portugal and Brazil, they are admired as unique pieces of handicraft.

I have a pair of toothpicks purportedly of this type but which I imagine were originally bought, probably in the late 1950s or 1960s, at some South American, Caribbean, or other tourist destination gift shop—and perhaps more for fun than for use. The picks are over four inches long and are decorated for a fair distance down their shaft with carvings so delicate-looking that, should I use one of these picks at all, I would be hesitant to touch the ornamentation. The picks are surmounted with gilt balls that are connected by light chains to small his-and-hers medallions. The chain on the "his" pick is threaded through the corner of a postage-stamp-sized booklet promising to cover the "Care and Social Significance" of the "Famous Hand-Crafted Golt-Tipped Tooth Pick" designed "for those who have everything else." The apparent misspelling of "gold" may well have been deliberate, lest some dissatisfied buyer find that no precious metal had been used, for the text of the booklet is in generally correct but intemperately corny English. It tells us, among other things, that owning these picks means that no longer will you have to "pick your teeth with your finger or sneak out one of the old fashioned, plebian toothpicks." The booklet concludes by identifying the picks as "hand carved out of Wango Wango trees by the natives of El Tiredo Woodchoppo Province of far away Portugal."[18] I did not expect any of my dictionaries to tell me anything about a Wango Wango tree, but the picks seem definitely too dark to be of orangewood.

Some Portuguese toothpicks have elaborately handcarved shafts, such as these examples evidently made for the tourist trade.

Lesser but perhaps more authentic examples of fancy toothpicks might sometimes have been used as rather extravagant cocktail picks. They are much more elegant than the common American "party picks," which are but ordinary (though often somewhat elongated) ones topped with frills of colored cellophane, perhaps conceived in vague imitation of the wooden curls of *palitos especiales.* Elaborately carved wooden toothpicks may have been on the mind of Cervantes when he wrote of Don Quixote's thinking about making small things. According to one wag, "Don Quixote thought he could have made beautiful bird cages and toothpicks if his brain had not been so full of ideas about chivalry. Most people would succeed in small things, if they were not troubled with great ambitions."[19]

But even plain wooden toothpicks can be appreciated for their exquisite simplicity. In a story about California frontier life, a raccoon dinner was followed by the traditional final course of toothpicks, which were described as "splendid" and remarkable. The cook commented, "What magnificent toothpicks we have for this season of the year."[20] He may have been referring to the absence of discoloring due to sap leakage, a perennial problem faced by toothpick manufacturers as they began to mass-produce the little sticks. No matter what the season, hand-carved orangewood toothpicks are always splendid. Although they were not likely used on the California frontier, they began to be imported from South America into New York and Boston by the middle of the nineteenth century. The first shipment into Boston may have been arranged by a man named Charles Forster, who worked in the import-export business in Brazil.

CHAPTER SEVEN

Charles Forster in Pernambuco

TOOTHPICKS AND THEIR use seem to evoke stereotypes and promote prejudices, but this is not a new phenomenon. In the sixteenth century, royal court receptions were customarily prepared with "salads for the Italians, porridge for the Englishmen, and picktooths for the Spaniards."[1] The Spanish reputation for this predilection has persisted over the centuries. According to one account:

> The educated Spaniard undertakes after each course at the table a very lengthy treatment of his teeth with the toothpick and keeps it in his mouth a very long time after meals. This decoration is said to be very becoming to the young ladies; they seldom put it away in the daytime; it serves to protect their virtue if some one approaches them to steal a kiss. In Sevilla the dancers during the public performances have a toothpick in their mouth as an indispensable part of their costumes and are said to develop particular grace in the manner they hold it between their lips, similar to the grace shown by the Parisian girls when smoking cigarettes.[2]

Seventeenth-century European literature, especially that from Latin countries, abounds with "numerous references to tooth-picks being essential in promoting oral hygiene" and personal image.[3] In *Don Quixote*, the pride of a "poverty-stricken aristocrat" kept him from letting it be thought that he was going hungry. Like a poseur, he appeared on the streets "with a toothpick," even though he had not eaten anything requiring its use. Ironically, had he encountered someone intent on inviting him to dinner, the invitation likely would not have been extended since it would be assumed that he had already eaten.[4]

In 1834, when toothpicks were not known to be used in polite company in the United States, one traveler summarized the state of affairs he encountered abroad:

> Customs vary in different countries, and in some parts of Europe, particularly among the genteel classes in the city of Lisbon, *picking one's teeth* is considered a ceremony necessary to be observed at dinner at the end of every course.—So much so, that tooth picks, fabricated of pieces of tough wood, are always furnished by the host. They are fancifully stuck into a neat little machine, resembling a pepper box, placed in the centre of the table, and which is regularly handed round to the company, who supply themselves and proceed to business. It is somewhat difficult for a Yankee to accustom himself to this singular habit, and we confess that we were not a little astonished, perhaps shocked, when at the first dinner which we had the honor to eat in Lisbon, a young lady, the daughter of the host, with a most engaging smile, pointed to an image resembling a Hindoo idol, with its head bristling with small splinters of wood, and asked me to have the goodness to hand her a *tooth pick*! But we heard aright—she selected one from the lot, and throwing herself back in her chair, and opening her pretty mouth, commenced operations in the most graceful manner imaginable.[5]

Almost half a century later, the toothpick was still very much the topic of discussion of some visitors to Lisbon. One delegate to the International Congress of Prehistoric Anthropology and Archaeology, which was held in that city in 1881, reported attending a ball and elaborate supper hosted by the king, at which "there were no plates, knives, forks or other appliances of civilization, nothing but large wooden toothpicks." Although the visitors were hesitant to approach the delicacies without the proper implements,

> the court, nay, royalty itself, unhesitatingly took a toothpick, dug it into the chosen morsel, poised it a moment in the air, and it was gone. Thus emboldened, all possessed themselves of these handy instruments, and dug in their turn, roving and sipping like bees, though all with inward misgivings as to whether they had been spirited away suddenly to China or some other

Eastern haunt of the primitive chopsticks. On after inquiry it was learned that in all large court assemblies these toothpicks were put in requisition, as it was feared that silver forks might be pocketed by the guests.[6]

No doubt the picks were also used to dig into the teeth after serving their purpose as eating utensils.

The association of the Spanish and related Latin cultures with toothpicks has continued, and not without reason. A dentist traveling through Mexico found himself in a small souvenir shop in San Luis Potosí, where he came across "a men's leather wallet containing a long, narrow, sewed pocket." When he asked what the thin pocket was for, he was told that it was for "the toothpick."[7] The point was made further evident by a passenger on a Portuguese steamer, which in the late nineteenth century was making its way from Boston to the Azores and then on to Lisbon. One of the passengers found the dining experience remarkable:

> Perhaps it was the abundance of meat, perhaps it was owing to the fact that the majority around our table belonged to the Latin races, who are all born with toothpicks in their mouths, but at all events these little wooden implements were flourished with a vigor and openness that was almost alarming to the Americans. These latter showed their comparative unfamiliarity with them by a cautious reserve. They held their napkins or their hands in front of their faces with a rather aggressive air, as of superior breeding. But the Portuguese and the Spaniards employed them in the open. They routed the offending particles out . . . with the zest of a hunter after game. Some even paused every now and then to examine their bag, and then returned to the chase with fresh energy and interest.[8]

Not only did the Portuguese have a long tradition of making and using toothpicks, they also began early to export them. In Brazil, where toothpicks were used even after breakfast, when they were employed over coffee on the veranda, they had been imported from the time of the country's occupation by the Portuguese.[9] And their use has continued into the present. When two schoolteachers from Kankakee, Illinois, took a trip to South America in the mid-twentieth century, they observed the Brazilians' "mandatory" use of toothpicks and reported that it was "really something to see portly gentlemen and stout dowagers conversing while they probe every nook, cranny and crevice of their teeth."[10]

Toothpicks had long been "used by every one in Brazil, from the Emperor to the lowest tradesman." It was not uncommon to see on the streets slaves with toothpicks "stuck behind the ear, where clerks sometimes put their pens when not in use." Also, dining tables were commonly set with accessories, such as the "neat little machine" and the "Hindoo idol," that held the toothpicks: "All repasts are wound up by pushing round [a] paliterio, a fanciful device for holding the picks, and often forming an item in a family's silver plate. With those who do not smoke, palitos are equal to cigars in promoting conversation, besides being cheaper and more durable."[11]

It was in this milieu that Charles Forster, a young man from Massachusetts, found himself working in the middle part of the nineteenth century. Forster was born in 1826 in Charlestown, just across the river from Boston, into a family described as "old and aristocratic" and that some believed "retained the early, and what was considered the better, mode of spelling the name" so commonly rendered "Foster."[12] His great-grandfather was Jacob Forster, who graduated from Harvard in 1754—in the same class as John Hancock—and soon became pastor of a church in Berwick, Maine, and later served briefly as a chaplain in the Revolutionary War. Charles's grandfather, another Jacob, was a cabinetmaker who started the Forster & Lawrence furniture business in Charlestown, which was eventually taken over by his son, whose name was also Charles.[13]

This Charles Forster followed his father, Jacob, into the furniture business and became a senior partner in the expanded firm of Forster, Lawrence & Co. Eventually he "failed in business," but he went on to serve as a deacon in the Unitarian Church in Charlestown and for many years was superintendent of the Sabbath school there. He recovered financially, perhaps through inheritance, and after a time paid his creditors in full.[14] He appears to have lived comfortably and prominently, owning an estate in the Winter Hill section of nearby Somerville while holding "many positions of trust" in Charlestown. He was known as an "ardent temperance advocate" and especially as a "philanthropist in the truest sense of the word." He befriended the poor, advised the wayward and unfortunate, comforted the mourning, and helped the weak:

> Mr. Forster would give a dollar to any man who said he was poor, whether he had ever seen him before or not. A peddler or tramp who called at his door at noon was often invited to dinner, and he would order the servant to prepare him a second table. He gave away money in great quantities to the poor, and the older residents of Somerville will never forget his freeheartedness. He was so charitable that he came near leaving his own family in poverty when he died.[15]

Forster served as a member of the school board of Somerville for six years, and in recognition of his service and philanthropy one of the largest grammar schools in the town was named after him. Some years after his death, the Forster School Alumni Association was established, and one of its goals came to be securing a portrait of the school's namesake to be hung in a conspicuous place in the building. Prominent residents of Winter Hill led the effort, but their task did not prove to be easy. No likeness of Forster was to be found, and so a crayon portrait was commissioned. Among those donating to the fund were to be the 250 pupils of the school, who would contribute ten cents each, in recognition of which their names were "to be written in a neat and artistic manner on the back of the picture." The portrait was finished in 1886, and a ceremony was planned to present it to the school.[16] But one must wonder how true to life could be a portrait created two decades after the death of the elder Charles Forster.

The entire Forster family seems to have been averse to having their portraits painted or their pictures taken. An otherwise thorough accounting of the Forster family pedigree does not contain a single engraving of a likeness.[17] This aversion to the image appears to have continued into the next generation, at least, for not a single portrait of the younger Charles Forster has been located. Neither has one of his wife, nor of their children during their youth.[18]

In any case, if the younger Charles Forster did not grow up among likenesses of his ancestors, parents, or descendants, he evidently did grow up amidst wood and woodworking tools, or at least

life, accumulated what he called his "tooth-pricks" in "boxes, tubs and pails" that he kept in a cupboard. They were of a "quantity sufficient to supply a Boston boarding house for years." He also whittled clothespins, but he did not sell any of his creations, having made them solely for "family use and his own amusement."[29]

Nowhere did there seem to be more wood for making toothpicks than in the still-young country, where one wag imagined that a fictitious St. Jonathan, the "personification of the Yankee nation-universal and particular," was consecrated in 1851, on the anniversary of the landing of the Pilgrims. Among the prominent characteristics of this "giant in the bloom of youth, graceful yet stalwart, dressed in all the peculiarities of rustic and fashionable costume," were his huge hands, in which were "the ever-accompanying jack-knife and piece of pine, which, with untiring industry, he manufactures into articles of every fashion and use, from a fancy tooth-pick to the model of a steamship."[30]

The likes of St. Jonathan were embodied in the more general "myth of the Yankee Whittling Boy," which promoted the idea that "American inventions of the nineteenth century came from youthful practice with a pocketknife." The essence of the myth is captured in the opening lines of a poem by the Reverend John Pierpont:

> *His pocket-knife to the young whittler brings*
> *A growing knowledge of material things.*[31]

Not everyone saw youthful Americans and their prowess with knives as models of wholesomeness, however. In 1874, a letter to the editor of the *New York Times* from "A Bachelor, who has been looking for fifteen years for a wife," lamented the difficulties of living in the city on a limited income. In response, a resident of the Fifth Avenue Hotel asked readers to consider the wisdom of investing in an apartment house to hold sixty bachelors. Among the worries about such an enterprise was this: "Isn't there some probability of their cutting monograms and eagles all over the wood-work, and carrying mantel pieces away by piecemeal in the shape of toothpicks and cigar lighters?"[32]

Such fears may have been fed by stories of soldiers running roughshod over private property. In one, John Bull, the personification of the Englishman, walked into an American house "and having eaten the dinner which he found ready cooked, split up the mahogany tables for tooth picks, and then set fire to the house."[33] A form of revenge would later be found in the hardwood toothpicks that would be manufactured from American logs on American machines—the product of American ingenuity—and eventually exported around the world. But first the machines had to be invented and developed, the appropriate wood had to be identified, and the right innovator had to bring them together into a money-making enterprise. In addition to the technological and entrepreneurial obstacles to be overcome, there were also the cultural.

The idea of paying even pennies for a supply of ready-made wooden toothpicks had been foreign to New England sensibilities, as suggested in an 1849 news item, of which variations appeared several times in Boston and other Massachusetts newspapers:

> *Novel Importation.*—A vessel lately arrived at this port from Rio Janeiro, having as part of her cargo, 55,000 *wooden tooth-picks.* Are Yankees done whittling?[34]

No, they were not. The transition from a handcrafted item to a machine-made substitute does not occur overnight. As late as 1874, the Yankee whittling boy was being held up as a paradigm of individual industry. One moralistic tale embodied him in a shy lad who wished to keep a store "to make some money for his mother." He dreamed of filling the store with things he would make: "I can whittle better than I can do anything else. I can whittle out of pine-wood a donkey and a sheep, and my knife makes the best kind of whistles. I can make wooden toothpicks too." He sold the toothpicks for five cents a bunch.[35]

In 1853, a correspondent for the *New York Times* described visiting a crowded Mechanics' Fair, which was being held in Faneuil and Quincy halls and was "presenting a favorable specimen of every man's manufactures in Boston and the neighborhood." The reporter admitted that it would be "rather impertinent" to "talk of small matters to people who have got a Crystal Palace," as New York did at the time, but he found the Boston exhibition "very rich and cheerful," giving many visitors "more gratification than they would obtain in a walk through the Louvre or the Vatican." Though he may have been playing a bit to the presumed greater sophistication

of his New York readers, the correspondent did go on to describe
what was on exhibit:

> Here are New-England notions, contrivances, fabrications of all
> sorts and sizes to suit the tastes or necessities of the million—
> from tooth-picks, rosemary for the hair, and cases of delicate
> bonnets, which Aphrodite herself would be delighted to wear,
> to an organ—
>
> > *Like a golden gate of Heaven*
> > *On its hinges, angel-driven;*
>
> and a huge buzzing and jarring machine for planing timber
> blocks.[36]

Toothpicks were being shown in good company, indeed, and the
context suggests that they were being offered as products of commerce. As such, they may have caught the eye of the New York
reporter more than they did the local fair-goers, for New York
appears to have developed into a market for commercially produced
toothpicks long before Boston did.

In 1860, an informative report appeared in the *American Medical Times*, which was published in New York. It not only traced
Yankee prowess with a whittling knife back to shaping toothpicks
but also told of a new source of the indispensable implements:

> Where do the toothpicks come from? It is supposed that the
> Yankee, when he first felt the necessity of cutting a stick of tim-
> ber in order to provide himself with a toothpick, gained the
> knowledge of whittling, and has since kept and improved upon
> the lesson. A New Englander will produce a toothpick with his
> knife from almost everything except a bar of iron; but with all
> his inventive genius it has remained for the natives of Chili to
> supply this toothpicking nation with a large proportion of the
> instruments for gratifying their habit or necessity. The aged
> and decrepit and the young of both sexes of Chili are engaged in
> preparing those little orange sticks. . . . These they whittle out
> with astonishing rapidity, at the rate of five or six hundred in an
> hour. The sticks are then packed in bundles of a thousand each,

and sent to this city; being imported expressly by a lady in Division street, whose son superintends their manufacture in Chili. Here the toothpicks are sold for twenty cents a thousand.[37]

Is it possible that the writer, in addition to misspelling the name of Chile, confused it with another South American country, Brazil? Or was he deliberately misinformed by the importer, who wished to conceal the true source? Chile is, of course, on the Pacific side of South America, and importing anything from there to New York or New England would have taken a much greater effort than did importing a similar item from Atlantic ports. However, by this time a railroad had been established across the fifty-mile-wide Isthmus of Panama, thus enabling an expedited transshipment of goods from ocean to ocean. It may have been that whoever was bringing toothpicks from Brazil into Boston had secured exclusive rights to do so between the countries, and as a result, a New York importer would have been forced to find a less convenient source. In any case, it seems clear that the importation of hand-crafted toothpicks may already have been established around the time that Charles Forster began to think about mechanizing their production—and underselling even the natives.

Well before he left Brazil for good, Forster appears to have had plans to make wooden toothpicks in America—and to have been talking to at least one other person about his hopes. The time was propitious, since it was a period when Yankees were devising machines to mass-produce everything from simple straight pins to intricate tapestries. In Boston, Forster would become associated with Benjamin Sturtevant, an inventor six years his junior, whose story deserves a chapter of its own. With the benefit of Sturtevant's inventiveness and generosity of spirit—not to mention his dire financial need—Forster would soon "embark on a venture that would socially and culturally change the habits of a country."[38]

CHAPTER EIGHT

From Pegs to Riches

B ENJAMIN FRANKLIN STURTEVANT may have been destined by his given names to be a great inventor, but that was about all the help his parents could bequeath him to launch his career. He was born into poverty in 1833 at Martin's Stream, in Norridgewock, Maine. The family's circumstances and his father's ill health required young Benjamin to work on a farm to support the family, a situation that did not allow him to acquire much formal education. In fact, he attended "a district school only a few weeks one winter."[1]

At six years of age, like many a young farm boy in the area, he spent evenings cutting out wooden shoe pegs by hand for the local cobbler. At fifteen, Sturtevant began working in the shoemaking trade in Northbridge, Massachusetts, and later worked at it in Skowhegan, Maine. Although he did not serve a formal apprenticeship, he developed considerable skill as a shoemaker, the trade he practiced until 1856, when he sought to make a better life for himself and his family, for he had married in the meantime. His many years hunched over a cobbler's bench had evidently given him plenty of time to think about how shoe pegs might be made and driven with less expenditure of human effort, and he was determined to devise a machine that could produce boots and shoes just as good as those crafted by hand, but could make them faster.[2]

As with improvements in technology generally, the idea of shoe pegging was rooted in attempts to address the shortcomings and outright failings of preexisting technology. In the early nineteenth century, shoes were made by hand and the methods used to fasten the bottom of the shoe, including the sole and heel, to the upper shoe body could be the weak link in the design. Hand sewing was time-consuming and imperfect. The wax that was used on the thread to reduce friction between it and the leather clogged the eye of the needle; dampness could rot the stitching thread, causing the parts to separate. Pointed wooden pegs, designed to be driven like nails into holes pre-formed with an awl in the shoe parts, thus provided an attractive alternative. Wood pegs could give shoes a longer life; in fact, the pegs swelled when damp and so could hold the leather with even more tenacity.[3]

The pegged shoe, which was worn generally by the "laboring classes," was developed by a Massachusetts mechanic and inventor named Paul Pillsbury, who had been approached in 1810 by a shoemaker who wished to have a machine for attaching heels.[4] Pillsbury saw an opportunity to develop also a machine for attaching soles to uppers by means of pegs. Since they were typically spaced at about six to eight per inch (or about the width of a peg apart), it took scores to finish a single shoe. Pillsbury saw another opportunity in supplying the pegs themselves, and so he established a peg mill in 1815. He became so successful in making and promoting the use of pegs and the pegged shoe that he was known throughout Massachusetts as "Peg Pillsbury."[5] By the late 1860s, pegs were used in almost 90 percent of all boots and shoes made in America.[6]

It was a much more efficient operation to employ pegs instead of stitching to attach shoe bottoms to uppers, and "one pegger could do the work of two or three stitchers" in a shoemaking operation.[7] According to one recollection, "Shoe pegging proved to be a fast technique and fast workers could finish four pairs of 'pegged shoes' a day using wooden pegs made cheaply by machine."[8] Benjamin Sturtevant came to believe that "a man or smart boy" using a machine also to drive the pegs could be much more productive than that. He would estimate that a very fast pegger could drive as many as thirty pegs per minute, thus making in a ten-hour day from twenty-seven to fifty-seven pairs of shoes, depending on whether they contained a double or single row of pegs.[9] Calvin T. Sampson, a manufacturer of "women's, misses' and children's lace-boots" in North Adams, Massachusetts, claimed that his operators of pegging machines could on average peg almost five hundred pairs of the smaller shoes a day.[10] This is consistent with the claim that a "pegging machine could peg a whole shoe in 10 seconds." Since the cost of pegging was about a fifth that of sewing, there was clearly much to be gained by making efficient pegging machines and the pegs they would drive.[11]

Patents for attaching shoe soles by means of pegs were issued as early as 1812. Many subsequent inventions, some patented as early as 1825, concentrated on speeding up the process of making the pegs themselves. Pillsbury's method " 'plowed and cross-plowed' a piece of maple to form the points, split it into combs and then into pegs."[12] At midcentury, the most advanced techniques for cutting pegs seem to have worked on a principle not unlike Pillsbury's. A "saw-toothed peg-card" was separated from a grooved plank of wood as thick as the pegs were to be long, and the card was "*split* in the machine" into individual pegs.[13] As early as 1847, steam power was driving such operations.[14]

Other manufactured items also depended upon the subdividing of wood into even more slender pieces, often referred to as splints. Among the most important products made from splints were cigar lighters and wooden matches of several kinds, including those with fanciful names such as "lucifers" and "loco-focos."[15] At the time that Sturtevant was developing machines to make shoe pegs, there already had been numerous patents issued for machines for forming the body of wooden matches and other splint products. Some might even be said to have suggested the solution Sturtevant was to take. Among these was a patent issued in 1840 to Norman Winans and Thaddeus Hyatt of New York, who employed a method that formed splints by passing wood veneer though opposed grooved rollers. The inventors argued that the attendant compression of the wood even lengthened the time that the matches made from it would burn.[16]

Most early machines for making match splints and shoe pegs continued to form them by splitting the wood along the grain, leaving them irregularly shaped. Such pegs tended either to jam the machine or to ruin the shoes being worked on. Newly developed pegging machines used prepared strips of wood (referred to as pegwood or peg blanks) that were split into individual pegs by the action of the machine itself. However, the machines used to prepare the pegwood tended to produce strips that varied in thickness and contained cracks, which consequently yielded imperfect pegs that could damage the shoes into which they were driven. According to the manufacturer Calvin Sampson, "If a lot of shoes are injured, in appearance even, by the use of a poor quality of pegwood, it depreciates their value in the market all the way from five to fifteen cents a pair; and if the pegs are so poor as to materially injure the shoes, they become altogether unsalable; so that the profit of the manufacturer is largely affected."[17] It would be Benjamin Sturtevant's signal accomplishment to develop not only an improved pegging machine but also the pegwood to feed it.

Before he began these tasks, Sturtevant is said to have had no knowledge of mechanics and no experience with making machinery, but he may well have been aware of some of the machines and patents that did exist to make shoe pegs and related articles, and he did have determination. After a month or so of effort, he produced a model of a machine that could take logs and make them into shoe pegs, not by splitting but by cutting across the grain, thereby producing a more uniform product. Although Sturtevant's machine had the bugs to be expected in a new device, it also held the promise of success. With barely enough money to get him and his model to Boston (he arrived there at age twenty-three with only twenty cents to his name), he set out to make his fortune.[18] Those who lived and worked closely with him during his early years in Boston recalled that Sturtevant's "pecuniary means were extremely limited" at the time and that no "inventor ever worked harder or under more discouraging circumstances than he."[19]

Starting in 1857, Sturtevant had begun securing patents for pegging machines and for preparing wood blanks for shoe pegs for such machines.[20] Existing machines could be worked only as fast as the pegwood could be fed into them, and they stopped whenever a knot or other imperfection was encountered. To improve the length and quality of the pegwood blank, and thereby the number of good pegs that could be realized from a single loading of the magazine, Sturtevant devised a new method. First, he sawed the raw lumber into planks of roughly two-by-two-inch cross section and cut out any bad sections. The pieces of clear wood were then glued together to form a plank of defect-free wood, and the plank was sawed lengthwise into slices only as thick as the pegs had to be long, which was no more than three-quarters of an inch. From these were cut pegwood blanks that were then assembled into groups held together with glued paper, not unlike the way a line of metal staples is held together with adhesive today. The assembly could then be loaded into the pegging machine to provide it with a relatively long-lasting supply of pegs.[21]

But protecting intellectual property that resulted from all his hard work would come at a steep price. In exchange for a guaranteed weekly stipend of twelve dollars, which had enabled him to file for patents and continue his development work, Sturtevant had assigned half the rights to his inventions to Elmer Townsend—variously described as a Boston auctioneer, a wholesale shoe dealer, and a manufacturer of wax-thread sewing machines—and also relinquished to him complete control over the remaining half.[22] Townsend was already familiar with the area of Sturtevant's patents, and he was also in the business of acquiring patent rights to pegging machines.

A few years earlier, Townsend had been assigned rights to a handheld pegging machine invented by George Wardwell of Andover, Maine.[23] He was also approached by another inventor, John Greenough, of New York, who was granted the first American patent for a sewing machine.[24] Greenough brought not a request for a stipend but the threat of a lawsuit. In 1854, Greenough had patented a "machine for pegging boots and shoes," which he claimed was being infringed upon by the Sturtevant machine.[25] A lawsuit did result, which Sturtevant ultimately lost on appeal, and Townsend withdrew his support.[26] Townsend went on to secure his own patents for pegging machines, which he naturally hoped would not be challenged in court, and he retained all rights.[27] Townsend became the "biggest pegging machine manufacturer," and was also the "largest waxed-thread producer," thus having a large hand in both means of shoemaking.[28]

There were, however, inventions that Sturtevant had not shared with Townsend, and these addressed the fact that the speed of the pegging machine continued to be limited by how fast and reliably the pegs could be fed into it. Hence the problem was reduced to one of preparing and feeding the pegs themselves. To address the preparation problem, Sturtevant devised a lathe attachment for cutting veneers, which enabled the use of a continuous strip of uniformly sized wood to make shoe pegs.[29] Patents for his inventions were secured in 1859, and Sturtevant retained all rights. The use of veneers had been proposed by other inventors, but the method of forming them had led to frequent splitting of the wood, reducing the strength of the veneer for applications. Sturtevant's analytical approach to the problem was evident in his understanding of the situation:

I think it proper to divide veneers into two classes,—thin veneers used for the outside finish of furniture and cabinet work, and thick veneers, commercially termed back stuff and box stuff. The thin veneers, being glued on to other woods, are usually sawed or cut from twenty-seven to thirty-two thicknesses to the inch; that is to say, from twenty-seven to thirty-two veneers, when piled up would make an inch; but, if these veneers are sawed from the log, it takes nearly two inches of wood to make one inch of veneers. As these veneers are generally sawed from most expensive kinds of wood, and oftentimes worth from three to ten cents a square foot, the value of the wood cut away by the saw is a very considerable item.

This led to the invention of machinery for shaving them off with a sharp knife, in which case there would be very little or no waste of the wood; and, as the veneers were glued on, the crooked, warped, and shattered condition of the veneer was, and is now, of but little consequence with many furniture manufacturers. Whereas, in the manufacture of thick veneers, such as are nailed up into cigar boxes, and fastened to the backs of looking-glass and picture frames, it is an object to have the veneers straight, both lengthwise and crosswise, and of uniform soundness, which would not be the case with them, if they were cut off from the log with a knife. Even if they were cut with my patented machine, and retained nearly all the original soundness of the wood, they would be more or less inclined to warp, because the wood has to be cut green; or, if seasoned timber is cut, it has to be boiled in hot water or steamed, and the veneers afterwards dried in the air, which process has a tendency to twist and curl them out of shape.[30]

The veneer that comprised what were variously termed blanks, strips, or ribbons from which individual shoe pegs could be cut fell into the thick category, and Sturtevant solved the splitting problem by introducing the use of a "presser bar" that exerted considerable pressure near where the cutting knife acted on the log being veneered.[31] (His was claimed to be the first device to "successfully cut veneers from round the logs" as opposed to slicing them off flat.)[32] He also made improvements in the use of peg strips themselves, being credited with "a method of compressing the peg strip

The proper placement and adjustment of the knife blade and presser bar on Benjamin Sturtevant's lathe attachment were crucial for cutting a continuous uncracked ribbon of wood veneer from a rotating log.

between hot rollers, so that the moisture was withdrawn from the wood and the peg reduced in size. When driven into the sole it absorbed the moisture of the leather and expanded, making a secure fastening."[33] In his first veneer patent, Sturtevant made it clear that he was concentrating on providing the pegs for machines rather than on the machines themselves. This distinction freed him from Townsend's control and would enable Sturtevant to go on to form a successful niche business.

Sturtevant "conceived the idea of making long strips or ribbons of wood" in the spring of 1859 and began developing his machines for doing so that same year.[34] The long coils of pegwood ribbon could be fed into a shoe-pegging machine the way flexible clips of cartridges might be fed into a machine gun. The scheme might also be likened to loading a stapler or nail driver with an eight- to ten-foot-long supply of flexibly connected fasteners and having the machine cut off and drive a single fastener with each percussion. Sturtevant's invention of the "first practical pegging machine" and the pegwood ribbon that fed it revolutionized the business of making pegged shoes.[35]

In the summer of 1859, Benjamin Davis, who had previously worked sawing peg strips for Elmer Townsend, began operating Sturtevant's first "machine for cutting ribbon for blank pegs."[36] Sturtevant set up his business on Sudbury Street in Boston and soon had seven or eight men in his employ. He got his raw material—probably birch, maple, or poplar—from Maine and New Hampshire, but it was costly to transport it to Boston. So, after a few years operating in that city, in 1864 he rented a mill in Livermore Falls, Maine, and relocated much of the peg-blank-making machinery up there. Not only wood but also labor expenses were "less when conducting business in the country." After about three years, the wood supply was becoming dear in the vicinity of Livermore Falls, and so the machinery was again moved—to a new mill that Sturtevant had built in Wilton, Maine. Three years later, in 1870, he added another factory, in Oxford, Maine, but after only a couple of years the dwindling wood supply in that vicinity led him to discontinue operations there. Even when sufficient quantities of trees were standing in an area, getting wood to the mill could be difficult. This happened in 1873, when extremely heavy winter snows and the subsequent spring melt and mud cut off the wood supply to the Wilton mill. The situation forced Sturtevant to buy a mill in Bethel, Maine, and to outfit it to make peg blanks in sufficient time to fulfill orders.[37] Generally speaking, it was more cost-effective to relocate an entire factory than to haul wood over great distances.[38]

Sturtevant was not the only manufacturer of shoe pegs. According to one report, there were "numerous factories in the Eastern States turning out from fifty to one hundred bushels" of pegs daily. The cumulative output of shoe pegs was so great that one contemporary observer speculated that even "if all the people in the world were shoemakers, they must be overstocked with pegs." Still the demand did not diminish, leading to the generalization that "anything in universal demand even if individually the demand is small, must foot up large in the aggregate for the civilized world."[39]

Sturtevant certainly knew this, and it was not only in Maine that he had located pegwood factories. As early as 1865 he had rented a building in North Sandwich, New Hampshire, and turned it into a peg mill. Unfortunately, one of the very real risks of running a woodworking factory anywhere is the danger of fire, and the North Sandwich mill burned down in 1870. Sturtevant also built a large mill at Conway, New Hampshire, which became one of his principal plants.[40] Multiple manufacturing sites and pyrotechnic risks would also come to characterize the toothpick industry.

Even with the demands of a budding business, Sturtevant continued to think of improvements for preparing shoe pegs for use in machines. Like most inveterate inventors, he was often occupied with a wide variety of new ideas virtually simultaneously. In an 1862 patent, for example, he described an alternative to the veneer scheme whereby individual shoe pegs were to be fastened to a strip of paper that could be wound into a spiral form and, like a ribbon coming off a spool, fed continuously into a pegging machine.[41] During the Civil War, he would be engaged in inventing improvements in fuses for explosive shells and projectiles for rifled ordnance.[42] It is not clear whether Sturtevant's forays into weapons work were driven by a deeply felt commitment to the Union cause or by more mercenary motives.

By his own account, it was because "expenses considerably exceeded the receipts during the years 1859 and 1860; and being wholly without pecuniary means" at the outset of his enterprise, that Sturtevant had found it necessary to part with some of his patent rights in order to go forward with his nascent shoe-peg business. Thus, for an initial loan of just three hundred dollars, he had given Francis Brigham the exclusive right to use his peg blank patent in the town of Marlborough, Massachusetts. Furthermore, at least initially, the peg blanks were to be sold to Brigham at the very low price of two and a half cents per roll, a price that would later be negotiated upward.[43] Sturtevant had made a similar deal with the firm of A. & A. B. Keith, granting it exclusive rights under the patent within the town of Raynham, Massachusetts, and also in the county of Plymouth. Thus, "he parted, for a loan of only seven hundred dollars, with the right to all profits arising from the sale of all peg-wood in a single large shoe town and county." The money loaned by the Keiths on the first day of December 1859 enabled Sturtevant to pay the application fee and other expenses associated with prosecuting his patent for the lathe attachment, which was granted less than four weeks later.[44]

Early in 1860, he sold to Harrison Parker and Jonathan C. Sleeper for the sum of six thousand dollars (to be paid over two years) the exclusive license to use the machinery covered by the new patent for producing veneers from a large variety of expensive and exotic woods, including "rosewood, mahogany, black walnut, zebra wood, birds-eye maple, ebony, satin wood, cocoa wood, Spanish cedar, holly, butternut, horse-chestnut, cedar, apple, pear, hickory, cherry, chestnut, figured white wood, and the woods known among mahogany dealers as curly maple, figured oak, figured ash and figured birch; and also all woods not grown in North America." He also gave Parker and Sleeper the right to cut "from birch and maple woods such veneers as are commonly used for cabinet and furniture wood."[45] Sturtevant was confident that the business of manufacturing peg blanks would be sufficiently large so that he could afford to give up such wide-ranging rights, as long as he kept the shoe-peg application as his own.

Benjamin Franklin Sturtevant, who had struggled financially in the late 1850s and early 1860s, when he was inventing and patenting machines and methods for making shoe pegs and toothpicks, prospered later in life by manufacturing industrial fans and blowers.

He was correct in his expectation that the shoe-peg business would grow. Soon Sturtevant had to confront the time-consuming necessity of keeping accurate account books, so it is understandable that he thus required the services of a bookkeeper. He found one in Caleb King, who held the position from the late spring of 1862 to the spring of 1867, at which time he appears to have become a salesman for Sturtevant.[46] King was succeeded as bookkeeper by Edward W. Simmons. Both of them were assisted by clerks or other bookkeepers, one of whom may have been Charles Forster, though the relationship between Forster and Sturtevant would eventually become much more complicated than one of basic accounting.[47]

In the course of developing his pegging machine, Sturtevant had also devised a means to remove the wood dust produced in forming the pegwood, and this led to his development of pressure blowers and exhaust fans. He began the manufacture of blowers in early 1863.[48] That activity would eventually become the mainstay of the very successful B. F. Sturtevant Co., whose products would in time include such large industrial devices as those used in the ventilation system of New York's Holland Tunnel. This first significant underwater vehicular tunnel would not have been safe without a proper means for removing the voluminous exhaust gases resulting from heavy motor traffic. The Sturtevant Co. provided the expertise to design the proper equipment.[49] But the commingling of manufacturing resources for shoe-peg blanks and blowers in the early stages of the business had confused and complicated the way profits were attributed to the separate products.

When Sturtevant began to switch his focus from shoe pegs to blowers, they were used industrially principally to provide blasts of air for furnaces and forges; at the time the use of exhausting fans was virtually unknown.[50] His first patents for improvements in fan blowers, which can be thought of as a kind of air pump, were issued in 1869, but even before then his product had developed a sound reputation.[51] At that year's Fair of the American Institute, "only a few" blowers were exhibited, and one report sandwiched a mention of Sturtevant's between those of two other inventors. However, his product was described as "a good blower, running without great noise and performing good work." No details were given because they were considered unnecessary, the device having "been before the public so long" and having been "so favorably known."[52] Sturtevant was to live another two decades after that exhibition, but according to a reporter writing without much elaboration or apparent grasp of the facts about "trifling suggestions that have won riches for lucky thinkers," he "went crazy later on."[53] There is no other indication that he did. His contributions to technology were eminently sane, which is the way he would be remembered.

In 1872, Sturtevant had advertised himself and his eponymous company as "patentee and sole manufacturer of pressure blowers & exhaust fans."[54] He wisely had kept all the rights to himself. Sturtevant's name became inextricably associated with the blowers and fans the company produced in Jamaica Plain, which is now part of Boston. The works he established there in 1878 became, by the time of his death, "the largest of the kind in the world and one of the most extensive industries in New England." But in addition to the fan company, he also continued to remain known for his earlier work, sometimes even being credited with the very invention of the shoe peg itself, and the idea was said to have "brought him millions of dollars."[55] Whether or not making shoe pegs was that lucrative, the profit from it must certainly have been enough to enable Sturtevant to get into the fan and blower business, which in the early twenty-first century would still be thriving as the Sturtevant Division of Westinghouse Electric Co.[56] But that was a long way off in the 1870s, when Benjamin Sturtevant was fighting to retain his monopoly in the shoe peg business.

Though the lenders of the money that enabled Sturtevant to have the funds necessary to patent his lathe attachment themselves eventually received payments of about sixty-four thousand dollars, they would not be satisfied with their return and they sued for more. According to George L. Roberts, Sturtevant's counsel, Brigham sued to recover some twenty thousand dollars, which he calculated was the amount he was overcharged for pegwood that he had expected to procure at two and a half cents per roll. Arza Keith threatened a similar suit to recover a like sum.[57] With legal fees and other expenses, the final cost of the seven-hundred-dollar loan could have been of the order of a hundred thousand dollars. It was in the midst of such significant, complicated, and distracting money matters that Sturtevant sought an extension of his patent on the veneer-cutting lathe attachment. His patent for the pegwood blank itself had already been extended, and so he must have been optimistic that he could achieve a similar outcome for the machinery that made the blank.

In 1873, Sturtevant filed for the extension of the patent on the lathe attachment, asking for another seven years of protection beyond its scheduled expiration near the end of the year. He based his case on the argument that his device was novel at the time of its invention and that it remained "useful, valuable and important to the public"; that he had used "due diligence" in exercising his rights and privileges under the patent; and that he had yet to receive a "reasonable remuneration for the time, ingenuity and expense" he had contributed to its success.[58]

The granting of an extension was opposed by Brigham, Arza Keith, and Theodore H. Videto, who was Keith's bookkeeper and who had acquired an interest in the rights assigned to A. & A. B. Keith. They argued that Sturtevant's invention was not novel, naming American and English patents from the 1840s that they said had its essential features; they also claimed that the inventor had been adequately rewarded for his efforts. Testimony in support of the opponents delved into the minutiae of how Sturtevant's presser bar was designed and functioned and also into the accuracy, completeness, and fairness of the account books. Among the questions asked of the bookkeepers was how Sturtevant's employees divided their time between various aspects of the veneer-cutting and blower-making sides of the business, something that was not at all clear from the books themselves.

According to the summary of book accounts presented as part of his application for an extension of his patent on the lathe attachment, almost five million rolls of pegwood were sold between 1859 and 1873, at prices ranging from two and a half to twelve and a half cents per roll. Over a half million rolls were sold in each of the districts controlled by Keith and by Brigham, at prices never exceeding seven and a half cents. The cash receipts to Sturtevant over the fourteen-year period were just over a half million dollars. However, after deducting expenses, apportioning half of the profits to his earlier patent for the shoe-peg strip itself and half of what was left to the machinery used in making the pegwood rolls, Sturtevant calculated that he realized a reward of less than sixty thousand dollars total for his essential lathe attachment patent.[59]

In fact, after his initial statement was submitted to the Patent Office, it was brought to his attention that during the accounting period under consideration there was indeed some additional income attributable to the lathe attachment patent. This included receipts for machinery sold for the purpose of making pegwood blanks in Germany, as well as machinery for making blacking boxes. There was also some income from the sale of veneers made for purposes other than shoe-peg blanks. These sales included veneers produced from woods not controlled by Parker and Sleeper. Other separate items included shanking, the "thin sheets of wood inserted between the outer and inner soles, in the shank part of the shoe," which had been manufactured since 1862.[60] The total receipts for all additional products was less than eighteen thousand dollars, and overall profit just under eleven thousand. Sturtevant added this amount to his previously calculated profit, which brought his reward to just over seventy thousand dollars over the life of the patent. This still paled in comparison to the benefit that society realized from his invention. He calculated that, from the operation of pegging machines that used his product, shoe manufacturers and the public had profited to the extent of no less than three million and as much as ten million dollars.[61]

Shoe manufacturers submitted affidavits in praise of Sturtevant's ribbon pegwood. Calvin Sampson swore that the quality of the product made by Sturtevant was so good that he would "regard it as a serious disadvantage to the boot and shoe trade, and to the public purchasing pegged boots and shoes for wear, if the manufacture of this ribbon pegwood was thrown open to others, not having the experience nor exercising the care and skill of Mr. Sturtevant, and allowed to become deteriorated by competition." Sampson estimated that his own business alone had realized savings of almost ten million dollars due to the use of the reliable Sturtevant pegwood. Alfred H. Batcheller, a North Brookfield, Massachusetts, manufacturer of boots and shoes, believed that Sturtevant's product "contributed more than any other invention to the practical success of the pegging machine" and that the profit Sturtevant realized from his invention was "trifling in comparison with its real value and importance to the public, and but a small portion of the actual amount of money saved by it to the manufacturers of machine-pegged boots and shoes." Batcheller praised Sturtevant's "first-class article, so free from defects as to bear evidence of great care in the selection of the wood, and skill in its preparation, all of which is of the utmost importance to the manufacturer of pegged boots and shoes, the finish and stability of which are intimately dependent upon the quality and structure of the pegwood." In Batcheller's mind, it was clearly advantageous to both manufacturers and the public that Sturtevant retain exclusive patent rights to "prevent the production of inferior articles by injurious competition." In the end, the acting commissioner of patents ruled for Sturtevant and granted him the requested extension.[62]

Sturtevant, impoverished of money and education as a youth, became a philanthropist in his mature years, eventually donating a quarter of a million dollars to worthy causes. His money made possible Sturtevant Hall at the Newton Theological Institution, and among his other charities was the Home for Little Wanderers. His religious affiliation was Baptist, and his political one Republican. During the 1888 presidential campaign he affiliated with the Prohibition party, and the next year he ran for lieutenant governor of Massachusetts under its banner. He died the following year, of a relapse of a "slight stroke of apoplexy," from which he had appeared to have rallied. His obituary naturally credited him with the development of a shoe-pegging machine, ribbon pegwood, and blowers and exhaust fans, as well as the "first machine to manufacture wooden toothpicks," which was described as one of his "minor inventions."[63]

In fact, Sturtevant was also one of the first wooden toothpick manufacturers in America. Among the non–shoe-peg-veneer activity that his amended accounting revealed was the manufacture and sale of toothpicks in 1865 and 1866.[64] That he took up the work reluctantly and that he abandoned it shortly thereafter suggests that he did not see wooden toothpick manufacturing as having a business benefit equal to its risk, or promise a profit worth the effort. Still, his minor invention would be the foundation for an industry that might properly be counted among his legacies. Why Sturtevant engaged in toothpick making at all has to do with the persuasiveness and perseverance of Charles Forster.

CHAPTER NINE

A Family Affair

A MONG THE THINGS about Brazil that had most impressed
Charles Forster were the beautiful teeth of the natives. These
he attributed to the "whittled slivers of wood" that boys made and
sold in the streets.[1] He also became acquainted with the "large
hand-made toothpick whittled with a jack-knife in Portugal from a
Spanish Willow. These toothpicks were about five inches in length,
about one-eighth of an inch in diameter, and packed twenty picks in
a box, sold in that country for the equivalent of fifteen cents of
American money."[2] They were "boxed and sold cheaply as disposables." Finding the price right and the concept attractive, Forster is
said to have purchased some and sent them back home. Though one
version has the recipient being "his wife in Boston, who offered
them round to her guests," another has it that mistakenly "a hotel
man received the package, and he ordered a box for himself."[3] Neither of these stories appears to be exactly true.

It is unclear how Charles Forster first became acquainted with
Benjamin Sturtevant, but it may have been that Forster deliberately
sought out Sturtevant to learn more about his shoe-peg machinery
and how it might be adapted to mass-produce wooden toothpicks to
compete with those made by hand in South America. The operation
of the veneer-cutting machine, which took a log and turned it into a
ribbon of wood, and the peg-pointing machine, which beveled
opposing sides of one edge to form the point of each shoe peg that
would be cut from the ribbon, would likely have fascinated Forster,
who would have seen not shoe pegs but toothpicks being produced
by the same apparatus. All that needed to be done, he must have
thought, was to make the pegs longer and point them at both ends.
If a machine could produce as many toothpicks in a minute as a man
could whittle in a day, they could be sold competitively. He even
envisioned exporting machine-made toothpicks to South America

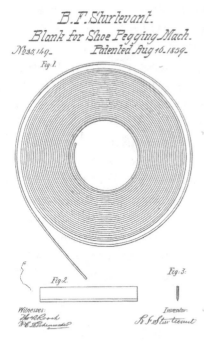

B. F. Sturtevant.
Blank for Shoe Pegging Mach.
No.25,149. Patented Aug 16, 1859.

When one edge of a coil of veneer
was beveled on opposing faces, the
coil provided a convenient blank
from which pointed shoe pegs could
be cut and driven by machine. This
blank was also suggestive of how
wooden toothpicks might be made.

to compete with the handmade ones, including those imported from Spain and Portugal.[4]

According to the hagiography surrounding Charles Forster, he tried to persuade Sturtevant to modify his machinery to accomplish the new task and to join him as a partner in the enterprise. Sturtevant, so the story goes, seeing no need or market for mass-produced toothpicks, ridiculed the idea of making and selling "slivers of wood."[5] But that was not likely the only reason Forster and Sturtevant did not form a partnership; in the late 1850s and early 1860s the aspiring manufacturer of shoe-peg blanks was looking not for untried new products to make but for money to file for protective patents and to finance his fledgling shoe-peg operation.

During that same time period, Forster was establishing a partnership of another kind. He was courting Charlotte Messer Bowman, of Somerville, Massachusetts, whom he would marry in Cambridge in October 1861. She seems never to have traveled to Pernambuco, either to visit her beau or later as his wife to live with him there, yet it does seem likely that she heard much about the place and its products from Charles. While he was still courting

Charlotte, he must have shared with her his impressions of how the Brazilians made and used toothpicks, including their practice of passing a container of them around the table after a meal, and confided in her that he saw in the mass production of the splinters a business opportunity of international potential. His enthusiasm was infectious, for even before they were married and while Charles was still away in Brazil, Charlotte made in her maiden name a deal with Benjamin Sturtevant that would ultimately give her and her future husband a monopoly for making toothpicks by machine.

In late March 1861, Sturtevant granted Charlotte M. Bowman in exchange for the sum of two hundred dollars exclusive rights to using his patented "improved lathe attachment for cutting veneers" for the sole purpose of "making strips of wood to be cut up into toothpicks."[6] According to the Patent Office's digests of assignments, the agreement was registered two years later with the following stipulation on what she was to get for her money:

> One machine under said Patent for cutting sheets or strips of wood from around the log not less than 2½ inches wide and not more than ⅟₁₈th of an inch in thickness for making toothpicks only, and any improvements thereon for same purpose.[7]

Not only had Sturtevant sold Bowman exclusive rights to making toothpicks, but he also sold her one of the machines that he had constructed to cut the veneer. In fact, it was only the third such machine that he had made, the first two being inadequate to the task. The first, which had been designed to accommodate a log of wood two feet long, was used for only two months, it proving not to be strong enough for the purpose. The second machine also "proved not to be perfectly adapted to the work." In the meantime, Sturtevant continued to make improvements in his veneer-cutting machines, and he also worked on "the pointing or chamfering machines," which he found "even more difficult to perfect" because of the "finer and more delicate" work they needed to perform in shaping the edge of the shoe-peg strip. The problem was that the bevels on either side of the strip "had to be of equal length, so as to bring the point of the peg exactly in the centre."[8]

The manufacture of veneer strips for making toothpicks was in a way simpler, in that only one side of an edge had to be beveled. At

same way, but having to tur
would not have been practic
ally be achieved by "changir
as was pointed out in *Kn*
Though *Knight's* was pub
might have had to wrestle w
cle strongly suggests that th
for some time.[13] As early as
a plant in Richmond, Virg
heated rolled iron by means
change the cutter angle as d
der on whose surface were i

B.F.
Manufactui
N? 38,768,

Fig 2

Fig 1

Witnesses
R. H. Eddy
G. P. Hall Jr

When both edges
but on one face o
machine that cut
picks in alternati

the same time, the operation was different, in that both edges had to be so cut. There was another difference between what Sturtevant wanted to accomplish and what the Bowman-Forster team wished to do. Sturtevant was basing his business on supplying strips of pointy-edged veneer to shoe manufacturers, who would feed the strips into their pegging machines, which would cut or slice off a single peg as it was needed. Bowman and Forster wished not only to make the double-edge-pointed veneer strips but also to cut them into toothpicks themselves. Whereas the peg strips and the pegs were only a means to an end (making boots and shoes), the toothpicks cut from the toothpick strips would be the end product.

If Sturtevant himself had not already been thinking about the similarities between making shoe pegs and manufacturing toothpicks when he and Forster first met, he soon took a significant step in declaring his own patent right to do so. Shortly after the two-year-old deal with Charlotte Bowman—now Forster—was recorded, Sturtevant applied for a patent for an improvement in the manufacture of a veneer strip for making wooden toothpicks, admitting in the specification that his toothpick blank was "in some particulars analogous to the shoe-peg blank" of his 1859 patent. Within a matter of weeks, the Patent Office turned down his application, stating that the shoe-peg invention was "considered as fully anticipating" the toothpick one. Merely "chamfering the other edge and widening the blank" to make toothpicks instead of shoe pegs was declared not patentable.[9]

Through his attorney, R. H. Eddy, Sturtevant requested a reconsideration of the decision. His argument rested on the fact that, unlike shoe-peg blanks, those intended for making toothpicks would be beveled at both edges. The change was stated to be "one of much importance and value as it enables the tooth picks to be made in a manner very different from which they have heretofore been made and to be supplied to the community at a very much cheaper rate."[10] Admitting that, if issued as first submitted, the toothpick patent could be interpreted to encompass that of the shoe-peg one, thus effectively extending its term of protection, Eddy enclosed an amendment from Sturtevant explicitly stating that in this new application he did not claim the shoe-peg invention but only the double-beveled toothpick one. He emphasized that the former was "insufficient for the purpose of making such tooth-

picks" as were described ir
tions, the patent was issue
succinct paragraph of the s

The nature of my invent
cut in a spiral from a log
at its opposite edges, th
chamfered band may be
a brad or sprig cutting r
such machine may be re

Thus, he in fact patentec
band of veneer that woulc
his agreement with Charl
new invention.

If Sturtevant had read
ful Arts, which was publisł
connection between the r
and other kinds of nails an
his veneer strip—and so g
dent of any prompting by
toothpicks. Tapered spri
because they resembled s
by shoemakers," among o
iron. Using shears tende
using a punch produced
ping the iron bar over afte

Clearly, single-pointe·

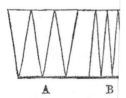

*The way nails were cut
toothpicks could be cut j
the nails designated A d
both used by shoemaker:*

The inventor Sturtevant would not have needed a bookkeeper's eye to see the manufacture of toothpicks as just another technical application of his veneer process, but someone with a bookkeeper's instincts had to apply a sharper pencil to show it to be valuable. That someone appears to have been Charles Forster. Less than a year after it was issued, Sturtevant's patent for an improved method for manufacturing toothpicks was reissued to him in a significantly strengthened form.

A patent could be reissued when the original was defective in some way, such as when it covered less than the inventor was entitled to claim. A reissued patent would supersede the first and protect the invention for the unexpired part of its term.[15] In the case of Sturtevant's toothpick patent, originally issued in 1863 and reissued in 1864, the protection would extend for seventeen years from the earlier date, or to June 1880. Whereas in the original he had claimed a new way of making the chamfered (beveled) band of veneer, from which toothpicks could incidentally be made, in the reissue he claimed a "new article of manufacture," namely, the machine-made wooden toothpick itself, thereby securing that much broader right.[16]

One of the witnesses for the reissued patent was "C. Forster, Jr.," a form of his name that the younger Charles Forster would cease to use after the death of his father in 1866. But regardless of how he identified himself as a witness, the text of the reissue patent that dealt with motivation for the invention suggests the hand or at least the experience of one who had lived in a country of hand-crafted picks, such as Brazil:

> Previous to my invention wooden tooth-picks, as articles of commerce and manufacture, were made abroad by hand-labor, and were imported into the United States. These were necessarily irregular in shape, and were expensive.
>
> My invention consists in a new article of manufacture—viz., machine-made toothpicks, when these are made from a band or strips of wood of uniform cross-section—the characteristic of the invention being that the form of the tooth-pick when viewed one way is that of the cross-section of the band or strip of wood from which it is made, while its form when viewed the other way is determined by the action of the mechanism by

which it is cut from said strip, each tooth-pick cut being a fac-simile of that preceedingly formed. Such tooth-picks differ from any made prior to my invention, in that they are perfectly uniform, each with the others, owing to which fact they can be packed in extremely small compass, and also in the fact that they can be produced very many per cent. cheaper than any heretofore made.[17]

The patent states that the invention "further consists in a band or strip of wood, cut in a spiral form from a log of wood, and having chamfers or bevels at its opposite edges," but it clearly was also the toothpick made from the band of wood that the patent as reissued had been amended to protect.

Sturtevant's reissued patent also makes clear that the invention had been reduced to practice, presumably by Sturtevant, Forster, or others in their employ, for according to the inventor of record, referring to the drawing sheet:

In cutting the picks, I generally introduce the strip A between two feed-rollers, which carry the end onto and projecting be-yond a horizontal bed, and so as to be presented to the action of a rotating knife or cutter projecting from a rotary disk plate. After such knife severs the pick from the strip, and before it or an opposite knife again reaches said strip, the feed rollers receive a rocking or vibratory movement, which swings the blank or strip around so that each successive cut is on an angle to the last cut, the effect being to give the tapering form of the picks. . . . This operation, as will be seen, utilizes all the mate-rial, each successive piece cut off forming a perfect pick and being a counterpart of the one previously cut.[18]

The process employed was thus a modification of that used by the nail-making machinery described in *Knight's Dictionary*. However, the repeated "rocking or vibratory movement" of the feed rollers would naturally take its toll on the equipment. In time, a less violent method of cutting out the toothpicks would have to be developed. This would be accomplished by mounting the rotating knives at alternating angles to the fixed orientation of the veneer strip being fed beneath them.

The fact that Forster witnessed the reissue patent places him in face-to-face proximity to Sturtevant in 1864, whether as an assignee, employee, or protégé—or all three. Whatever their relationship, Sturtevant had years before sold to Forster, albeit through his fiancée, the right to use the veneer-cutting process to make wooden toothpicks, which were "cut from none other than a widened peg-strip beveled on both edges." Whether as part of this deal or in exchange for bookkeeping or other services, Sturtevant, who at the time was concentrating on establishing a "ribbon peg-wood manufactory" in Maine, was being lured into the toothpick business.[19]

The delicate operation of getting the point just right on a shoe-peg strip might not have been as touchy when making a strip from which to cut out toothpicks, for now only one side of the strip on each edge had to be beveled. However, this would have been a problem that Charles Forster faced when he returned from Brazil and began to employ the Sturtevant machine in toothpick production. The basic machine only cut the veneers; it did not point them or form them into individual toothpicks. Forster may have tried to devise his own methods for doing these finishing operations, which may have involved some handwork. In any case, according to Sturtevant, although Forster was "at first successful, at length he met with reverses." These setbacks would likely have been due to the fact that the early-generation machine would not have continued to work as smoothly as when it was new. Forster, who was evidently no inventor or mechanic, could not have been expected to maintain the machine to the degree that it required or to fix it as its parts aged and broke.[20]

After a period of frustration, Forster approached Sturtevant with a proposal that he take back the machinery and operate it as part of his veneer-manufacturing business, giving a royalty to the Forsters for all the toothpicks he made and sold. Sturtevant agreed to the arrangement, even though "the machinery was in bad condition, needing extensive repairs." Thus, in April 1865 the Forsters' interests in the patent and machinery were reassigned back to Sturtevant.[21] In addition to the veneer-making machine, the inventory included others that beveled the edges of the veneer strip and still others that cut the beveled strip into individual toothpicks. At the expense of his own business, Sturtevant had his machine shop repair what machinery it could and make new equipment where necessary.[22]

After getting everything operating, Sturtevant's company produced toothpicks "for several months," beginning late in 1865 and continuing into 1866. According to the account books, the gross receipts for this activity was under twelve hundred dollars, some of which naturally went to the Forsters.[23] Sturtevant, who had found that "the expenses were nearly equal to the receipts," declared the toothpick business "unremunerative, and gave it up to Forster again." Thus, Forster was back on his own making toothpicks.[24] But if Sturtevant could not make money at it, then how could Forster? He would have to find ways to produce toothpicks for less and at the same time sell many more of them.

It may have been at this point that Sturtevant put some of his machines and space on Sudbury Street in Boston—and one of his most talented young employees, Charles Freeman—at Forster's disposal. The machinery was evidently still under the control of Sturtevant, to whom it had been assigned, but in 1869 he entered into an agreement with Charlotte Forster to assign back to her the machinery and rights to the toothpick patent and its reissue. The arrangement made it possible for Charles Forster to set up shop independent of Sturtevant in the Boston area with the intention of making slivers of wood on his own. This time, no doubt with the continuing help of Freeman, he evidently succeeded. According to a Forster family genealogy published in 1870, at that time Charles Forster was "largely engaged in the manufacture of toothpicks at Cambridge, Mass."[25]

Forster may have traveled back and forth between Boston and Pernambuco during the early 1860s, but he appears to have left Brazil for good to return to the Boston area no later than about 1865 or 1866, when he was approaching forty years of age. By that time, Charles and Charlotte Forster had also started a family. Their two daughters, Charlotte Bowman Forster and Annie Eaton Forster, had been born in Massachusetts in the late fall of 1863 and 1864, respectively, meaning that Forster had spent at least some of his time in the first half of the 1860s in New England with his wife. Their last child, a son named Maurice Webb Forster, would be born in 1873, in Maine, where the toothpick business had caused the family to relocate.[26]

CHAPTER TEN

Going Where the Wood Is

G IVEN OUR KNOWLEDGE of how toothpicks were made by hand
in Portugal in the late twentieth century, there is no reason to
doubt that a nineteenth-century South American native could produce as many as six hundred toothpicks in an hour.[1] This would
equate to a per capita output that could easily exceed five thousand
daily. According to his company's own lore, while he was in Brazil
Charles Forster had purchased a significant quantity of such handmade toothpicks: "75 cases at $50.00 per case which he brought
back to Boston as an experiment." Representing a $3,750 invest-
ment, this would have been an expensive experiment indeed, and
its outcome should not have been encouraging, for "it was only
with extreme difficulty that he could sell them, and even then, at a
sacrifice."[2]

Since the wooden toothpicks said to be imported from Chile
were at the time reportedly selling for twenty cents per thousand
in New York, we can expect that Forster should have had to pay
no more than about half that rate in Brazil.[3] Thus, each of his
experimental cases would have held at least five hundred thousand
toothpicks, and seventy-five cases would have amounted to over
thirty-seven million of the splendid splinters. These were the kinds
of numbers he would have had to be thinking of as he contemplated
the mechanization of toothpick making. And in spite of what might
be called a failed experiment at importation, Forster seemed determined to forge ahead with his plans for domestic production. If he
could figure out how to make toothpicks efficiently by machine and
how to convince people to buy them, he stood to reap a fortune. At
least that is what Forster the bookkeeper must have calculated, and
he would also have been able to calculate what kind of production
rate and economies of scale he had to achieve through the use of
machinery to do so.

As good a numbers man as he may have been, Forster was no mechanic. This was attested to by the condition of the machines that he turned back to Sturtevant. Forster may have been able to strategize, scheme, and set goals, but coaxing toothpicks out of shoe-peg and other machinery appears to have fallen ultimately on the shoulders of the hardware specialist Charles Freeman. He was still "a young man, but known as an ingenious mechanic" who became "a staunch and loyal friend." Freeman, "who believed in Mr. Forster, stood by him even when the business outlook was dark [and] did much to develop the methods of manufacture."[4] Evidently, whenever they were not fulfilling whatever obligations they may have had to Benjamin Sturtevant's shoe-peg operation, Forster and Freeman worked at developing the strategies and machinery needed for the new task. Though Freeman's essential role was often to be forgotten in later recollections, it was no doubt he who eventually did the successful mechanical design and heavy lifting. Still, Forster maintained the vision and the hope, even when "machine after machine was built at large expense, only to be broken with a sledge-hammer when it was found impossible to cut the picks as he had hoped."[5]

Although wooden toothpicks may have been made by machinery as early as 1860, that is unlikely to have been the case to any great extent before about 1865, when Sturtevant was briefly directly involved. At that time, they were reportedly being made by "patent machinery," but more likely it was 1867 or 1868 before a machine was developed that worked efficiently enough to produce a profitable item.[6] By December 1869, "a machine was finally built which was almost exactly like" machines that would continue to be used into the 1930s.[7]

But some time before Forster and Freeman had gotten their machine to work properly there was potential competition. In 1864, the inventor J. C. Brown, of Brooklyn, New York, received a patent for "an improvement in machines for cutting tooth-picks, match-splints, &c." His invention consisted of "a series of knives fixed into the surface of a cylinder and arranged in combination with a mechanical contrivance for revolving a block of wood, against which the said knives are pressed with sufficient force to cut the surface of the said block (revolving) into narrow splints, and combining therewith a cutter to shave the splints from the block." In

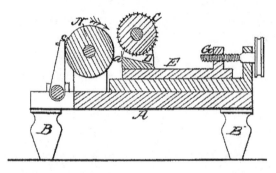

A machine patented in 1864 by J. C. Brown was capable of making toothpicks and a variety of similar wooden items known as splints. In this figure, the cylinder marked C was fitted with knives that when pressed against the log rotating in the direction indicated by the arrow made incisions that outlined tapered flat toothpicks. These were then to be shaved off the log by the blade at the tip of the wedge below cylinder C.

other words, rather than cut toothpicks from a coil of veneer, as the Sturtevant process did, Brown's first incised the outline of the splints directly on the rotating log, what he called a "round block," and subsequently shaved them off. *Scientific American* praised Brown's achievement, stating that "perhaps in all the arts there is no other machine that multiplies the product of labor to a greater extent than the simple little machine recently invented by [Brown]. A block of wood is placed in the machine and the splints pour out in a constant stream, or cataract, like the pouring of corn from a half bushel."[8]

Unfortunately for Brown, just a little over a year earlier Benjamin Sturtevant had been granted his patent for the machine-made toothpick itself. Since that patent claimed not only a process for making toothpicks but "a new article of manufacture, a machine-made tooth-pick," Brown could not legally apply his own machine to toothpick manufacturing.[9] Though it might have been easily argued that Sturtevant's claim covered only toothpicks made from a veneer cut first from the log, it is also easy to imagine that Sturtevant and Forster, who had witnessed the toothpick patent and who had acquired the rights to its protection, could have bullied Brown into restricting his machine to producing matchsticks and other kinds of splints. After all, Sturtevant had seen Elmer Townsend, his early backer of a pegging machine, drop all support at the threat of a lawsuit. At the same time, Sturtevant himself may have shied away from toothpick making because he feared Brown might sue him. In any case, Brown seems not to have pursued the manufacture of toothpicks the way Forster did. Indeed, Forster's fervor appears to have been so intense that he had been able to persuade Sturtevant to help him to the considerable degree that he did.

By 1870, it could be stated that the disposable wooden variety of toothpick had "to a considerable extent superseded the gold, horn, ivory," and other kinds of personal toothpick that had previously been the object of one-time purchase. And Forster held a virtual monopoly on production and sale of wooden toothpicks. He controlled the patent on the toothpick itself, and had successfully fended off challengers and potential competitors.[10] Eventually, he reportedly "expended $50,000 on his patents in litigation," but, by controlling the rights to make, and thereby distribute, toothpicks, all orders had to go through his hands. By 1869, the manufacture of wooden toothpicks using patent machinery was reported to have been going on for four years and was "principally, if not wholly, carried on at an establishment near Boston," which likely meant Cambridge, where Forster relocated the manufacture of toothpicks after Sturtevant ceased it.[11] In July 1870, when the decennial census was taken, Forster was residing across the river in Brighton.

Another potential challenge to Forster's advantage came in 1872, when Silas Noble and James P. Cooley—Granville, Massachusetts, drum manufacturers—were granted two patents for improvements in toothpick-making machines. Indeed, it is an often-repeated "fact," or at least the answer to a trivia question, that the first toothpick-manufacturing machine was patented that year by Noble and Cooley. Not only does this ignore the machine patented by Brown, and that invented by Sturtevant and operated by Forster and Freeman, but also it ignores the word "improvement" in the titles of their patents, which clearly signals that Noble and Cooley were not absolute pioneers.[12] That is not to say that their machines were not new and original. The double-pointed and beveled toothpicks that they produced might have been preferred to the ones made by

S. NOBLE & J. P. COOLEY.
Improvement in Tooth-Pick Machines.
No. 123,790.   Patented Feb. 20, 1872.

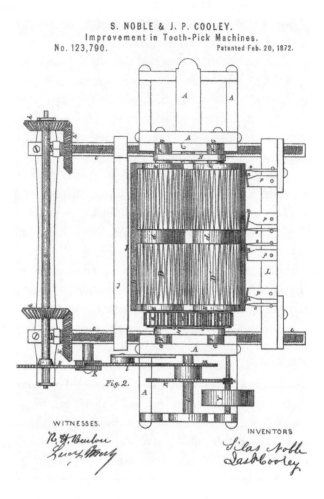

WITNESSES.

INVENTORS

In 1872, Silas Noble and James Cooley patented an improved toothpick machine. It operated in a manner similar to that of Brown's, but made flat toothpicks with a lenticular shape and beveled ends, which were produced by the auxiliary knives mounted on the bar L located on the right in this plan view.

Forster, but the Sturtevant patent rights that he controlled prevented the Noble and Cooley machines from being used to produce wooden toothpicks for sale. They could, however, use the machines to make "matches, lamp-lighters, and other similar articles from a ribbon of wood, which do not require subsequent pointing."[13]

In the course of developing the domestic toothpick industry,

Forster is said to have "whittled the original box of Yankee tooth-picks" to fulfill his first order. While he may have learned the skill as a Yankee lad or from South American natives, it is unlikely that even they would have been able to keep up with the demand that was to develop. In the first year of machine manufacture, Forster is said to have sold sixty-five cases of a quarter million toothpicks each, for a total of over sixteen million. By 1869, according to some reports, machine-made toothpicks were selling in the United States at a rate as high as five million per day, which would have kept quite a few South American natives constantly busy.[14]

But making toothpicks in great quantity did not necessarily mean that they were of the best quality. As they must have been in Forster's time, hand-crafted toothpicks made of orangewood in Portugal today are slightly irregular in size and shape but are smooth to the touch. Forster's first machine-made toothpicks were of willow, but the wood "was very small, crooked and hard to secure." Then he used maple, but it had "too much fiber."[15] Neither of these varieties would have fared very well in comparison to orangewood, but that was to prove not to be a viable option in the Northeast.

White birch was identified as the wood to be tried next, most likely at Sturtevant's suggestion. Unfortunately, there were not sufficient supplies of it available in the Boston area. However, by this time Sturtevant had a shoe-peg mill operating at Wilton, Maine, where the tree grew abundantly, and Forster got in touch with the mill's manager, a W. H. Chamberlain, who had a cord of white birch shipped to Boston. Observers who saw it arrive and who knew that the whole load of wood was to be made into "slivers to push between the teeth" thought it enough "to supply the entire world." When tried in the machines, the birch worked perfectly, and the cord of wood was made into fifteen hundred boxes, each containing twelve hundred toothpicks, for a total of almost two million.[16] This is believed to be the first time white birch toothpicks were manufactured in the United States; thereafter the overwhelming majority of American toothpicks would be made of white birch from Maine.[17]

In the wake of the Civil War, there developed a general sense that it did not pay to engage in farming in Maine. Not only was the growing season short, but farming in the state involved hard work for comparatively low compensation. Simultaneously, there was a "general awakening throughout New England to the advantages and necessities of more and varied manufacturing enterprises." Driven by these perceived givens, towns began to hold out inducements to manufacturers who would relocate to Maine.[18] Sturtevant had known firsthand the difficulties of farming and how little it rewarded his father. Thus he must have welcomed the opportunity to engage in a more agreeable occupation in Maine, and one made all the more attractive by the conditions of the times.

The favorable economic climate may also have helped attract Charles Forster to move to Maine, but he did so principally to have his manufacturing operation closer to the supply of wood that worked best in the machines. Where that wood was, of course, was "out in the sticks." In 1870, most likely in the latter part of the year, Forster set up a factory in Sumner, Maine, which is located in the sparsely populated hills of the west-central part of the state, halfway between Augusta and the New Hampshire border.[19] This mill, at which Charles Freeman supervised production, burned down the next year, and the operation was moved to the nearby town of Canton.[20]

In early 1872, the Canton mill was described as a "thriving toothpick factory." At midyear it was employing thirty workers, and was soon providing "more work than all other business in town."[21] Over one hundred cords of poplar, a less expensive alternative to birch and therefore used for making cheaper picks, were hauled to the mill for use that season, and each of its machines was said to be capable of making five thousand toothpicks per minute. The factory was also reported to have had an average daily output of 180 bushels, but boasting of making toothpicks by the bushel was not conducive either to offering for sale an exact quantity of the product or to conveying a sense of the sanitary conditions under which it was manufactured.[22] It was preferable that a factory's output be counted by the box, carton, and case, each containing a stated amount of clean and neatly arranged toothpicks. Boxes would be packed in cartons and then those in shipping cases, and they would add up to millions of toothpicks per day. But producing vast quantities of anything is no guarantee that they will be bought, and so Forster had to project from year to year how much wood he would need during the next season to keep warehouses full but not too full.

Forster had also to determine how long he would operate the mill at Canton, which was, after all, established in a hurry after the fire at Sumner. As the year 1873 drew to a close, he decided to continue operations there for at least another year and so had to secure a supply of wood. He thus entered into a contract with Otis Hayford for five hundred cords of poplar, a move that made news from Boston to Atlanta. The use of the wooden toothpick was becoming widespread, and by the mid-1870s it had "almost entirely displaced the sharpened match."[23]

The plant at Canton remained in use until 1874 or 1875. In the meantime, Forster had decided to relocate his operations to the town of Mexico, on a site just across the Webb River from Dixfield and so referred to locally also as Dixfield. Like his earlier mills, this Forster factory at Dixfield was in the heart of the "birch belt." This region began in the Berkshires in Massachusetts and extended north into Canada and west through northern Minnesota. Very little birch grew farther west, and so no toothpick factories would be established there.[24] Where to locate a toothpick mill was principally influenced by the availability of wood and by access to a means of transporting cases of finished toothpicks far and wide. Thus, Forster had to find the right combination of proximity both to plentiful stands of birch and to roads and railroads. The factory at Dix-

In the late nineteenth century, Charles Forster's toothpick mill at Dixfield generated enough excess electricity to power lights that would be added to the nearby bridge over the Webb River.

field was in place as early as 1873, but it may not have begun operations immediately. It would develop into one of two major plants in
the Forster enterprise.[25]

Soon another plant was established at Andover, Maine, but perhaps because there was no railroad nearby, it was moved in 1883 to
Strong, where it was set up in an old starch factory.[26] The machinery brought from the Andover plant consisted of "five toothpick
lathes and a cigar lighter lathe." The toothpicks were "blunt on both
ends and thick in the middle"—a type that would be designated
no. 2. The cigar lighters were made from poplar veneer formed into
strips "4 inches long, ¹⁄₁₆ inch thick and ½ inch wide, wrapped with
wire in bundles of 50."[27] They were a kind of friction match known
as a locofoco, believed to be "so called from a self-lighting cigar,
with a match composition at the end, invented in 1834 by John
Mack of New York, and called *locofoco cigar* in imitation of the word
locomotive, which by the uneducated was supposed to mean self-
moving." Members of the antimonopolistic wing of the New York
City Democrats came to be referred to as Locofocos when they held
a meeting by the light of candles and locofocos.[28]

Though both the Locofocos and locofocos themselves would
eventually fall out of fashion, Strong would continue to be considered the best location for a toothpick plant. It was well situated for
a future supply of white birch. It was also on the Sandy River Railroad line, which terminated at the Maine Central about ten miles
away at Farmington.[29] By the end of the century, with the purchase
of the old J. W. Porter mill, in which clothespins, excelsior, and croquet sets had been manufactured, a second Forster toothpick plant
was added in Strong. As the Forster business continued to grow,
"additions sprouted out all over the main building to accomodate
new machines," giving the structure the look of something designed
by committee.[30]

It was the factories at Dixfield and Strong that would remain the
principal Forster toothpick mills well into the twentieth century.[31]
The early Dixfield mill, which was located where the Webb River
joins the Androscoggin, was at one time known as "the Power
Plant." Its water-powered dynamo had the capacity to light 360
incandescent lamps, which not only provided light for the mill but
also lighted the nearby bridge and the highway running in front of

the mill.[32] Eventually a newer plant was constructed at Dixfield, which, according to a 1907 newspaper feature, was a "town built out of toothpicks," and their manufacture by a workforce of nearly two hundred had made it "one of the wealthiest of the villages in New England."[33] Still, Strong would eventually supplant Dixfield as the town most known for toothpicks.

For a time, there was also a plant at Buckfield, which was once identified as "the only and original factory,"[34] but that latter honor, at least, surely belonged to Sumner's plant. Nevertheless, the evidently well-planned move into Buckfield provides insight into Forster's methodical approach to conducting business generally and establishing factories in particular. In 1871, water still provided the preferred power source for running a mill, and so it was imperative that access to such power be secured before planning a new factory. Shortly after moving his operations from Massachusetts to Maine, Forster acquired the water privilege at a location near Buckfield. At a town meeting held the following month, he was "exempted from taxation for a term of 5 years, 'on any mills, machinery and stock he may put upon the mill privilege at Shaw's bridge.' " Forster, who would be identified as "a large manufacturer of toothpicks in Buckfield," would also set up residence there, which may have led to the later misconception that he was born in the town. In any case, it was likely in Buckfield in the late 1870s that he first met Oscar H. Hersey, a young attorney who had opened his office there.[35] (As we will see, Hersey would move to Portland late in the century and play a central role in running the toothpick business after Forster's death.)

The tax advantage offered by Buckfield and taken by Forster was not unlike the kinds of incentives communities use to this day to attract industry and its attendant payroll and other economic benefits to their area. Though Forster may at first have employed workers numbering only in the dozens, that many jobs was significant for west-central Maine towns whose population may have been only in the hundreds.[36] Of course, in addition to workers a factory needs a ready supply of raw materials. In every case, Forster located his plants so that a good supply of birch and poplar logs was within an easy sled or rail haul, and later a reasonable truck ride, of where the wood was to be processed into toothpicks.

In the nineteenth century, when horses and oxen provided the principal hauling power, only wood within a radius of about ten miles was considered readily accessible. It is thus no accident that early toothpick factories were located in towns that are no more than about twenty miles from each other. With the introduction of trucks, the acceptable radius could reach over a hundred miles from a mill, but as late as the 1940s there was concern for the future of toothpick mills that were farther than that from abundant sources of wood. Still, most factories continued to be located in or near Maine's Oxford and Franklin counties, "where the soil and climate are especially well suited to the growth of the birch tree."[37]

The choice of wood was important for more than technical reasons, as was pointed out in an item in *Scientific American* titled "A Fortune in Toothpicks":

> It seems that it was not the invention of the wooden toothpick, *per se,* that netted the inventor $50,000, but the idea of making the toothpicks out of soft, brittle wood. It is said that, when first brought out, the toothpicks were made of hard, fibrous wood; but the inventor soon found that this would not pay, as the picks lasted too long, and he went to pine. It now takes four sound picks to get the broken end of one out from between the teeth; and it is the latter discovery that is said to have realized the inventor his fortune.[38]

The story, as embellished in a British journal—and then retold in *Scientific American*—identified the original hardwood as hickory. Forster seems never to have used either hickory or pine for his toothpicks, so the story appears to stem from a confusion of manufacturers or from a misinterpretation of a brand name. The "hickory" toothpicks may have been Baker's Old Hickory brand, which were evidently being distributed from Brooklyn in the 1850s and thus predated Sturtevant's and hence Forster's patent protection. The label carried an image that resembles Andrew Jackson, "Old Hickory" himself.[39]

But whatever other toothpick brands may have been on the market through the 1870s, it was the output of Forster's mills that overwhelmingly dominated, due largely to the patent control he had acquired from Sturtevant. However, patents do expire, and the seventeen years' protection came to an end in 1880.[40] This led to the speculation that "whether competition will prove the life or death of the toothpick remains to be seen."[41] It was no doubt in anticipation of such competition that Forster had long before mounted a very clever and aggressive marketing campaign.

CHAPTER ELEVEN

An Article for the Million

M AKING MILLIONS OF wooden toothpicks by machine was one thing; selling them box by box to individuals was another. Although regular, if not ritualistic, wooden toothpick use was a long-standing tradition in homes in countries like Spain, Portugal, and Brazil, as late as the 1870s it was far from universal in the United States. Here, where the toothpick had traditionally been an ad hoc implement, whittled and used when needed, it was said to be an "unknown adjunct to the dinner table."[1] Getting Americans to buy toothpicks by the box and to use the newfangled mass-produced splinters at home was no easy task, at least if the stories told about Charles Forster the marketer are true.

In spite of published reports that as many as five million wooden toothpicks were being sold every day, millions more were being pro-duced.[2] Even in the earliest days of operation, there appears to have been considerably more production capacity than demand, and to make any significant amount of money on articles that sold for so little per item, as many as possible naturally had to be sold. Accord-ing to one account, in the spring of 1870 the machinery that Forster and Freeman had developed was working "perfectly," but "not a single sliver had been sold in this Country."[3] This clear contradic-tion with contemporary published reports is typical of the hyper-bolic Forster lore that developed long after the fact. Certainly there was some market for wooden toothpicks, for Benjamin Sturtevant evidently had sold modest amounts even in the mid-1860s.

There were in fact two distinct markets: the large-volume buyer (including hotels, restaurants, and retail outlets) and the individual (or family) toothpick consumer. At the time, Forster's toothpicks were generally packaged in boxes of twelve hundred each, the boxes packed in cartons, and the cartons in cases. The hotel and restaurant market, which can be expected to have bought by the carton, if not the case, seems to have been well established by 1870, but not according to Forster lore, as we shall see. Although it could be argued that this should have been viewed as the primary market for which the mass-produced wooden toothpick was invented, Forster invested a lot of marketing time and energy into reaching retail outlets.[4]

According to repeated reports, Forster's early attempts to sell his wares to merchants were "met everywhere with jeers and laughter." However, rather than being discouraged by a cool reception to his product, the man who had the patience to spend years investing his time, money, and hopes in machinery to manufacture toothpicks was "determined to create a market" for them. How he evidently proceeded to do so shows that his inventiveness "equaled the genius of any modern chain-store magnate."[5]

Forster does not seem ever to have wanted to sell directly to the individual consumer. Rather, he wished to sell wholesale to retail establishments such as stationers and other dealers in small goods. Yet he appears to have had a difficult time getting started. The first problem that he faced in placing his product in retail stores was that the proprietors were not used to carrying so frivolous an item as a disposable wooden toothpick, let alone boxes holding large quantities of them. To overcome this resistance, Forster is reported to have conceived of a rather elaborate scheme that involved well-coordinated planning, organization, execution—and patience.

Since it became evident that he was getting nowhere as a salesman, Forster stopped making the overt attempt and let the story circulate that his efforts had proven to be a failure, leaving the impression that he had given up on selling toothpicks altogether. In the meantime, he had prepared a long-term plan:

> He then quietly hired more than a dozen men and women from different parts of Boston and gave each a list of stationery and novelty stores they were to visit and seek to purchase wooden toothpicks. Day after day these men and women, one after the other, called at the different stores and sought to purchase wooden toothpicks. For more than a month this attempt to purchase toothpicks was kept up. Each day one or more of these men and women would call upon the various store-keepers trying to purchase wooden toothpicks.[6]

Though this account suggests all the repetitive charm of water torture, it captures in that way the relentlessness with which Forster attacked, albeit vicariously, what he saw as his potential market. When the targeted retail-store buyers were sufficiently softened up and primed, Forster visited the same stores. He naturally was able to sell boxes of toothpicks to the proprietors. Then, "no sooner than these boxes were sold were the men and women, hired for that purpose, calling at the stores to purchase these toothpicks, at a profit to the dealer. These toothpicks were immediately returned to Mr. Forster. All through the Summer and Fall of 1870, Mr. Forster was selling and buying back his own toothpicks over and over again."[7] The retail stores had been convinced of the value of carrying Forster's toothpicks.

At first, Charles Forster may have been alone in his scheming, but if he was, he soon had an associate and accomplice in Levi L. Tower. Tower was born in western Massachusetts in October 1826, the same month and year that Forster was born in the eastern part of the state. It has been said that one way the young Forster traveled to Brazil was by captaining a ship owned by Tower and others.[8]

In 1870 Tower was residing in Somerville and Forster in nearby Brighton, both of which border Cambridge. Tower was a serious and imaginative stationer, at one time patenting the design for an ink or mucilage stand that resembled a beehive. He was also the assignee of rights to an improvement in suspension rings, devices that could be attached to a calendar or business card to make it suitable for being hung on a hook or nail.[9] In association with James M. Cutter and others, Tower was engaged in the firm of Cutter, Tower & Co., a partnership registered in New York.

One day in 1871 that business was dissolved, and on that same day a new one with the same name was established, but without Levi Tower as a partner. The new enterprise dealt in the wholesale and retail stationery business, as well as in book, job, and color printing.[10] Tower and Cutter themselves formed another partnership, under the name Cutter-Tower & Co.—with a hyphen rather than a comma and often subsequently abbreviated C.T. Co. This business was evidently established to focus on the wholesale toothpick trade, and it did so very successfully. Tower's brief obituary in the *New York Times* would identify him as the "dean of the stationery trade," and he was also "said to have been the first man who introduced the wooden toothpick into this country, and is said to have realized a fortune as the result of the introduction."[11] It very well may have been Tower and his partners who had been importing into New York and Boston handmade wooden toothpicks from South America while Forster was still down there dreaming of making them by machine.

The success of Cutter-Tower and its prominence in New York, as well as Levi Tower's position as president of the firm, likely provided an entrée to him to serve on the board of directors of the Greene Consolidated Copper Co., of which he also became president. Greene Consolidated was apparently no small enterprise. According to a stock offering, the company owned "the Vast and FAMOUS CANANEA COPPER MINES in Northern Sonora, Mexico, containing an inexhaustible supply of Self-Fluxing COPPER ORES of high grade, from which copper can be produced at a lower cost than ANYWHERE ELSE IN THE WORLD." Tower also served as president of the United Cotton Gin Co., but his anchor remained Cutter-Tower & Co., which dealt in "novelties, patented goods, typewriters, and wood toothpicks."[12]

Although he was deeply engaged in business in New York, Tower's base of operation was in the Boston area. He bought an estate in Newton known as Mt. Ida, and in 1902 Mr. and Mrs. Tower would celebrate their fiftieth wedding anniversary there.[13] The Cutter-Tower Co. was among the contributors to the Fifty-fourth Regiment during the Civil War, reportedly donating a box of stationery to the colored unit. He was also active in the First Methodist Episcopal Church of Somerville, having served as a superintendent of its Sunday school.[14]

The story of Forster and Tower's deception in seeding the toothpick trade is an engaging one, evoking simultaneously disdain of the charlatans and admiration for their cleverness. Although Forster and Tower seem to have worked together very closely through at least the fall of 1870 to establish a market for wooden toothpicks in the Boston area, they soon went their separate but integrally connected ways.[15] Forster would become exclusively a manufacturer and Tower, at least until after Forster's death, a wholesaler of much of what he would make.

Over half a century after the fact, Levi Tower's role in marketing schemes was usually mentioned only in passing, if at all, at least

within the Forster toothpick family. It was the role of Charles
Forster in promoting his toothpicks that was recounted with gusto.
It was told with especial emphasis in 1931, at the dedication cere-
monies for the Forster Memorial Building (in honor not of Charles
but of his son, Maurice) in Strong, by Frank W. Butler, a Maine
lawyer who by then had become a judge, a state senator, a trustee of
the Forster estate, and eventually a partner in the Forster firm.[16]

Whether the story Butler told was literally true or an embellish-
ment of something told to him by Maurice Forster (and possibly
embellished by him) or by toothpick-plant employees, it does provide insight into the psychological makeup of Charles Forster and
his capacity for deception. It is difficult to imagine that such a story
could be or would be fabricated out of whole cloth by a third party.
Unfortunately, there are other parts of Butler's speech, as reported,
that are clearly in error, such as the assertion that "Forster was an
Englishman" who as a young man "entered the employ of a large
English exporting house." Evidently Butler did not know Forster
or his heritage well enough to know of his family's deep Boston
roots and mercantile prominence.[17] It is entirely likely that the two
men never met.

The story of the Forster marketing scheme was repeated, almost
verbatim, by George P. Stanley, who at one time carried business
cards that identified him as manager of the Stanley Manufacturing
Corp. of Dixfield, maker of "high grade tooth picks."[18] In 1943,
when he was seventy-four, Stanley wrote that he thought it fitting to
record some history that he had "gained by experience and learned
from others" in the toothpick industry. According to Stanley, his
earliest recollections dated from 1881, when at the age of twelve he
ran the toothpick-cutting machinery known as "choppers" in
Forster's mill in Dixfield. It is possible that it was close to then that
Stanley heard the stories he later related.[19] It is also possible that his
oral reminiscences might have been the source of Butler's stories
(and some of his in turn the source of Stanley's written account).
But Stanley also makes some errors about Forster's life. For example, he has him not as an Englishman but as a "Buckfield native."
Mainers make sharp distinctions between natives and those "from
away," and Forster may have promoted, or at least did not discourage, the idea that he was born in Maine. It certainly would not have
been inconsistent with his deceptive marketing practices, and it

definitely would have given him a business advantage in the state, although it is impossible to imagine for a moment Forster fooling true Buckfield natives.[20]

He does appear to have fooled Boston businesspeople, however. And it was not only stationery and novelty stores in which Charles Forster wanted to place his toothpicks. Hotels and restaurants should also have been large potential customers, but "the restaurants he approached were not willing to give this new item a chance."[21] This can only be exaggeration at best, for by this time many restaurants had indeed given the toothpick a chance, offering it free to patrons and no doubt feeling it necessary to do so to remain competitive. And so the story—or, again, stories—about how he tricked such establishments into buying them must be taken with a grain of salt. Nevertheless, they are stories that bear repeating, for their relative amusement if not for their absolute truth, and also for the further insight they provide into the character of Charles Forster, who most likely was their ultimate source. According to one version, which was related within a generation of the event, Forster created a demand "in a most ingenious manner":

> In his circle of acquaintances was a man of remarkably fine personal appearance, but utterly lacking in all other qualities necessary to earn a living. Forster at once took the fellow into his service and dressed him up to the height of fashion. The man was supplied with money and directed to go to all the best hotels and restaurants and get his meals in turn. After eating he would come out and ask the manager of the place for a toothpick. Of course, no such article could be furnished, and then the fellow would express his regrets and assure the landlord that he was far behind the times. By a prearranged coincidence Forster would soon turn up with samples of his goods and solicit orders. In this way the article soon came into common use and now they are everywhere, both in public and private places.[22]

In another version, Forster hired "Harvard scholars" to eat at local restaurants and then after their meal to ask for toothpicks, and to do so loudly enough so that they could be heard by other customers. After this occurred repeatedly, the restaurants were in a vulnerable position when approached by a toothpick salesman. On the occasion of the self-declared centennial of the corporate descendant of Forster's toothpick business, Senator George Mitchell of Maine repeated the story, with adjectival embellishments, in remarks introduced into the *Congressional Record:*

> Forster hired several suave, successful-looking young men to visit prominent Boston restaurants where, after finishing an elegant dinner, they would ask for disposable wooden toothpicks. When they were told there was none to be had, they would demand to see the manager and protest loudly. To quiet the indignant gentlemen, restaurant managers would ask how they had learned of these toothpicks and where they could be purchased.[23]

This story, in all its variations, reinforces the impression of Forster's being quite capable of using deception to market his product. It may be a corruption of the tale about selling to stationers, or vice versa; it also may be simply apocryphal.

Whatever the truth or falsity of the near legend about Forster's use of Harvard students to place toothpicks, their introduction into restaurants is commonly associated with an eating establishment that such students might well have frequented at the time. The Union Oyster House claims to be "the oldest restaurant in Boston and the oldest restaurant in continuous service in the U.S.," its doors having been open since 1826. Its Web site makes a sweeping assertion and repeats the toothpick story:

> The toothpick was first used in the United States at the Union Oyster House. Enterprising Charles Forster of Maine first imported the picks from South America. To promote his new business he hired Harvard boys to dine at the Union Oyster House and ask for toothpicks.[24]

A quick reading of the Oyster House's Web page may give the impression that the toothpick was introduced in the United States in 1826, but tying the implement to Forster, who was just born that year, puts toothpicks in the restaurant much later. This version is also in conflict with the prevailing Forster story that it was his machine-made toothpicks that he was trying to place in the restaurant and so establish the expectation that they be available to diners. Nevertheless, however the story is told, adherents to the Forster mythology continue to think that it was a result of his ploy that "soon restaurants were offering toothpicks on small silver trays."[25] Even were we to believe that toothpicks were not commonly used in the Boston area before being introduced there by Forster, they certainly had been widely used elsewhere in the United States before that time.

If there were geographical and national differences in what food was eaten where—and there certainly were and are—there obviously also could have been regional differences in the availability and use of toothpicks. We do know that wooden toothpicks from South America were imported into both Boston and New York well before Forster returned to the United States for good.[26] Very likely it was by Levi Tower and his associates, one of whom may have been Charles Forster. If we believe the Forster lore, to what uses the toothpicks might have been put in Boston might remain a mystery, but there can be no doubt as to how eagerly they were consumed in New York and elsewhere in the country.

In 1859, an article in the *National Era,* a Washington, D.C., newspaper, noted that toothpicks, an "article of table furniture naturalized on Continental Europe since Greek and Roman times," had "but recently been admitted into our restaurants and hotels." And it may well have been that they were introduced by Henry Stearns Mower, whose 1917 obituary described him as "a hotel proprietor for sixty years and who is said to have introduced toothpicks to the American public," but did not say exactly when or where.[27] The humble toothpick appears to have had many fathers.

Regardless of who introduced them, the use of toothpicks in—and out of—restaurants was soon something to be remarked upon. In a story about the completion of the first Atlantic cable, in 1858, a reporter described how a party "with tooth picks came out of a restaurant" in New York and how the "knights of the tooth picks" began celebrating the technological achievement.[28] Much such reporting of toothpick use was, perhaps appropriately, tongue in cheek, but that does not diminish it as testimony to the then-recent appearance of the sliver of wood into public use. Even if ridicule, the reporting should indicate all the more the pervasiveness and prominence of the device. One wag, writing humorously about etiquette for gentlemen, indicated that the proper use of the toothpick required study and models. To find these, he wrote, "We recommend the steps of any fashionable hotel as a studio, between 3 and 7 p.m., where lessons are given, gratis, by distinguished masters."[29]

By 1860, the availability of wooden toothpicks appears to have been widespread in America. They were referred to as "those little orange sticks that one finds at every restaurant and hotel in the city and country," and they were "scattered all over the country— placed in the restaurants and hotels, and in the hands of every toothpicking Yankee in the republic." Furthermore, "to such an extent is this traffic carried, that the proprietors of the Astor House alone purchase eight or ten barrels of each importation, and retail them among the country hotels. A restaurant with a good run of custom will consume about twenty thousand toothpicks in three weeks," which amounts to about a thousand a day.[30] The "little orange sticks" would, of course, be the hand-crafted toothpicks imported from South America. Forster also attempted to manufacture such items for a while, but he found the orangewood not to be suitable to being worked by machine.[31]

No matter of what wood or how made and marketed, wooden toothpicks soon became ubiquitous. In 1870, *Manufacturer and Builder*, which was published in New York, described where exactly the toothpicks were placed: "Every eating-house visitor in this city and other leading cities of the Union has doubtless noticed a small tumbler of wooden toothpicks upon the counter of the cashier, for the use of customers."[32] It would not be very long before such ready accessibility of the "familiar toothpick which we take up as we leave a restaurant" would lead to quips: "Why is it that a man cannot see a bundle of toothpicks without helping himself when he does not need them at all?"[33] It appears also that a certain kind of man could not see a hotel without availing himself of a space on its front steps. The resulting scene had become a cliché. One observer asked in 1872, "Will anybody explain where all the men come from who lounge about the hotel-entrances and stand three-deep on the steps at the hour of the ladies' promenade? The oldest inhabitant will tell you that they do not belong to the hotel, and that they are never known to get a meal there, though you will always find them on the steps, tooth-pick in hand, after dinner."[34]

The public places, and hence the public itself, had finally been hooked, and by 1874 more orders for toothpicks were coming "from hotels and restaurants than from all other sources combined." Forster was manufacturing toothpicks in Maine, but they were sold through "the Boston house which contracts the whole business in this country," which would likely have been Cutter-Tower. The number sold had reached five hundred million and was on the increase. Wooden toothpicks were selling for twenty-five cents per box of two thousand, a small fraction of the price of the same number of quill picks. By one account, "So great has been the popularity of cheap wood toothpicks in this country that the experiment of exporting them is soon to be tried."[35] Charles Forster would seem to have achieved his goal, but he continued to develop real and imaginary marketing schemes, presumably to sell more and more slivers of wood to push between the teeth. He appears to have been the kind of person who is never satisfied and thus is always on the lookout for new opportunities.

According to another Forster legend, the Centennial Exhibition held in Philadelphia in 1876 provided one such opportunity. That year Forster engaged in a new effort "to educate the public" about and to promote the toothpick.[36]

[He] prepared for a grandstand display of his goods. First he shipped a full carload of picks to Philadelphia, and then went to that city himself. Upon his arrival he hired a large, high, beautifully decorated wagon, drawn by four snow-white horses. The driver was clad in a uniform of bright scarlet. Mr. Forster, attired in a dress-suit and wearing a tall hat, mounted the wagon and was drawn through the streets of Philadelphia. These streets were lined with people, and as the wagon proceeded on its journey, Mr. Forster himself threw boxes of toothpicks into the crowds of people standing in the streets watching the procession pass by. The newspapers of Philadelphia printed pictures and cartoons of Mr. Forster "throwing away his slivers," all of which were paid for, later, by Mr. Forster, at regular advertising rates. This unique and original idea of advertising was kept up at frequent intervals during the entire time the Centennial Exposition was open.[37]

Pictures of Charles Forster on his wagon have been as elusive to come by as portraits of him and his family. Nevertheless, the report may well be true, for such a spectacle would have been in keeping with the goings-on in Philadelphia that year. On May 10, the opening day of the exposition, the city had "reduced flag flying to a science." But the events of that day were outdone on July 4, when the streets of Philadelphia "were crowded all day with people who turned out in throngs." That evening, there was a torchlight parade, with the Northeast Division bringing up the rear. Representatives of Maine could be expected to have marched in this division, which had a "smaller quantity of advertising wagons." One of them may well have borne Charles Forster and his toothpicks. The crowd along the parade route was reported to have been "immense, numbering thousands, and could be counted by tens, twenty and thirty thousands. Indeed, the great sweltering mass of humanity reached into the hundreds of thousands, including men, women, and children, from the lowest to the highest in the social scale." That would certainly have been the kind of audience to which Forster would have loved to throw boxes of his wares.[38]

The procession culminated in a midnight "ceremony of ushering in the new century" of the nation, and the next day's *Philadelphia Inquirer* carried an extensive report of the events.[39] However, that newspaper was not heavily illustrated at the time, and no image of a toothpick wagon is to be found in its pages. (Had one appeared, it would likely jump out at the reader of microfilm carrying column after column of virtually unrelieved text.) In general, the advertising wagons were not individually identified in the reports, which is understandable. Newspapers are in the business of selling advertising. It is certainly possible that another Philadelphia paper did carry—for an advertising fee—an illustration of Forster on his wagon, but such an image is not known among students of World's Fairs. Although it has been said by propagators of the Forster myth that "only a very limited number of toothpicks were manufactured and sold prior to 1876," the purported Philadelphia stunt would presumably have resulted in a greater demand being created for the "Forster toothpick." In any case, the annual demand would rise and rise in the coming decades to reach well into the billions.[40]

At just about the same time that Charles Forster was working to mass-produce wooden toothpicks and convince people to buy

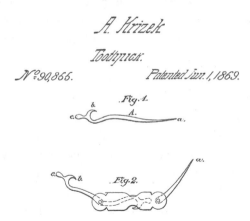

Numerous variations on the basic toothpick have been patented. Alphons Krizek's invention combining straight and hooked pick points with a small spoon could be made out of a variety of materials, either all in one piece or in a folding version.

them, other inventors were also trying to strike it rich with tooth-pick variations of their own. Some of these "improvements" would in fact be retrograde, employing older materials and older ideas. Among the inventors was Alphons Krizek, of Philadelphia, who in June 1869 was issued a patent for an "improvement in toothpicks." His reusable instrument incorporated, in addition to the usual more or less straight pointed end, a pointed hooked end and a small spoon-shaped projection. The hook enabled "the teeth to be cleaned from the inside in a manner which cannot be accomplished by means of a straight instrument," and the spoon could be used "to clean out hollow or sensitive teeth into which the sharp point . . . could not be inserted without giving pain." The spoon could also serve to "cleanse the teeth of tartar at points adjacent to or beneath the gums." Furthermore, according to Krizek, the device could be made in one piece "of ivory, bone, hard wood, or other suitable material," or it could be "made in sections, jointed to and arranged to fold within a suitable handle."[41]

Krizek's patent, which consisted of multiple tools and did not infringe upon the Sturtevant patent that Forster controlled, was assigned to two other Philadelphians besides the inventor. One was Thomas Richardson, who within weeks of the issuance of the patent took out a classified ad in *Scientific American,* which read:

> FOR SALE at a Bargain, entire, or by State Rights, Pat. No. 90,855, for an Improved Toothpick. An article for the Million. Samples furnished.[42]

The word "million" was all that anyone interested in getting rich should have needed to read. It was a word that Henry Ford certainly was to respect. On the occasion of a dinner celebrating the one millionth Model T to come off the production line, Ford was asked to give a speech. He obliged with, in its entirety, "Gentlemen, a million of anything is a great many."[43] In fact, the future market for toothpicks was in the multiple millions.

It was thus the likes of Forster, Krizek, Richardson, and a few others who recognized the potential for ultimately large profits in small, trivial things such as toothpicks sold by and for the millions. According to a casual glance at the Patent Office index, through 1873 fewer than a half dozen patents were issued for toothpicks and machines to make them, including Sturtevant's method for manufacturing the simple wooden kind and Krizek's combination device.[44] But there were plenty of other patented combinations not indexed under the letter *T.* Among them was John Sturdy's "Combined Case for Pen, Pencil, Knife, Toothpick, &c.," Richard Cross's "Watch-Key, Toothpick, and Toggle Combined," and Henry Graham and Richard Child's "Improvement in Combined Watch-Key and Toothpick."[45] Such gadgets were common in the mid-nineteenth century, but the only examples that have tended to survive are those made of materials with some intrinsic value, that is, those that were not melted down as scrap. Before the widespread availability of wooden toothpicks, these combination devices were as popular as Swiss Army knives were to become in the twentieth century.

As with patented inventions generally, the word "improvement" was commonly included in the title of the patent, and the concept of improvement was prominent in the text of the specification. To establish such an advance in the state of the art, the problems with the prior art were typically enumerated. Thus, in his 1875 patent for an "Improvement in Tooth-Picks," the New Jerseyite Garret Boice acknowledged that the "thinness and great strength and elasticity of pieces of quill eminently adapt them to serve for tooth-picks," but, he implied, when put directly into the pocket they were lacking in cleanliness. To improve upon this condition, he invented a holder "of precious metal, or other rich material," which consisted of parts that could be rotated relative to each other and thereby extend or retract the toothpick, thus keeping it clean when not in use. He pointed out that toothpicks made of rubber, whalebone, ivory, and metal could also be used in his device, but ordinarily he believed that quill would "generally be preferred."[46] A subsequent team of inventors, the New Yorkers Howard Stephenson and William Bennett, also patented an improvement in toothpicks, putting forth many of the same arguments as Boice. However, their case was to be "comparatively inexpensive, and yet have all of the advantages of cleanliness, convenience of carriage, and facility of manipulation possessed by the costly article made of precious metal." They referred to Boice's device not as a toothpick case but simply as a "pocket toothpick," for it was not combined with any other tool. In addition to the split quill that it was designed to receive, Stephenson and Bennett's case could accommodate any "like curved form from wood, rubber, tortoise-shell, bone, or other material flexible or strong enough" to be inserted into the cylindrical holder and serve its purpose.[47]

Other inventors looked not to improving or complicating toothpick-containing mechanical devices and their operation but to supplementing the basic toothpick in other ways. In 1870, William Blake of New York City patented an improvement in wooden or other toothpicks that consisted of applying "a coating of any approved medical composition, that may be taken after meals for the affections of the voice, mouth, and throat, as well as to aid digestion, &c., said medical composition being overlaid with a coating of perfume for the breath, and the latter coating overlaid with gold-leaf, sugar-coating, or their equivalents," thus producing "a neat and ornamental toothpick."[48] A few years later, George Clark, Jr., a Bostonian, also patented an improvement in toothpicks that were "artificially charged with a substance which either flavors it to the taste or perfumes it." The patent describes his manufacturing process as either steeping the raw wood or the toothpicks made from it in a suitable liquid or putting it into a vacuum chamber into which the liquids were injected, to be absorbed by capillary action. Among the oils he used were those of checkerberry, lemon, and peppermint, sometimes mixing in also cottonseed, olive, or rape-seed oil. He also used sassafras bark and other substances to make the toothpicks "more attractive" and capable of "perfuming the breath."[49] Such could be the basis for a fortune, as it would be in the 1931 film *Finn and Hattie*, which was based on the flip and quippy Mr. and Mrs. Haddock novels of the 1920s, he being "the wealthy inventor of a combination toothpick and mint."[50]

But the toothpick design that would be the true article for the millions, if not the billions, was neither fancy nor complicated—no matter how complicated and temperamental the machinery to make it by the millions and billions might be. Even before Charles Forster moved his operation to Maine, he was reportedly producing four million to five million toothpicks per day, though that may well have been more than he could sell then. By the mid-1870s, five hundred million a year were supposedly being distributed, but that was still well below the reported capacity of his machinery. To a manufacturer, the business of toothpicks had not mere millions but, literally, "billions in it."[51]

Who was reaping the profit remained as difficult to pin down as who was to be publicly thanked for the mass-produced and mass-marketed toothpick. A story that first appeared in the *New York Tribune* in the mid-1880s did not credit Forster's New England enterprise for the introduction of the wooden toothpick but attributed its success to the efforts of an unnamed firm in New York City.[52] It is unlikely that the firm alluded to was any other than the Cutter-Tower Co., or possibly the Tower Manufacturing and Novelty Co., a New York agency whose toothpicks were later advertised as made in Maine of birch and poplar, though then they were likely made in Forster's mills. According to the *Tribune*'s source—an anonymous member of the anonymous business—the firm began to make wooden toothpicks in 1870 and was the sole manufacturer of them for ten years, after which it began to "share the wholesale trade."[53] Charles Forster's curious shunning of the limelight, at least when not putatively riding on a wagon in Philadelphia, sometimes left him without the recognition that he surely deserved.

In 1885, at least three billion toothpicks were being made annually in the United States, and two years later domestic production was up to "five thousand million," that is, five billion, with Forster accounting for 60 percent of them. At the end of the nineteenth century, a single factory in Maine was making five hundred million *packages* each year.[54] The United States' population then was only about seventy-five million.

CHAPTER TWELVE

Advice and Dissent

I N ADDITION TO the many manufacturing and marketing issues associated with the increasingly widespread availability and use of the wooden toothpick in nineteenth-century America, there developed a host of considerations relating to propriety and etiquette. The resulting difference of opinion was nothing new, for it had existed across the ocean centuries earlier, when "tooth-picks were a symbol of gentility and not in general use in England. Those persons, who possessed them, were invariably of the higher ranks of society."[1]

It is likely that the kinds of toothpicks then in use were made of a variety of materials. In the fifteenth century, the Italian Giovanni Arcolani, who wrote his *Practica* under the Latin name of Johannes Arculanus, "advocated the use of thin strips of wood (preferably those of a bitter or styptic nature, such as aloes or pine), in order to remove food particles: and thus minimize the risk of dental caries." Contemporaneously, metal toothpicks were also in use, and gold was the choice of those who flaunted their position and wealth. Such implements are believed to have been imported into England at least as early as the mid-sixteenth century, for at the time there existed an import duty of a penny a dozen on both toothpicks and ear picks.[2]

Wherever there is a class structure, an accompanying behavioral structure serves to distinguish the classes. The proper and improper use of the toothpick seems always to have been the subject of much opinion, assertion, and declaration. Petronius, who satirized the excesses of first-century Roman society and who has been called "the Emily Post of his day," took exception with the emperor Nero for "entering a banquet hall with a silver toothpick in his mouth."[3] A fifteenth-century "book of courtesy" advised, "At meal cleanse not thy teeth or pick / With knife or straw or wand or stick."[4] Another admonition on what to do and what not to do at the dinner table was captured in a verse from the latter part of the sixteenth century:

> *Pick not thy teeth with thy knyfe*
> *Nor with thy fyngers ende,*
> *But take a stick, or some cleane thynge,*
> *Then doe you not offende.*[5]

And there was also contemporary advice, full of mocking imagery, about how not to retire from a meal: "When the table is cleared, to carry about your tooth pick in your mouth like a bird going to build his nest, or to stick it behind your ear, as a barber does his comb, is no very genteel custom."[6] Even the Dutch humanist Erasmus weighed in: "If something has remained between the teeth, do not fetch it with the fingernails as the dogs and cats do, nor with a napkin, but either with a mastic staff or a quill from a chicken or a rooster."[7]

In *Much Ado About Nothing*, Shakespeare used the willingness to secure a toothpick from a distant land as a means of demonstrating loyalty. Thus, the soldier Benedick says to Don Pedro, the prince of Aragon,

> *Will your grace command me any service to the world's end?*
> *I will go on the slightest errand now to the Antipodes*
> *that you can devise to send me on;*
> *I will fetch you a tooth-picker now from the furthest inch of Asia.*[8]

In *King John*, the character Bastard speaks of toothpicks and toothpicking in a soliloquy full of double entendres:

> *. . . Now your traveler,*
> *He and his toothpick at my worship's mess,*
> *And when my knightly stomach is sufficed,*
> *Why then I suck my teeth and catechise*
> *My picked man of countries . . .*[9]

And in *The Winter's Tale*, Clown assesses the character of a stranger and warrants that he is a great man: "I know by the picking on's teeth."[10] That such references to the simple object and its use and significance would have been understood by Shakespeare's audiences speaks to the familiarity with the toothpick in his day.

In Elizabethan times, gallants carried "elaborate cases of tooth-picks, and after a custom introduced from abroad, used it osten-tatiously at meals and other times by way of distraction."[11] Not everyone was believed to have had the same manners or sensibilities, and "toothpicks were believed to be used only by the upper classes." In the early seventeenth century, a list of "directions for health" included "cleaning the teeth with a toothpick," but with the proviso that it be done in the privacy of one's own chambers. Another giver of advice declared that "teeth were cleaned with a silver instrument and the gums cleaned with a wrought handkerchief."[12] Fancy and valuable toothpicks were considered part of a woman's dowry.[13]

An oft-translated assertion of Archenholz in his *Tableau de l'Angleterre* became an oft-repeated filler in newspapers and magazines of the nineteenth century. According to one version of this quip, "An Englishman may be discovered anywhere, if he be observed at table, because he places his fork on the left side of his plate; a Frenchman, by using the fork alone, without the knife; a German, by planting it perpendicularly into his plate; and a Russian, by using it as a toothpick."[14] Evidently, the Spaniards, who properly used a toothpick as a toothpick, were not invited to the table. Cardinal Richelieu is said to have so disliked the use of the knife point as a toothpick that he ordered knife tips ground down, thus instituting a fashion that persists in sterling flatware to this day.[15]

The English had forks for two centuries before the items were familiar to Americans, and when the Americans began to use the new implements they often confused their function. One early etiquette writer hoped to forestall faux pas by holding their use up to ridicule: "To give any thing from your own plate to another to eat of, shows great good nature and amiableness of disposition, particularly if on the point of a fork with which you have been picking your teeth."[16] As she or he did any dining utensil, a refined lady or gentleman saw a toothpick as having a proper purpose. To use it otherwise was anathema. On one occasion, a British gentleman, deathly ill in America, was having rhubarb and mint water prepared for him by his servant, who stirred it with a toothpick. A second gentleman present, who knew of the other's penchant for an elegant table, called the servant a "slovenly dog" and ordered him to bring a spoon, but death came first to the ill man.[17]

As much as Charles Forster and others might have asserted that they had brought the toothpick to America and introduced it to Americans, the implement as a symbol of good breeding, contentment, and inspiration preceded by decades, if not by almost a century, the era of mass-production. In 1745, George Washington set down his "rules of civility," the one hundredth maxim of which cautioned against using a knife or fork to remove stuck food, advising rather that one should "let it be done with a pick tooth,"[18] a pair of words that in time came to be hyphenated and then spelled as one and when used as an adjective carried the meaning of idle, indolent, easy, leisurely—qualities that would be associated with the use of a picktooth or toothpick, as in "the pick-tooth carelessness of a lounger."[19] But at the same time, the toothpick could also be associated with refinement and good taste.

In 1778 the baggage of a plenipotentiary—the "first Commissioner"—setting out for America was reported to include, among other things, "half a dozen opera glasses," "forty boxes of pearl coloured powder for the teeth," and "twelve dozen best toothpicks."[20] But at the same time, the eighteenth-century *Treatise on the Disorders and Deformities of the Teeth and Gums* cautioned against the use of toothpicks with reasoning that might be employed by a dentist today: "They tended to injure the gingival tissues; causing them to recede and the interstices to widen. Consequently, additional food lodged between the teeth; and the use of the toothpick then became more necessary."[21] The misuse or overuse of a good thing can turn it into a bad thing. Ambivalence about toothpicks and their use would persist.

At the beginning of the nineteenth century, one commentator noted that in that "age of political innovation, it is curious to observe the great veneration for antiquity which prevails in all our dresses and fashions." Among developments that he considered remarkable was the "total abjuration of the female pockets," for fashionable women had come to carry everything in a "workbag." In the bag, money and needlework intermingled with handkerchief, watch, and "tooth-pick-case."[22] But even if fashionable women always had a toothpick with them, they presumably used it discreetly. The same could not be said for gentlemen.

An 1812 edition of the *American Watchman* gave the following advice: "Let any gentleman, who is a gentleman, step into a public hotel, take up a *Delaware Statesman,* call for a glass of water, an American segar, an elbow chair and a toothpick, and if he does not enjoy what the poets call '*a feast of* REASON *and a flow of* SOUL,' he has no taste for good living." That such an item caught the eye of readers was made evident when the *Watchman* apologized for printing it first with a typographical error, which left out the "not," an unfortunate omission that caused a "derangement of idea." The paper assured its readers that the error "was unintentional, and not caused by a wish to convey a blunt, uncivil insinuation."[23]

Criticism of picking the teeth in public came from many directions. Among the disagreeable practices that the English novelist Frances Trollope reported observing on her visit to America in the late 1820s was the habit of military men using their table knives to shovel food into their mouths, the fork not yet having been fully adopted in the New World. Since by that time most such knives were blunt- if not bulbous-tipped, they could not serve as a toothpick too, and so the diners had to employ a more sharply pointed implement. This they found in their pocketknife, which was always readily available.[24]

Most gentlemen did not need to employ a pocketknife or call for a toothpick in early-nineteenth-century America, for they usually carried their own personal toothpick in their pocket. Speaking of the implement then typically referred to a more or less permanent and reusable possession of some value. It was often a gold or silver item, with or without a case, and typically classified as "fancy goods" by importers, who regularly advertised having received fresh stock for gentlemen. It was no doubt such a toothpick that a satirist had in mind when he reported on April Fools' Day of once having seen a gentleman with "a fashionable air, and his very horse seemed to say there are few people like me and my rider." While riding by, the gentleman unknowingly dropped a piece of paper from his pocket, and it was scooped up by the writer, who found it to be a list headed "Resolutions for Conduct in Life." Among the items listed was, "Must not forget to carry my toothpick."[25]

Gentlemen and toothpicks were generally considered inseparable, the implement to be surrendered only in the most unusual of circumstances. One such occasion arose when a gentleman wished to make the acquaintance of a woman in a green silk cloak: "I would give my gold tooth-pick, my silver-mounted snuff box, my seal ring, and a Canton segar case—making up the sum and substance of my ornamental treasures—to find out the wearer of that green silk pelisse."[26] In another fantasy, two women who were raised on an island after being shipwrecked as infants went fishing. Upon catching a "man-fish," they threw it back, thinking it to be a monster. When their mother learned what they had done, she explained that it was "one of the most tame, domestic animals in the world" and much more interesting to play with than their squirrel or monkey. Upon hearing this, one of the daughters said she would fish for him again, catching him whenever she pleased by using as bait "three tooth-picks to one pound of snuff."[27]

Regardless of the material of which it was made, using a toothpick as a toothpick in public was roundly condemned in the 1830s. Once, on a boat on Lake Erie, "a well-dressed youth was parading the deck with an air of self-satisfaction, and amusing himself by poking a large silver tooth-pick in his mouth." Another passenger was offended. He grabbed the youth's arm and said, "Young man, if your mother's sugar plums have rotted your teeth, and you must pick them, go below and get a quill of the steward; or beg a pen from the clerk, and cut it into the required shape. Use it privately. To clean your teeth in public is a sign of vulgarity—but to torment your gums with a metal spike, when you can treat them to the softness of a quill, is sheer stupidity."[28]

Even when properly used, toothpicks had to be of the right kind for the right effect, whether real or imagined. According to a Scottish superstition, "A rusty nail from a coffin will cure the toothache if used as a toothpick."[29] In Maine, a folk belief that lingered into the twentieth century had it that "a bit of wood cut from a lightning-blasted oak and used as a toothpick" would prevent the user from having a toothache. The validity of the practice was attested to by a septuagenarian, who "never had the toothache until he lost a toothpick that he had carried in his vest pocket many years."[30]

Some 1834 advice regarding the toilet noted that "the use of metallic toothpicks, pins, forks, &c., with which people are in the slovenly and thoughtless habit of picking their teeth, ought to be studiously proscribed."[31] One newspaper story considered "picking one's teeth in company or at table" to be "a breach of every rule of propriety." The story went on to explain that customs did vary from country to country, and that picking one's teeth was not considered improper at all "in some parts of Europe, particularly among the genteel classes in the city of Lisbon."[32] Such a story, sometimes but not always attributed to its original source, was commonly repeated, usually verbatim, in countless newspapers across America, at a time when copyright was apparently interpreted to mean a right to copy. The practice meant that the dissemination of opinion on everything from politics to manners was easily accomplished, albeit without regulation. Wit and humor, jokes and riddles, were also freely cribbed from one source to another. One gem, called a "good conundrum," ran in its entirety, "Why is a newspaper like a tooth pick? Do you give it up? Because every man should use his own and not borrow his neighbor's."[33]

As much as the use of a toothpick in public was frowned upon, however, the private use of one was encouraged. Indeed, the person who did not clean food from between his teeth was looked upon with derision. Again, it was satire that got the point across, in the form of a description of Job Doolittle, a man "remarkable" for his poor breeding, lack of initiative, and slovenliness: "He thought combing his hair a great waste of time; and, for the most part, dispensed with the use of buttons in his dress, from the needless labor they occasion every morning and night. His favorite food was small potatoes, placed very few in a pile. Toothpicks he never used."[34]

Increasingly, toward the middle of the nineteenth century, the use of a toothpick in the company of others began to be mentioned in a more neutral tone, with neither shame nor disdain. A club member, having dined on mutton chop and porter, admitted to "picking my teeth very leisurely to give William time to get me my pint of Sherry."[35] In another instance, a "literary gentleman" and Nicholas Nickleby were portrayed as engaged in a discussion about changing attitudes toward adapting material for the stage. At one point, Nickleby is interrupted by the gentleman, "leaning back in his chair and exercising his toothpick."[36] Sometimes, identifying the toothpick user's background or origins seemed to be enough of a comment on the practice. Thus an academic's view of Gotham described people he observed on the street after breakfast, among them "a Southern merchant with tooth-pick in teeth, hurrying from his hotel to Pearl or Wall-street."[37]

Class, or at least breeding distinctions, became emphasized in toothpick use, often in apocryphal but effective stories. One that may or may not be literally true seemingly involved the English actor and playwright Charles James Mathews:

> Mathews once, at a public dinner, sat opposite to a gentleman who, after using his toothpick, put it by the side of his plate; on seeing which, his next neighbor took it up and did the same. Mathews, horrified, said quietly, "I beg your pardon, but do you know you are using that gentleman's toothpick?" "Oh! yes, yes," was the cool reply, and in a few minutes more he repeated the dirty trick; when Mathews, unable to contain himself, bellowed out, "Sir, do you know that you are using that gentleman's toothpick?" "Well, sir, suppose I am, I mean to give it him back again!" was the answer of the offended citizen.[38]

Wanting to borrow another's toothpick continued to be among the attributes of the fool, according to *Vanity Fair*.[39] In a raucous dialog between a Yankee peddler and an Englishman who criticizes the peddler's dirty fingernails, the Yank explains that he lost his jackknife and proposes to the Englishman, "You lend me your knife to dig out my nails, and I'll lend you my tooth-pick."[40] The idea of using another's utensil was also commented on by an English visitor who observed in a betting parlor that the serving fork on a plate was used to convey the food directly to the mouth of various individuals and that bare fingers were thrust into a pot of hot chowder, which prompted him to note the adage that "fingers were made before forks," suggesting that the offender was not fully evolved, socially at least. He opined that the use of common dishes so literally in common was perhaps derived from the Declaration of Independence. As a final application of the principle, he noted that "it is by no means an unusual thing for a Yankee to offer you his tongue-scraper or his tooth-pick."[41]

Among the most remarked-on practices relating to toothpick use was the propensity of young men, especially, to employ them while congregating on the sidewalk, "lounging along with toothpicks."[42] The practice was recorded as early as 1836 by the Canadian humorist Thomas Chandler Haliburton, whose *Clockmaker* series of "anecdotes and cautionary tales" about contemporary foibles featured "the sayings and doings of Samuel Slick, of Slick-ville," Connecticut. Sam Slick, who regularly traveled to Nova Scotia to sell the clocks he made, was intended to be an archetypal Yankee peddler.[43] In one *Clockmaker* episode, "a shabby-genteel fellow, with not a cent in his pocket, is represented lounging or 'loafing' in the evening at the portal of a fashionable hotel, holding a toothpick in his mouth meanwhile, to give persons the false impression that he had just been dining there."[44] Indeed, even outside fiction, it was commonly held that "the man who carried a toothpick in his mouth did so in order to convey the false impression that he had dined on meat instead of coffee and cakes."[45]

The image of the contented toothpick chewer was invoked during a debate over a statue of Abraham Lincoln by the sculptor George Grey Barnard. The original had stood unchallenged in Cincinnati, but controversy arose when a duplicate was to be erected in London in commemoration of a century of peace between England and America. In the bronze, Lincoln was represented with his hands "grasped over his abdomen in a gesture strongly suggestive of pain," a stance to which opponents, including Lincoln's son, objected. The *New York Times* editorialized, "The humble origin of the man, the uncouthness his enemies found in his personality, need not be suggested in a statue intended to personify for all time the triumph of the democratic principle." Proponents retorted that the position was "one of the most natural assumed by Mr. Lincoln." While this was granted, the objectors still felt that it was reminiscent of the way the "Middle-West and middle-class public man of the late fifties stood on the hotel porch and plied his toothpick."[46]

If the hotel porch was where the toothpick users congregated, whether lounging about or holding up their bellies, the rear gate was where the young toothpick sellers huddled in the cold, waiting for their chance to enter. Although they were not looking for a simple handout, these children were classed with beggars by reformers, who feared "their exposure to the contraction of vicious habits."[47] At midcentury there were estimated to be on the order of ten thousand such homeless children in New York City alone.[48] The situation must have grown intolerable, for in 1869 the streets were "to be cleared of juvenile match and toothpick venders."[49] Adult vendors could also be annoying, especially when they approached someone at an inopportune time. In an 1870 cartoon, a sidewalk merchant offered a bundle of toothpicks (for three cents) to a gentleman coming out of a dentist's office. The aggravated gent retorted that he had "just bin and had" his last tooth pulled.[50] Even extractions did not stop a man on Guam from using a toothpick. The dentist who had taken out the man's last tooth came across him in a restaurant "picking his new dentures!!"[51]

Street peddlers seemed to be ubiquitous in the 1860s and 1870s. Here, one offers a bundle of toothpicks to a man exiting a dentist's office, where he had just had his last tooth pulled.

Whether children or adults, toothful or toothless, among the characteristics of street vendors of all kinds was the "extraordinary cheapness of their wares," including, of course, their toothpicks.[52] Perhaps it was some of the junior salesmen who would grow into the men who would be called "an ignoble class of nondescripts, neither peddlers nor beggars," who hovered about the edge of society. According to one observer, "These characters make a semblance, in the streets or public vehicles, of giving a *quid pro quo*, but offer wares so preposterous that buying is out of the question, and the only transaction performed is giving." Among the wares proffered were "many-pronged toothpicks," whose ends were likely frayed from much prior use. The characters were not believed to expect to sell the toothpicks but rather "to effect a gratuity instead of a purchase."[53] Who might want to buy their wares was the subject of jokes and cartoons, like the one about the street merchant behind a tray of toothpicks he was selling three for a penny: "Pick and try 'em, before you buy 'em!"[54] A similarly smart-aleck remark came from a prisoner appearing before the Paris Correctional Court. When asked how he made his living, he responded, "I sell second-hand toothpicks."[55] There would perhaps be more respect for the inmates of New Jersey's Trenton State Prison, whose residents spent their free time making toothpicks "and other trifles," which they were allowed to sell to visitors and then send the money to their wives and families.[56]

In fin de siècle New York, there was a class of peddlers on the Lower East Side who made a living selling small items such as pins and needles from a basket on their arm as they went door-to-door on regular routes. Their stock was obtained from "pack peddlers" who worked out of "dark, unwholesome holes in the ground" or some "ill-smelling rear apartment." A soldier of the army of basket peddlers was described as he exited such premises:

> He laid in a generous supply of ribbons, cheap laces, buttons, collar buttons, glove buttoners, tooth picks, garters, suspenders and green goggles.
>
> These he carefully placed in his basket, letting some of the ribbons and laces trail gracefully over the side, till they dangled half way to his feet.
>
> He was a member of a house to house peddling coterie. His basket usually contains $4 or $5 worth of goods, which he sells at seldom less than $12 or $13. To make $2 or even $3 a day is considered ordinary work, even when trade is not especially good.[57]

The toothpicks offered by peddlers and freeloaders were certainly not of the fancy-goods kind. At best they would have been goose quills. At the other extreme were the gold and silver toothpicks, peddled not by children and deadbeats in back alleys but by jewelers and silversmiths on the finest streets in New York. They dealt not in pennies but in fortunes:

> A gentleman in California has ordered from this City a splendid service of gold plate. . . . It consists of three hundred and fifty pieces, comprising dinner, dessert and tea services, from massive centre pieces, vases, tureens, &c., to cigar stands and tooth picks. All the pieces are elaborately chased, representing characteristic scenes among the mines. The whole service is valued at $15,000.[58]

So was the disparity of wealth encapsulated in a toothpick. It could be bought overpriced for a penny from a cold and unwashed hand or thrown in almost as an afterthought among hundreds of pieces of gold plate. But in the middle of the nineteenth century there existed strong feelings among some that everyone, no matter what their station in life, should do something "useful and honorable—no matter how small or how lowly—in the great workshop of social and intellectual organization." Although the life of the mind seemed clearly to provide an advantage to achieve a place in front rather than behind the hotel, nothing was considered to be without merit: "If you cannot contribute to intellectual pleasures—lasting forever—contribute to corporeal ones, though they be ephemeral. Peddle matches and toothpicks; cry out, with stentor lungs, the issues from the press; cart clams; be apprentice to a knife-grinder or an itinerant tinker, sooner than be idle."[59] But there was an inconsistency in it all, for the dandies on the hotel porch were the idlers, and neither Benjamin Sturtevant nor Charles Forster could be counted among them.

CHAPTER THIRTEEN

The Nasty Instrument

A LL THE WHILE Charles Forster worked to realize his obsession with making and marketing toothpicks, American society continued to debate, praise, ridicule, and generally remain ambivalent about their value and use. It was hard to deny that using a toothbrush, no matter how primitive, was good for keeping the teeth clean and thereby reducing the chance of decay, but in 1870 one could also read that "one good toothpick is worth an armful of toothbrushes," making the little sticks all but essential then.[1]

In one instance, young teachers were urged to "save the cost of all needless luxuries, that you may procure a good supply of the best of educational books." Among the luxuries were hiring a horse (when hoofing it would do), candy and peanuts, rings and gewgaws, and tobacco. At the same time, the teachers were advised to "keep up an intimate acquaintance with soap and water, brushes and toothpicks," so that they might "become as perfect as possible in health, person, mind, and morals."[2]

Though a young teacher might be able to get along with quill or wood toothpicks, rather than silver or gold, the precious metals continued to symbolize and represent wealth, achievement, and status. And it was not only men who carried the pocket trophies. In 1864, a woman leaving Albany on a train was the victim of a pickpocket, and among the contents of her purse was—along with eight hundred dollars in greenbacks, three thousand dollars worth of diamond rings, her railroad ticket to Indianapolis, and other valuables—a gold toothpick.[3]

In the latter part of the nineteenth century, Christmastime gifts for gentlemen included "books for those who are fond of reading, and toothpicks for those who are not."[4] Gold and silver toothpicks were advertised as suitable gifts, and a "splendid tree" might be decorated with red books with dazzling gilt edges, gold rings, gold pencils, and gold toothpicks. (Out-of-the-ordinary toothpicks and their accessories would also become appropriate gifts for special occasions. At a "brilliant wedding" in Mont Vernon, New Hampshire, a "silver tooth pick holder" would be among the presents.)[5] But the class given to ostentation and even to benevolence also came in for criticism of its "let them eat cake" insensitivity: "There are persons who would show their liberality to a starving man by sending him a costly toothpick instead of bread." The sentiment was frequently repeated in magazines that might scarcely be considered domestic, as it was in the 1873 Christmas number of *Forest and Stream:* "A gold tooth-pick can be hardly regarded as a suitable Holiday present for a starving man."[6] Nor could one of wood or quill be considered so.

Whatever the material of which it was made, the toothpick of a Victorian man (or at least one with teeth) typically was carried in a vest pocket. Getting the slender little rod out could take some fishing, especially since the index finger and opposable thumb entered the pocket most naturally and easily in a position aligned along the stick rather than across it. This often led to fumbling. A medical doctor writing of the physiological anomaly of having six fingers on the hand pointed out that the extra digit could be viewed as an asset or liability, depending on one's attitude toward it. Some families wore the genetic trait on their sleeve as a sign of distinction, but "Anne Boleyn kept her sixth finger bent downward upon the palm," and another afflicted with the trait complained of the cost of gloves. One person bragged a practical advantage, however, in that "his double thumb enables him to take a pencil or toothpick from his vest pocket without using the fingers."[7]

Those with a normal hand of fingers should nevertheless have known how to use their toothpicks discreetly, which would have included taking them gracefully out of the pocket. But there remained a plethora of those in any class who engaged in all manner of annoying habits with their toothpicks. In the early 1870s, on the occasion of court proceedings relating to a ring of thieves who had dealt in altered vouchers, one of the defendants, a functionary named Hagerty, "dressed in the latest mode, with an enormous diamond . . . blazing on his shirt-bosom, sat or rather reclined in a chair. . . . At times he would appear excessively nervous and downcast, and clutched spasmodically with one hand an ivory-headed cane, and with the fingers of the other snapped constantly a quill tooth-pick."[8]

Others engaged in "the funny and disagreeably-looking practice of sucking away at a toothpick." While perhaps also silly-looking, the habit was said to have made the practitioner calm and contented. Still, even a state of contentment could be the source of a complaint. One writer considered it to be "an injustice to the men" that feminine figures always represented virtues such as faith, hope, and peace. He wondered what would be wrong with employing the "figure of a man, with a toothpick in his mouth, to represent Contentment," even though it was not considered a classic virtue.[9] But regardless of who symbolized contentment and its corollary agreeableness, some believed that there were limits to the value of and tolerance for them. One commentator on manners observed that "in the United States, that quality of character expressed by the Yankee word clever, is always popular. Good-natured men and women are so heartily liked, that amiability is continually mistaken for good-breeding. A fellow that *means* well is at once assumed to have graduated in the school of manners; his cheerful or accommodating spirit covers a multitude of sins, and he is at once elected to the rank of a gentleman." But there were patterns of behavior of even the most agreeable characters that revealed them for what they really were: "An amiable fellow may so persistently masticate his toothpick while he is talking with you, that you would willingly accept a little acid in his composition in exchange for this unpleasant habit."[10]

One wag even invented a theory of toothpick chewing. The playfully sarcastic inventor asserted that "the leaders of the great anti-tooth-pick crusade" were mistaken in calling it a vice, in which a young man, "becoming addicted to the tooth-pick habit, sunk, step by step, until he would frequently chew tooth-picks at the public table or in the presence of ladies with a callous shamelessness that proved how completely the last vestige of self-respect had vanished from his breast." The theoretician did admit that the habit did indeed begin with "the seductive and costly quill" and escalated to an "unholy appetite" for "the coarse, fiery, wooden tooth-picks of the cheap restaurants," but he argued that this was in fact "a blessing to mankind." His satiric argument rested upon the observation that there was "an intimate connection between the mind and the teeth," as evidenced by the maddening effect of toothaches. Because of the proximity of the jaw to the brain, what affected one affected the other, or so it was reasoned. This led to the development of "the tooth-pick treatment for idiots and persons of weak mind," in which the stimulus of chewing a toothpick was sufficient to awaken the intellect. The treatment was said to be administered with great success in "our largest idiot asylums," and thus "nine out of every ten men who are seen in public in the act of chewing tooth-picks are discharged idiots, who are stimulating their minds in accordance with the system."[11]

The habit, no matter how arbitrarily embraced or rejected, was definitely not restricted to idiots. It seemed to be picked up by members of all classes and political persuasions. According to one man, who identified himself as an "observing and fastidious citizen who at noonday finds himself in the vicinity of restaurants":

> nearly every man he meets, whether of high or low degree, is industriously occupied in masticating his toothpick. Some are engaged in devoting the pick to its legitimate duty, others chewing the end as if they derived some sort of solace from the occupation, and others carrying between their lips this reminder of their recent refection, as if loath to part with it. Of course, he is only a slave to trifles who ventures a protest against this display. If gentlemen like to carry into the street reminiscences of their meals, if they disregard that social obligation which requires people to complete their personal adjustment before going into public, who shall deny them in a country where everybody asserts his fundamental privilege of doing, in all these things, just as he pleases?[12]

In 1876, the year Charles Forster purportedly threw toothpicks from a wagon along the streets of Philadelphia, chewing the wooden slivers was considered by some to be "in approved modern style."[13] But what was proper behavior was evidently not obvious to everyone. Letters seeking advice were frequently answered in the columns of ladies' and home magazines. But even just asking was often the subject of ridicule, as it was in the *National Police Gazette*, where it was noted that etiquette books did not necessarily help in the kitchen: "Alas! too true; and until they do fashionable women will be in doubt as to whether a silver spoon or a gold toothpick is the proper weapon with which to mash a cockroach."[14] Everyone wanted advice on the correct and proper use of a toothpick, and though the question was seldom made explicit, it was always clear from the answer. It certainly was in the response to a letter from one Blanche that appeared in *Arthur's Illustrated Home Magazine* in 1873, a time when streets were full of horses and their leavings:

> No, we do not think chewing a toothpick a mark of good breeding, nor the other practice to which you refer, as occasionally seen in young men, that of sitting with one foot over the knee. To carry a toothpick between the teeth and chew the end of it, is a disgusting practice; and not less so that of carrying a dirty boot across the knee, the sight and smell of which may sometimes be anything but agreeable to one sitting near. The toothpick should be kept out of sight as much as possible, and rarely, if ever, used at the table; and the boots kept near the ground, where they properly belong.[15]

Depictions of young gentlemen chewing toothpicks were published as early as 1857. The length of the toothpicks shown being used here suggests that they may have been imported from Portugal or Brazil.

While many women may have worried about the placement of feet and the proprieties of toothpick use, many young men evidently did not. In an 1879 issue, the *National Police Gazette* described "the man about town" as "a gentleman whose hardest task is to assassinate time." In a jocular exchange in the pages of *Puck*, the debate over whether women had endurance equal to men was considered to be moot "until women demonstrate their ability to undergo the fatigue of chewing a toothpick on a corner half the day and sit up playing billiards all night."[16]

The "gilded youth" of Paris were the touchstone and the toothpick the badge of the leisure class. However, there were geographical distinctions, and New York was said to be "more like Paris than London, where the stupid 'Crutch and Toothpick' gang are supposed to represent the class in male society whose members have nothing to do but to rise at noon, pass through a perfumed bath to the barber, and so to the 'Row,' to dinner, to the theatre, and finally to Evans's supper-rooms, or some faster place, where the night is given up to heavy carouse."[17] The production *Crutch and Toothpick* was a popular hit of the London stage in 1879, running for 240 nights. The play was written by the socially sensitive journalist George Robert Sims, who modeled it on a French farce. The phrase quickly entered the language, and "crutch and toothpick brigade" came to refer to "the dandies who affected [walking-] sticks with crutch handles, and held toothpicks between their teeth."[18]

Such types also became the subject of musical amusements. A typical member of the brigade was described in the opening bars of the song "Crutch and Toothpick," which was sung at surprise parties in the 1880s: "I'm an aristocrat, / Make no mistake in that; / I come of a line / Remarkably fine, / For troubles I do not care." Another song, "Toothpick and Crutch," which was suitable for accompaniment by banjo music, began: "I'm a dandy a nobby young fellow / The pride of the girls in this town / By jove they're in love with Adonis / The name I am known by around." The chorus of each song ended with words that were more physically descriptive of the swell: "Close cut hair; elbows square, with my tooth-pick and my crutch."[19]

Even when the likes of a crutch and toothpick brigade was on its best behavior, including keeping both feet on the ground, toothpick users could come in for unequivocal disdain, which sometimes degenerated into a rant. In 1874, an Englishman named S. Phillips Day had declared his agreement with "that public censor, *Punch*," which, he reported, had excoriated those " 'Savages in Clubs' who, dead to all feelings of delicacy, adopt the revolting and brutal practice of picking their teeth with sharp instruments while at table, and even in the presence of ladies." Day could "scarcely conceive a habit more ungentlemanly, offensive and abominable." He admitted that although the practice was "not followed so generally as it used to be, still the objectionable act [was] sufficiently in vogue to justify private animadversion and public reprobation." He agreed with *Punch* that "fashionable persons who persist in this odious practice should be ostracized from Club dining-rooms, and special chambers set apart for them, so that they may no longer inflict suffering upon others whose delicate organizations they cannot understand, and whose sense of decorum they fail to appreciate."[20] Just as there emerged a society-wide sensitivity to secondhand smoke in the late twentieth century, so there had developed an aversion to secondhand toothpicking in the late nineteenth.

After speculating that the toothpick was derived "from the *stecco* of the Italians and likewise formed the crude idea from which the two-pronged fork was drawn," Day went on to give his views of American toothpick use:

Some persons in America are particularly addicted to the foul practice of using toothpicks. In fact, not satisfied with the vigorous employment of such weapons during meals, they are said to carry them in their mouth out of the dining-room, and to keep digging at their teeth, or else twirling them between their lips for an indefinite period. This is an amusement equal to "whittling"; and a certain Yankee, as has been incisively observed, "can whittle a toothpick out of a pine log."

Nothing can well be more revolting to sensitive, cleanly persons than the habit of picking the teeth either at meals or afterward. The material of which the nasty instrument of torture is made, whether of wood or quill, does not render the practice less reprehensible. . . . The use of toothpicks should not be tolerated in civilized society, especially in what is termed "good society." Negroes do not need such things; then why should the white man? Savages can get on very well without such skewers; then why should the Christians patronize them?[21]

This tirade, reprinted in the *Missouri Dental Journal,* was followed immediately in the same publication by a response from an American dentist, who stated that American Negroes were "far nicer and more cleanly about their mouths than some Englishmen, inasmuch as they do use toothpicks very intelligently indeed." He further took the suggestion "that it is essentially English and genteel" not to pick one's teeth to have cleared up a mystery and have accounted "for the nasty, dirty condition in which American dentists usually find Englishmen's teeth." The response to Day repeated the widespread disapproval of "shabby gentlemen" who stood "on the steps of Boston hotels picking their teeth to induce passers to think they had just dined there."[22] And the phenomenon of picking one's teeth after eating was reaching from coast to coast, for even the occasion of the opening of a new hotel in San Francisco prompted the connection. It was to "accommodate 1,200 guests— quite a family to be supplied with tooth-picks."[23]

By the early 1880s, even smaller families needed a ready supply of toothpicks, and they were used by both men and women. According to an 1884 editorial in the *New York Times:*

The fashion of holding a toothpick in the mouth and chewing it in public has been adopted by ladies only within the last two years. Previous to that time the fashion was confined exclusively to men. At the present time no lady at a watering place hotel seems to regard her toilet as complete unless she carries a toothpick between her lips, and it is said that some ladies have become so addicted to the habit that they cannot feel at ease on rising in the morning unless they consume two or three toothpicks before breakfast.[24]

The editorial went on to muse over why "the human race is addicted to chewing toothpicks," and it focused, albeit hyperbolically, on the practice taken up by women:

Can it be that women imagine that carrying a toothpick in the mouth is a graceful and fascinating act? This seems hardly credible. A toothpick is sometimes useful, and even necessary, but the same may be said of a fine-tooth comb. Yet were a woman to carry the latter article in her hair and use it in momentary lulls of conversation she would scarcely attract admiration. The toothpick may be a useful auxiliary to the tooth brush, but no woman dreams of brushing her teeth in public. Whatever else the object of publicly carrying toothpicks in the mouth may be, it is incredible that any woman does it in order to add to her attractiveness. She might as well try to fascinate men by taking ipecacuanha or anti-bilious pills in their presence.

It may be suggested . . . that women carry toothpicks in order to protect themselves against unwelcome kisses. Undoubtedly the presence of a toothpick between lips presumably fair is admirably adapted to drive all thoughts of kissing from the minds of all but the most reckless men. Still, the fact that among women addicted to the toothpick habit are those who are ready to remove their toothpicks at the slightest prospect of a display of ardent affection forbids us to believe that toothpicks are carried in self-defense. Among the most ardent devotees of the toothpick are New England spinsters of advanced years and Emersonian views. To suppose that such women need any defense against sudden kisses is too preposterous to deserve consideration.[25]

Regardless of why women of the 1880s had taken up the wooden toothpick habit, it was not confined to the Northeast or to California. One could read of "an old sporting character" who was "plying a toothpick in the office of the Grand Hotel in Cincinnati, after a dinner of stalled ox and contentment, followed by a couple of fingers of the oil of joy."[26] Even when proper toothpicks were not available, toothpicking with improvised instruments was allowed. In Chicago, it was "not considered bad taste to nip the sulphur from the ordinary match" and use the matchstick as a toothpick.[27] And though in the post–Civil War decades there remained an impression that "the Southern people cherish such a deadly hatred to Yankees that they will neither smell, taste, touch, nor handle anything contaminated by Yankee hands," they were known to "call for Yankee toothpicks."[28]

Perhaps it was one of those that David Dickson, an Atlanta-area man, took to the grave with him in 1885. He had left with his nephew strict instructions about how he was to be buried, and as odd and enigmatic as his wishes may have been, they were carried out. Dickson's body was placed in an unpainted pine coffin covered with "common white alpaca" and buried in the garden of his home. Inside,

> the corpse was dressed in an elegant suit of black broadcloth and black silk velvet, but wore no shoes. The feet of the deceased were crossed, his right arm lay at his side, his hand clutched with the exception of the index finger, which pointed towards his feet. The left hand lay on his breast, and held a beautiful pocket handkerchief, and in the right pocket of his pants was a pocket-knife, a pocket-comb and a toothpick.[29]

But no matter where and how made, sold or cherished, it was the manner of using the toothpick in life that separated the well- from the ill-bred. According to one student of etiquette, writing in 1889, the instrument was properly used "simply without affectation and without obstinacy." He elaborated:

> At some of the best tables at which I have had the honor of sitting in Europe I found a quill toothpick laid at the foot of the wine glasses as being as indispensable a part of the couvert, or service, as a knife and fork. But unless I deliberately watched for a certain length of time, thereby losing the enjoyment of a part of the dinner . . . I never noticed guests using these toothpicks. And yet they did use them, certainly; but when doing so they did not hoist the white flag to call the attention of the whole table to the operation, as those do who try to hide their faces behind their napkin. This maneuver, so common among the Americans, is at best a false prudery, worthy only of the intelligence of an ostrich.[30]

In the last decades of the nineteenth century, toothpick use became a touchstone for judging a person's class and future. In a passage reminiscent of *Pygmalion*, two American women are talking about a mysterious stranger, trying to figure out his background. One of them says, "He'll be sure to have a title for us to know him by. He certainly is superior to all these creatures here; the bare look of him shows that, besides his table manners. He doesn't eat with his knife, or handle his fork and spoon like unaccustomed and inconvenient articles, or make his tooth-pick the prominent feature of his repast. He's an educated and eminently attractive being."[31] The other half of society was equally characterized by how it dealt with a toothpick. According to one late-nineteenth-century impression of New York, "At the news-stall you often see a little boy of nine or ten in charge; he has to stand on a stool and reach up to hand your change, but he already chews a wooden toothpick like his elders, and looks as disappointed and embittered as a man of fifty."[32]

But for those not embittered, the "gold-plated, everlasting toothpick" had become as much a part of a gentleman's "personal equipment as the key to his watch." When a residence in West Washington was broken into, the clothing was gone through for valuables, and among the items stolen was listed a "gold toothpick valued at $1.50," a not insignificant sum.[33] Though gold and silver toothpicks were declared "dangerous because the metal may scratch or chip the enamel of the teeth," they were also symbols of affluence and status, as well as "snobbishness." There was also the question of cleanliness. Ivory, in particular, was objected to by hygienists, since it is "absorbent and in the course of use becomes unclean." Those reading *Harper's Bazar* were advised not to carry a reusable toothpick "unless you are traveling in barbarous or over-squeamish countries." Otherwise, the advice was to "use a tooth-pick and throw it away afterward," meaning that it was wood or quill.[34]

The toothpick—of whatever material—so pervaded late Victorian life and sensibilities that over the course of that period it lost its hyphen as a noun and also became an adjective. By one anecdotal estimate, in 1888 three out of ten "average American men" would "sport a toothpick in their mouths in public."[35] But just as clothes styles and food preferences went in and out of fashion over the next century or so, so did the toothpick itself. In an 1890 farce, Mrs. Ambition plans to give a luncheon. When she says to her daughter, Kate, that perhaps Mrs. Splinter will loan her some toothpicks, Kate warns, "But, mamma, they don't use tooth-picks now. It isn't good form." The mother retorts angrily, "I don't care if *they* don't use tooth-picks—*I will*, anyhow, so there now! Do without tooth-picks, indeed! Kate Ambition, do you think your pa is a millionaire? What under the sun should we have for the fifth and sixth courses were it not for tooth-picks and water?"[36]

Mrs. Ambition was not the only one who continued to set her table with toothpicks. Fad and fashion are tidal, ebbing and flowing on a roughly generational time scale. It may well have been a contemporary of Kate's mother who was the author of an anonymous piece of verse entitled "The Pervasive Toothpick," which appeared in the *St. Louis Republic* in the early 1890s:

> *The tablecloth was fresh and neat,*
> *The china bright, the viands sweet,*
> *And slim and straight beside the meat*
> *Stood proudly up—the toothpick.*
>
> *Stood stiffly, as a toothpick ought,*
> *Which once was shunned but now is sought,*
> *For time has turned and forward brought*
> *To prominence the toothpick.*
>
> *The dinner done they passed it round,*
> *And none said "Nay," and no one frowned,*
> *But all, with dignity profound,*
> *Applied the nimble toothpick.*
>
> *Oh, other things of meaner sphere,*
> *Comb! tweezers! brush! The time draws near,*
> *Perchance, when each shall be the peer*
> *Of the promoted toothpick.*[37]

In spite of what her elders thought, young Kate Ambition still knew that toothpick use had fallen out of favor among girls.

But not everyone was a slave to fashion. In 1892, some men in a hotel writing room were discussing differences in European and American aristocracy, which was the subject of a new play by Bronson Howard, who wrote drama of social criticism. The conversation turned to how the American girl was "seldom treated seriously by the dramatist":

"She is always an extremely romantic, innocent, trustful girl," said one, "or a gushing, hoydenish, romping young female, who makes and breaks engagements with remarkable ease. Now, the typical American girl, I take it, is a modest, self-possessed, sensible woman, whose last wish would be to make herself conspicuous. Is it not so?"

The man who was addressed for answer waved calmly with his cigar toward the doorway. A picture was framed there, of which the center-piece was a young lady whose name figures near the head of the list in all official social events. She was leaning against the clerk's desk in a careless attitude chattering to two men, neither of whom was related to her, and in the intervals of her conversation she wielded with vigor a wooden toothpick of heroic proportions.[38]

She may have been an exception, admitted the defender of the American girl, but then there must have been a lot of exceptions. Not long before the men's debate, *Ladies' Home Journal* was offering guidance and responding frequently to questions about toothpick use. In a first-person item addressing young girls, Rose Cooke was firm: "Don't use toothpicks in public any more than you would a tooth-brush. I have actually seen these articles—toothpicks I mean—set on many a table in the country in sickening prominence, and I have seen girls and women, who would be very unwilling to be called anything but ladies, take them out after meals and use them as openly as they would their needle or their crochet-hook."[39] May S. was advised, "It is very bad form to hand around toothpicks." In a subsequent issue, the same magazine told H.E., "Toothpicks should not be put on the table, nor should they be used outside of one's own room."[40] Advice to the country girl away from home, who evidently was a composite formed from many letters to the *Journal,* took the form of a narrative that among other things warned, "If you are in a hotel, and the woman opposite you uses a toothpick and walks out of the room with one in her mouth, don't follow her example."[41] In the 1890s, the diminished position of the toothpick was evidently taking hold on the North American continent; a contemporary Canadian etiquette book, repeating what had appeared many times in magazines, advised mothers never to permit their children "to use a toothpick in public."[42]

In an 1894 issue of *Good Housekeeping*, an article on etiquette at the dinner table declared that the toothpick was "steadily falling into disuse, except in private." The writer hoped that the day of its demise might be hastened, for the toothpick's "free public use is absolute vulgarity, and for a person to go about with one sticking out of the mouth, after a meal, is disgusting!" And this barely a decade after the wooden toothpick was called "one of the adjuncts of civilization."[43]

In the meantime, toothpick use had become, in the words of an editorial in the *New York Star,* "an ugly habit." The sentiment was repeated verbatim by the *Los Angeles Times,* which reprinted the piece, in which a social critic mused,

> I wonder what saccharine or succulent qualities inhere in wooden toothpicks that so many persons cling so persistently to those unlovely little instruments long after they have performed the service for which they were designed. On any elevated railway train one is sure to see one or two men with toothpicks protruding from their lips, as if to advertise to the world a recent breakfast. Not all of those who make this exhibition are ill dressed or boorish, and one is left to conjecture in vain why a particularly private portion of the toilet is thus thrust upon public attention.[44]

Anonymous and pseudonymous arbiters of taste, ubiquitous in newspapers and magazines, were not shy about giving their opinions and advice. One asserted that toothpicks were "certainly never in evidence at good tables." Another, writing as Uncle Peter, was asked by a young student about the propriety of using in public the toothpicks that a college professor passed around the table. Uncle Peter agreed with the student that toothpicks were "only to be used in the privacy of a person's own room," and added gratuitously that he had learned that "too often college professors are surer about Greek roots, the mathematics of four dimensions, or the customs and manners of the antediluvians than they are about the rules of good society, or even English grammar." But even Uncle Peter acknowledged (parenthetically) that the wooden toothpick "is a most commendable instrument in its proper place."[45]

The presence of toothpicks on college campuses seems to have brought out intense feelings on both sides of the argument. A woman responsible for the commons at the University of Chicago once called toothpicks "vulgar relics of barbarism" and refused to furnish them to her boarders. In response to this announcement, a newspaper declared, "On the contrary, toothpicks are unknown to barbarians and are a mark of the highest civilization." Still, it was admitted that there was a proper place to pick one's teeth.[46] There was also a proper place for the toothpicks themselves, according to one playful observer who did not show a high regard for college boys: "The University of Chicago has abolished toothpicks, owing to the carelessness with which the students used them. Probably they wouldn't put them back" into the holder.[47] The homemaker had long been advised that there was a place and time for the location and use of toothpicks, and for houseguests that place was in the guest room, which should have a thoughtfully outfitted toilet table, complete with a supply of wooden toothpicks so that the guest could "make use of these necessary articles without offending your taste at the table."[48]

Although in literature "chewing the quill" had replaced "chewing the cud" as the metaphor for rumination and reflection, it was principally because of the quill's temporal priority over the wooden toothpick. In the mouth of reality, the latter had by now all but displaced the former, which remained "wholly imported" from Russia, Sweden, Germany, and France. Even with the introduction of hand-operated machinery for making quill toothpicks, the manual labor involved kept their price well above that of the wooden kind, rendering them noncompetitive.[49] At least one dentist saw no need for either wood or quill; he recommended "a small rubber band as the best toothpick."[50] No doubt he expected it to be used like floss in the privacy of one's room.

Whatever was thought about the toothpick and its use, by the end of the century it had definitely become so familiar an object that speculation could abound about its larger cultural significance:

The caricaturist of the future may perhaps represent the typical Yankee no longer as whittling, but going through the process called "chewing the quill." In nearly every hamlet, town, and city of the country the consumption of the tooth pick, both physically and commercially, has become a national characteristic. Few, indeed, are the hotels, restaurants, etc., where the invitation to "have a toothpick" is unknown; while no private house is considered well ordered in the absence of this edible.[51]

At the end of the nineteenth century, the wooden toothpick was indeed widely known and used, whether sanctioned or not. The future of the business of manufacturing toothpicks appeared to be bright, and new entrepreneurs were clamoring to get their foot in the door.

CHAPTER FOURTEEN

The Toothpick Man

CHARLES FORSTER MAY have established a broader market for toothpicks, but he could not keep it to himself forever—especially after the expiration of his patent protection. In the early 1880s, when the market was growing at a rate of about 8 percent a year on average, Forster went from being virtually the sole manufacturer to having competition that forced him to share the wholesale trade.[1] Toothpicks were being made not only by other manufacturers in Maine but also in Michigan, Minnesota, New Hampshire, and Wisconsin.[2] Though old factories were constantly closing and new ones opening, the establishment of any new facility was news, just as it is today, for with it came jobs for residents and benefits for the local economy. The news around eastern Massachusetts in January 1882 was that a factory capable of producing seventy thousand toothpicks per hour was going to be established at Brockton, about twenty miles south of Boston. It was an event remarkable even for the *New York Herald,* which expressed the "hope that the Boston ladies will not imitate some of the New York ladies who carry toothpicks in their mouths in stores and streets."[3]

But soon a headline was proclaiming, "An Immense Number of the Wooden Splinters Used in Boston." A reporter overheard one diner in a "somewhat tony restaurant" asking in a rather gruff and ungrammatical manner, "Where's your toothpicks?" And the answer was a disquisition of sorts:

"Well, the fact is, sir, it is almost impossible for us to keep a supply of these articles on hand. It may seem strange to you, but, if you would only believe it, there are no less than 5000 toothpicks used here every day. Oh, no, of course I don't mean to say that 5000 people dine here daily, but the people who do come here actually steal the toothpicks. Yes, sir; carry them off by wholesale. Other hotels experience the same trouble? Yes; it is a universal custom for people to carry off toothpicks, and, although it is a very small matter, still I venture to say that there is not a hotel man in the city but notices the rapidity with which his stock of toothpicks disappears."[4]

Prompted by this, the reporter inquired of a wholesale druggist how many toothpicks were used in Boston each year. The estimate he received was "at least a billion." He next asked to what use they were put and got a mouthful:

"Well, some people pick their teeth with them, others chew them, many dine off them, pensive people have a habit of whittling them away just for pastime, while others—and by others I mean the majority—carry whole pocketfuls about with them. These toothpick fiends you may observe anywhere and at all times. They pick their teeth at the table, in the parlor, on the street, in the horse-cars, in the hotel office, on the rotundas, and in fact everywhere you meet them; the mania is prevalent, and is increasing rapidly."[5]

In the 1880s the toothpick craze had been growing not only in Boston but everywhere else, feeding on and in turn feeding an increasingly fluid industry. According to an 1883 report in the *Boston Commonwealth*, a toothpick factory was established in the little town of Sebec, Maine, located near the eastern end of Sebec Lake in Piscataquis County. This was about as far to the north and east as the industry would reach, but the area was believed to have had "one of the best water-powers in the State." The Sebec factory was said to be doing a "large business," consuming annually more than a thousand cords of birch and poplar and "turning out a two-horse load of tooth-picks daily." The mill employed eight to ten men and between twelve and fifteen girls, but according to one source it gave "employment to as many girls as can afford to pack a hundred boxes for 25 cents, and feed themselves." Though this was considered "small wages for women" working in Boston, it was thought to be more than could be earned otherwise in rural Maine.[6] Another "flourishing" toothpick factory was located in Harbor Springs, Michigan, just across Little Traverse Bay from the now-thriving summer resort town of Petoskey. Then "one of the largest factories of the kind in the country," it was turning logs of white birch into about seven and a half million toothpicks per day, employing a process not unlike that developed by Forster.[7] Ribbons of veneer were fed into cutting machines, and out came toothpicks, "the perfect pieces falling into one basket, the broken pieces and refuse falling into another." There was a "wide market" for the toothpicks, which were "very neat and clean in appearance, sweet to the taste." They were packed into boxes of fifteen hundred each by "girls, mostly comely looking young squaws," and were sold at the factory for $1.90 per case of one hundred boxes—less than two cents per box, wholesale. The retail price of a boxful was five cents, which equates to three hundred picks per penny, "at which rate almost everybody can afford to take a fresh toothpick after each meal."[8]

In 1887, Forster was thought to be producing "three-fifths of all the wooden toothpicks made in the country." The business was shipping thirty thousand cases of a quarter million toothpicks each (for a total of seven and a half billion) annually and was averaging a hundred thousand dollars a year in sales, which was not bad for a business of its nature. However, the competition had grown so fierce that an industry trust was formed to protect the competitors from underselling themselves out of business.[9] In the 1880s in America, the term "trust" did not necessarily imply any form of ownership; rather, it was simply "a combination of manufacturers, engaged in the same industry, to kill competition and establish a monopoly."[10]

One wag joked about a Combined and Consolidated Association of Toothpick Manufacturers. In fact, it was the National Toothpick Association (NTA) that was said to manage toothpick production the way the Standard Oil Trust did petroleum: it "controlled the trade" by regulating output and price. At the time, the NTA had control over the production of five mills in Maine (one each at Strong, Farmington, and Canton, and two in Dixfield) as well as factories in Belmont, New York; Harbor Springs, Michigan; and Fond du Lac, Wisconsin. The association had "contracted for enough toothpicks to be made in Maine the coming year to load a freight train of fifty cars with nothing but toothpicks." In 1887, by June alone, five billion toothpicks would be shipped out of the state.[11]

The toothpick trust—what we would today call a cartel—did not control the entire industry, however, for there was "one small mill" in Massachusetts and another in Mechanic Falls, Maine, that were not members of the association. When one young Maine machinist and a partner were making a go of it alone, the trust attempted to "freeze them out." The machinist's response was to invent a "machine that would make a fabulous number of tooth-picks a day," securing orders for up to two years ahead. The slower machines used within the trust put its members at a disadvantage, and so they tried anew to force the upstarts out of business and again did not succeed. The young competitors continued to refuse to join the trust and so prevailed in their independence, but they were the exception. It is likely that the clever young machinist was Ernest Harris of Mechanic Falls, who in 1888 patented a machine that permitted the "passage of two or more parallel strips of veneer at the same time, and thus increase its capacity."[12]

With so much capacity, by the end of the 1880s the toothpick mills in Maine were said to be "suffering from overproduction," prompting the tongue-in-cheek question, "Have the young men who used to loaf around the hotels of the country gone to work?"[13] Still, in the early 1890s, new factories continued to open up else-where. A new plant in Chicago was said to be "one of the few out-ward signs of the World's Fair," a reference to the fact that the World's Columbian Exposition, which was to celebrate the quater-centenary of Christopher Columbus's famed voyage to America, was behind schedule and would, of course, open a year late, in 1893.[14] Amidst the proliferation of toothpick companies, desirable wood was getting scarce and manufacturers had to "send men into the woods now to hunt material up." While some observers may have recalled the quip that such a situation was "no reason why any man should chew a tooth pick for half an hour after lunch," it was no laughing matter.[15] As if growing competition and raw material shortages were not enough, the toothpick industry was, along with everyone, in the throes of a depressed economy. An 1894 report that a toothpick factory in Biddeford, Maine, had shut down operations prompted the comment that "so have a great many American teeth of late."[16] Still, the overwhelming bulk of all fin de siècle American wooden toothpicks were being made from white birch in Franklin County, Maine. At the time, there were also what were considered "principal factories" in Massachusetts, Pennsylvania, and western New York, but they were by no means locations that had the same association with toothpicks that the Pine Tree State did. There, independent factories continued to appear. One, the Hallett Toothpick Co., was established about 1897 in Mechanic Falls and was expected to have a daily output of a hundred cases of toothpicks.[17]

Years before there was a need to establish a toothpick trust, the Forster enterprise had also found itself in the midst of competition and circumstances that it had not previously known. Although it still dominated the market in terms of production, it wished to maintain that dominance and perhaps even increase it. But how? There was no more recourse in the courts after exclusive patent protection had expired. The only way to gain a renewed competitive advantage bordering on another monopoly was to gain new patent protection. The situation thus had spurred on invention. In 1881, Charles Freeman, then residing in Dixfield, filed an application for a patent on a machine that would produce not more of the familiar toothpick but an entirely new type. As inventors are wont to do, he made the case for his invention by criticizing the prior art, even if it was, at least in part, of his own creation. And even though it had been mechanized for barely two decades, Freeman wrote of the industry as if it were as old as whittling:

> It has been common to manufacture wooden tooth-picks by machinery; but the tooth-picks thus manufactured are somewhat rough, and have not sufficiently fine points. Tooth-picks made by hand are generally smoothed and provided with fine points; but both in the case of machine-made and hand-made tooth-picks they are, being made of wood, of a certain degree of softness and not sufficiently hard and firm to readily serve the purpose of a good tooth-pick.[18]

To solve the problem of soft, rough toothpicks with dull points, Freeman invented a machine that contained "devices for so manipulating the ordinary wooden tooth-pick as made on the machines now in use as will convert it into a polished, rounded, compressed, and pointed tooth-pick." To accomplish this, the ordinary toothpicks were fed into a progressively narrowing annular space between a rotating wheel and a stationary apron, both of which were "covered with fine-ground quartz, glass, or any other substance suitable for polishing wood." As the toothpick advanced through the mechanism, it was transformed into the promised product.

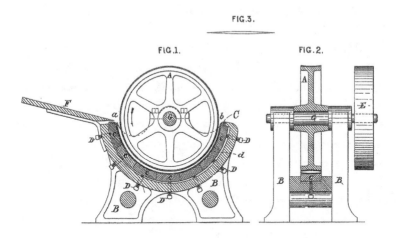

Toothpick-machinery inventor Charles Freeman patented this machine for compressing flat wood splints into double-pointed round and polished toothpicks, which came to be sold under the Worlds Fair brand name.

The new patent was issued in 1887, almost five and a half years after the application was filed, and it was assigned to Charlotte M. Forster, whose residence was identified as Buckfield. Just as she had obtained rights in her own name to Benjamin Sturtevant's toothpick patents while her fiancé, Charles, was away in Brazil, so Charlotte was once again at the center of the Forsters' intellectual property. At the time of the patent's issuance, Charles had recently turned sixty, and he likely was beginning to think about the fate of his business and how he might provide for his family. His wife, almost ten years his junior, was a logical successor to whom to pass control of the enterprise. The new patent over which she would have control could be expected to provide some degree of security for her and for their children. (In 1895, she would be assigned rights to still another toothpick-making device, this one invented by John C. F. Scammon, who directed the Forster factory at Strong. The application for the patent was made in 1886, which would have been during the time a new plant was being set up there, and Scammon would naturally have been looking for ways of eliminating problems that had complicated operations at earlier plants. The focus of Scammon's patent was an attachment in the form of a receiving tube with a narrow throat that constrained the picks to stay in a parallel arrangement, thus simplifying their subsequent handling.)[19]

Although Charles Freeman's new machine would be protected by a patent, Forster and Freeman no doubt realized that this would not prevent other inventors from coming up with different clever ways of turning an ordinary flat toothpick into a competing rounded, pointed, and polished one. The partners in spirit if not in fact, knowing how long it had taken them to develop an effective toothpick-making machine in the first place, could not count on any great period of time before significant competition surfaced. Whether it was Forster or Freeman who initially made this realization, almost a year and a half before the machine patent was issued an application was filed by Freeman for the product itself. But within four months the Patent Office examiner found the specification for a round, double-pointed toothpick to be "vague and indefinite" and the object itself not to be of "patentable novelty" because it had "long been common to cut tooth picks round and pointed at each end."[20]

Through his attorney, Charles Drew, Freeman sought to amend the specification to emphasize that he wished to patent a "new article of manufacture, a wooden toothpick, which by means of compression and polishing is made round and smooth." He added a drawing of the toothpick he described in words as being "rounded, compressed, and tapering gradually from its centre to each end." The minimalist patent drawing is perhaps the simplest ever produced.[21]

But the Patent Office was not satisfied with a drawing. It demanded that a model be submitted.[22] Though physical models were once required to accompany all patent applications, the problem of storing them had led to a relaxed policy. Nevertheless, the Patent Office had retained the right to ask for a model, which it did in this case. In response to the request, Drew forwarded "some of the toothpicks made in accordance with the specification in this application and embodying the invention."[23] The toothpicks were evidently accepted as proof of principle, but the patent examiner indicated that the claim for a "round compressed and polished tooth-pick tapering gradually from its centre to a point at each end" was "broader than the statement of invention" with which it was supposed to be in harmony.[24] In response, Drew submitted an amendment that referred explicitly to Freeman's 1887 machine patent.[25]

The claim was again rejected, however, with a new examiner asserting that "a tooth pick made of naturally hard wood would be the full equivalent" to what Freeman's machine produced from soft wood.[26] That assertion was met with an attorney's letter stating that no known "ordinary" wooden toothpick had "its longitudinal fibres compressed all the way to the point." Rather, ordinary toothpicks were pointed by "shaving or cutting" away at the fibers. To make the point, two sketches were incorporated into the text of the letter, which went on to note that "in a knife-sharpened pick, innumerable fibre ends, each of which is free to catch in and leave in the teeth, small woody threads or bits, more troublesome than the remains of food." The compressed toothpick did not have these disadvantages. The letter concluded with a new set of claims,[27] which the examiner finally accepted (more than five years after the initial application) and which were incorporated into the patent as issued.[28]

In the process of securing a patent on the round toothpick, Freeman's attorney argued in a letter to the commissioner of patents that compressing wood fibers was superior to cutting through them to form a point.

(No Model.)

C. C. FREEMAN.
TOOTH PICK.

No. 448,647.

Patented Mar. 24, 1891.

WITNESSES:
Chas. S. Gooding.
Chas. H. Swan

INVENTOR:
Charles C. Freeman
by Charles Drew
Atty.

The patent drawing for Freeman's improved toothpick was stark.

In his petition for the patent, Freeman had asked that it be granted to "Charlotte M. Forster, wife of Charles Forster of Buckfield," to whom he had assigned his interests.[29] However, patents can be issued only to the inventor. He could have assigned his rights to Mrs. Forster, but no assignee was listed on the patent as issued. Whatever the disposition of the ultimate rights to the compressed, rounded, polished, and double-pointed toothpick patent, the patent itself was as focused as that for Sturtevant's wooden toothpick. Freeman's stark patent drawing—a single view of a single toothpick, which sufficed because of the rotational symmetry of the object—had survived the examination process intact. The drawing of the toothpick was a small thing in a large blank field, but its implications were huge. Instead of the flat, wedge-shaped toothpicks that were stamped out of a ribbon of veneer, this new shape had a round cross section throughout its length and tapered down to a point at each end. Superficially, it resembled the toothpicks that Noble and Cooley had proposed to be made on one of the machines they patented in 1872, but those were just two-pointed variations on the flat toothpick.[30] The Freeman cylindrical toothpick had a more fully three-dimensional point and was almost ahead of its time as a designed object. This was later attested to by the erroneous claim that "the invention of the disposable round toothpick was one of the great multipurpose inventions of the 20th Century, ranking only second in this category to the paperclip," which in fact also dated from the late nineteenth century.[31]

Freeman, as the visible, residential head of operations of the mill in Dixfield, had become so closely associated with the toothpick enterprise there that a history of the town referred to the "Freeman (Forster) Toothpick Mill," but it was Forster himself who would be identified as "the toothpick man."[32] Nevertheless, it should not have been any surprise that it would be Freeman more than Forster who would invent new mechanical ways to protect their interlinked futures. Moreover, another distinguishing feature of Freeman's round-toothpick patent is the fact that one of the witnesses was the oldest Forster child, Charlotte B. Forster. She would have been twenty-one years old when the patent was filed, and that she served as a witness to this important document that would mean so much to the Forster business suggests that she was possibly being groomed to get involved in it, perhaps with her mother. There would soon be other indications that the business was inextricably connected with the Forster family.

Before that would play out, there were the new toothpicks to manufacture, market, and sell. They appear to have been offered as early as 1889, the year of the International Exposition in Paris, under the brand name World's Fair. Perhaps with that name Forster was evoking, at least to himself, the spectacle of his throwing out his boxes of flat toothpicks from the hired wagon pulled through the streets of Philadelphia during the Centennial Exposition. Later, in describing his use of the World's Fair trademark for toothpicks, he identified the words themselves as its "essential feature." However, they were then generally arranged over a pair of crossed double-pointed toothpicks, with the letters "C. T." above and the abbreviation "Co." under the intersection. "C. T. Co.," or Cutter-Tower Company, had at one time distributed Forster's entire output. As a cautionary note, the words "trade" and "mark" appeared to the left and right of the intersection, respectively.[33]

Although the letters and abbreviation and their placement were explicitly mentioned in trademark statements Forster would eventually file, he did not register "C. T. Co." No doubt the Cutter-Tower firm rightly considered that to constitute its own long-established trademark. Thus, the label on an old box of double-pointed flat toothpicks gives no hint of their manufacture by Forster, the only brand name being "C. T. Co." appearing within a ribbon-like border. Though identified on eBay as coming from the Civil War period, the box more likely dates from the latter part of the 1870s.[34] The designation for the grade and type of toothpick the box contained would later be trademarked by Forster, who would at that time assert that he had used it continuously in his business since 1875.[35]

Early boxes of the newer World's Fair "Wood Tooth-Picks" displayed prominently, in letters as large as those of the brand name itself, the four notable features of the new picks ("COMPRESSED · POLISHED · ROUNDED · POINTED"), with the words bordering the crossed-picks trademark. In other arrangements, the words "Velvet" and "Finish" appeared above and below the "C. T." and the "Co.," respectively, emphasizing the smoothness of the compressed and polished wood. Indeed, there was no comparison between how smooth to the touch the compressed picks felt and how rough were the flat ones stamped out of veneer like cookies cut out of dough. The New York–based Tower Manufacturing and Novelty Company, familiarly known as Tower M & N, devoted the entire first page of its 1899 catalog to the World's Fair Wood Tooth-Picks, which it advertised as Tower's Compressed Round Polished Wood Tooth Picks. As the catalog copy explained, "For many years we have been trying to get up a Toothpick to take the place of the low-priced, cheap pick, and the high-priced orange wood pick, and have at last succeeded. The World's Fair Wood Toothpick is compressed, rounded and polished. The point being compressed makes it very tough, so that it will not chip or break off as the common wood pick, nor will it bend as the orange wood."[36]

Tower's World's Fair toothpicks came in boxes described as "hotel size," containing fifteen hundred picks, and "family size," holding three hundred. The latter was offered by Tower in cases of twenty-five cartons, each carton containing twenty-five boxes, for a total of 187,500 toothpicks. Clearly, Tower was selling to the trade. At the time, Forster itself did not market toothpicks directly either wholesale or retail. Rather, it sold through its agent, Cutter-Tower. Some years later, Tower M & N and the Samuel Cupples Wooden Ware Co. of St. Louis would deal with Forster directly in distributing Worlds Fair toothpicks in boxes that bore the crossed picks but not the "C. T. Co." designation.

Early boxes of World's Fair toothpicks also bore the notice "Patent allowed." Later, boxes of "The Improved Worlds Fair Polished Wood Tooth Picks" bore the more explicit "Patented March 24, 1891. Made in U.S.A." It is possible that boxes so marked were produced before the patent had expired and thus served as a warning to competitors. It may be equally likely that they were produced after its expiration, there being no longer any reason to be coy about when the patent was "allowed" and to use the date to impress upon toothpick users that the brand was long established. The reminder that these picks were made in America was no doubt prompted by increasing competition from imported varieties that were not polished to so velvety a finish. Still, one end of the box declared that the World's Fair remained "THE MOST POPULAR PICK IN USE," without qualification.[37]

Charles Forster, who came to be referred to as the "veteran toothpick manufacturer," had achieved his objective. However, in the early 1890s business generally remained depressed. Toothpicks were selling for only one-twelfth the price at which Forster had first sold them.[38] Circumstances did eventually improve, of course. As the century drew to a close, his company was producing machine-made toothpicks from birch that arguably were superior even to handmade Portuguese orangewood ones. Forster's decades-old dream had become reality, but the septuagenarian was now ill and knew that he wanted to put his house in order. Not being an inventor himself, Forster had never had a patent issued to him, and so he could not pass that form of intellectual property on to his heirs. It was true that the 1887 Freeman patent had been assigned to Mrs. Forster, but it would expire shortly after the turn of the century. In 1900, the Forsters' son, Maurice, who in the meantime had become involved in the production end of his father's business, was issued a patent for a device to remove the dust and small splinters from a falling stream of newly formed toothpicks, with the rights assigned to Charles Forster.[39] In securing a patent, the son may have been trying to show the father that he was a logical and worthy successor in the business; in assigning the rights to the old man himself—as opposed to his mother or older sister or to the company—Maurice may have been demonstrating genuine filial piety or seeking to ingratiate himself with the old man. Or the old man may have demanded that the son put the patent in the family pool. Generally, however, there seemed to be no consistency in how patents associated with the Forster business were or were not assigned, nor to whom—which suggests that there were at best changing models of, if not arguments about, how the business should be run by and held in the family. One conclusion seems inescapable, though, which is that in the year 1900, when he was ill, Forster had begun to amass intellectual property under his own name.

Within a period of a couple of months in the fall of that year, Forster applied to the Patent Office for a series of ten trademarks, some of which he stated that he had used for twenty-five years. In addition to the World's Fair brand, he claimed protection for the names "Ideal" and "New Century," whose use he dated only from the late 1890s. He also trademarked the phrases "Flat Hard Wood Tooth Picks," and "Double-Pointed Wood Tooth Picks," in addition to the cryptic designations "1B," "2B," "1P," and "2P," whose use he dated from 1875.[40] The numerals 1 and 2 indicated relative quality. The 1B toothpick was a "second grade, birch toothpick" inferior to the Ideal, and the 2B was "not as high grade" as the 1B. Initially, the designation B and P may have designated toothpicks made of birch and poplar, respectively, but eventually the 2B "was made of both birch and poplar."[41] A box of "Double Pointed Wood Tooth Picks" bearing the designation 2P and the statement that they were "patented June 2, 1863, and April 26, 1864" advised the customer that "all genuine Picks have Patent Dates, as above, on each box."[42]

Some toothpicks made by Charles Forster were sold in plain boxes whose contents were distinguished by a paper label pasted on one end. This box, from the 1870s, contained flat toothpicks stamped out in a lenticular shape, which alone provided the pointed ends.

All ten trademarks registered to Charles Forster show his eldest child, Charlotte B. Forster, as a witness, further evidence that he was trying to involve her in the business, perhaps even thinking of leaving her in charge. As a young girl, Charlotte had evidently been sent away to school, but neither she nor her sister, Annie, appears ever to have married, and neither seems to have borne any children. Charles Forster's son, Maurice, did marry, but he also did not have any children. Hence, the patriarch Forster was faced with the problem of how to draft a will that would divide his estate, which other than his house and home consisted essentially entirely of his toothpick empire, among his wife and children without running the risk of having the business inappropriately dissolved after his death.

The declaration accompanying each of Forster's ten trademark registrations was notarized by Oscar H. Hersey. Oscar was the son of Levi Hersey, a Free Will Baptist clergyman who around 1870 moved his family to Buckfield. About five years later, Oscar "began reading law in the office of Hon. Geo. D. Bisbee" and was admitted to the bar in Oxford County in 1877, eventually becoming a partner in the firm of Bisbee & Hersey.[43] Practicing in this region of Maine,

Hersey likely had at least a passing familiarity with the toothpick industry and with the more general wood novelty industry that produced such things as checkers, clothespins, and other items for amusement and household use. It is also very likely that some if not much of his law practice had touched on related business matters. Indeed, around 1890, Hersey had joined six others in loaning twelve hundred dollars to the firm of Harlow & Son, which was then operating a steam-powered toothpick mill in Buckfield. The mortgagers wanted to encourage local industry, and the Harlow operation had been in need of capital. With the money loaned to them, the Harlows were able to buy new equipment for making wooden cutting blocks and meat blocks. Unfortunately, the new enterprise failed the next year, and Hersey took over the plant and sold the machinery to recoup some of his and his partners' money. However, those who sold the equipment to Harlow & Son understood that they were to retain title to it until the entire bill had been paid, and so they took Hersey to court to recover its value. The Supreme Judicial Court of Oxford County found some but not all of the machinery to have been properly disposed of.[44]

Hersey had established a successful practice in Oxford, Androscoggin, and Cumberland counties. Oxford County included, in addition to Buckfield, the towns of Sumner and Turner, the locations of Forster's earliest toothpick mills, as well as Dixfield, one of his most successful. Hersey, "a sharp and efficient business man as well as a lawyer," became involved with county and then state politics, always as a Republican, serving in Augusta as both a representative and a senator. In 1899 he moved to Portland, which is in Cumberland County, and with Judge Enoch Foster—"one of the brightest of legal minds" but of no apparent relationship to the Forster family, even though some Forsters did use the spelling Foster—established the law office of Foster & Hersey. The firm "grew rapidly and prospered," becoming the largest practice in Maine.[45] It is possible that Charles Forster became acquainted with Hersey during their mutual residence in Buckfield, or it may have been that Forster first established contact in Portland, where he resided during the last years of his life.[46]

The house that Charles Forster lived in since at least 1889 was an elegant freestanding Victorian on now-again fashionable Park Street.[47] The residence, number 85, was built at midcentury by a local jeweler and 150 years later still commands a dramatic prospect down Gray Street, over which it looks from the elevated levels of its living quarters. It is possible that Forster invited Hersey to this house to discuss his last will and testament and the disposition of the business he had built up from scratch. As any good lawyer would have done, Hersey could be expected to have asked Forster to spell out what assets he had, and it is likely that then the question of intellectual property was raised. There can be little doubt that there was a systematic discussion of the future of the toothpick business, for applications for trademarks began to be filed within days of Forster having signed his will, in the first week of October 1900.[48]

Since Forster had not registered trademarks previously, it seems very likely that it was Hersey's suggestion that this be done. Then the question would naturally have arisen as to how they should be recorded. Since one of the Freeman patents had already been assigned to Mrs. Forster, it might seem logical that the trademarks would have been assigned to her or registered in the name of the business itself. That they were not suggests that there continued to be some uncertainty as to how the business was to be controlled after the patriarch's death. It is possible that Cutter-Tower would wish to acquire it but that Forster might not wish to see his name subordinated to, if not completely absorbed by, that of his early associate Levi Tower.

At the time the trademarks were registered, in his name only, Forster had been ill for a couple of years, and the end may very well have appeared to be in sight. Mrs. Forster, who had suffered "many years of sickness" in her life, seems also to have been ill at the time, and she would die within fourteen months of him.[49] For whatever reason, Forster decided not to leave the business directly to her. Perhaps he sensed that there was some disagreement between his elder daughter, who had served as a witness for all of his trademark statements, and her mother, who had been assigned rights to the valuable Freeman patent, about what direction the business should take in the future or whether it should be liquidated. Or perhaps there was disagreement or uncertainty about whether the son should have a voice, if not a controlling hand, in the operation. In any case, it is likely that Charles Forster and his lawyer Hersey discussed these and other matters relating to the final disposition of the estate, and a will was drafted accordingly.

On March 9, 1901, Charles Forster died in his home, after being ill for nearly three years. His obituary read, in part:

> When yet a young man, he went to Pernambuco, Brazil, and became a partner of Henry Forster & Co., which was one of the largest commission houses in that country, and enjoyed a world-wide reputation for honest dealings. In about two years his health failed him and he returned to the United States and became largely interested in patents and patent-rights. In 1870 he moved to Maine where he has since been engaged in manufacturing, having at the present time large factories in the towns of Strong and Dixfield. . . .
>
> Mr. Forster was philanthropic and kind in all his dealings with mankind and all who knew him learned to love and respect him. In the great temperance reform movement of 20 years ago he was one of its warmest supporters contributing liberally of his time and money, and no worthy object ever went from him with an empty hand. In religion he was a Unitarian.[50]

Understandably, the man is presented in the best light, but how curious that there is no explicit mention of Forster's pioneering work in making and marketing wooden toothpicks around the world—his dream fulfilled. Yes, he "became largely interested in patents and patent-rights," but these were exclusively for making toothpicks and for protecting a monopoly to do so. His interest in controlling the rights to patents rather than seeking to secure patents of his own may have been prompted by his knowledge of how Benjamin Sturtevant, by all accounts a brilliant inventor, was exploited for his patents, including perhaps even by Forster himself. Of course, Forster also had no choice but to exploit inventors, given that he was no inventor himself. In any case, his interest in patents and patent rights seems to have centered on what they could prevent others from doing, since he had reportedly spent as much as fifty thousand dollars on "patents in litigation" over the course of almost two decades in maintaining his intellectual property advantage.[51] His first direct dealings with the patent office appear to have occurred when he was on his deathbed, when he wanted to protect the trademarks of his business from beyond the grave.

Could it be that his obituary portrayed Forster the way he himself wanted to be portrayed and remembered? Did his neighbors on Park Street and in Portland society know the source of his fortune? Forster was capable of deception and self-promotion, as the stories of his introduction of toothpicks to the Boston trade and Philadelphia fair-goers demonstrated, and he may well have been able to misrepresent what it was he did for a living. But how could he have hoped to keep hidden what must have been common knowledge in Maine, at least? Even though his name did not appear on the boxes of toothpicks that were sold exclusively through wholesalers, agents, and distributors, he was eventually to be referred to posthumously as the "millionaire toothpick king of Maine" and his product as "probably the greatest social innovation of the 19th century next to the spittoon."[52]

One thing is certain, and that is that his own family could not have been deceived, his wife and elder daughter being involved as they were in the intellectual property and his son in the production end of the business. But could it be that one or more of them, who may well have had the greatest control over the content of his obituary, wished that their fortune and their house on Park Street had come from some more dignified pursuit, such as exotic trade or fancy goods or patents for more glamorous inventions and processes?[53] Could it be that Charles Forster had glimpsed evidence of this in his survivors and wrote his will accordingly?

In his last will and testament, Forster left all of his "household furniture and personal effects of every name and nature" to his wife. Everything else—including the factories in the towns of Strong and Mexico (the jurisdiction within which the Dixfield plant was actually located)—was left "in trust" to the attorney Oscar Hersey, who was also named executor of the will. As trustee, Hersey was authorized to "engage in and carry on the manufacture and sale of toothpicks and cigar-lighters." Forster also authorized his trustee to use his judgment regarding making "any wood novelties or other manufactures."[54] In other words, Hersey—and Hersey alone—was given carte blanche to carry on and even expand the Forster enterprise.

Since Hersey had had some firsthand experience with the complications if not the operations of running a woodworking business and had some appreciation of the issues that could arise, he might have been a wise choice to run the Forster toothpick business after

Oscar H. Hersey is shown here in a portrait dating from about 1902, when he began his term as trustee of the Estate of Charles Forster.

its founder's death. According to Forster's will, out of the profits from the business were to come quarterly payments amounting annually to one thousand dollars each to his wife, his daughter Charlotte, and his son, Maurice. His daughter Annie, because of her unspecified "physical disabilities," was to receive a total of thirteen hundred dollars a year. If the net profits from the business fell short one year, the amounts were to be reduced pro rata.[55]

Forster, who later was credited with having "done more for the teeth of America . . . than any other man under the sweep of her eagle's wings," appears to have wanted to control his money posthumously.[56] In his will he directed Hersey to accumulate additional profits until twenty-five thousand dollars was amassed as working capital. Only after that happened were any additional net profits to be paid to the heirs in equal amounts. Forster also empowered Hersey, as his trustee, to buy, sell, and convey real estate and personal property as he judged "necessary for the welfare of the business." The trust established was to remain in effect until one year after the last member of his family died, and then what remained in the estate was to be distributed "according to the laws of descent then in force," which would mean to his surviving nieces and nephews.[57]

The Forster toothpick business was thus perpetuated through the creation of the Estate of Charles Forster, the legal name under which it would be known, operated, incorporated, and sued.[58] Trustee Hersey was given wide-ranging powers to "carry on the extensive business Mr. Forster had built up," a business that Forster evidently did not trust any of his heirs to carry on successfully.[59] Perhaps Forster sensed or feared that his survivors and heirs would have distanced themselves from the toothpick business the first chance they got. Perhaps he was right, considering that one or more family members were probably responsible for omitting all mention of his central involvement in the toothpick industry from his obituary. By creating the Estate of Charles Forster, he frustrated once and for all any attempt they might make to extricate themselves from the toothpick business. His will saw to it that his family would be reasonably well provided for, but as a result of the manufacture and sale of toothpicks and related articles—and not through the sale of his business.

CHAPTER FIFTEEN

Cords of Paper Birch

B Y THE EARLY twentieth century, 95 percent of toothpicks made in America were of birch. The white birch that Charles Forster had found so perfectly suited to making toothpicks by machine was more commonly known among foresters as paper birch (*Betula papyrifera*), the name referring to the fact that "its bark consists of several thin layers resembling paper," which could be "cut from the tree in long sheets." This property, plus its resistance to being damaged by water, made it ideal for American Indians to use for the "skin" of their cedar-framed boats, thus giving the wood the additional name of canoe birch.[1]

For canoes, the bark was split from the tree lengthwise, yielding a single sheet perhaps sixteen feet long and as wide as the tree's circumference, which might be as great as five feet. This was sufficiently large to cover an entire small craft. To do so, the sheet of bark was laid flat on the ground, with its white outside up, and the cedar frame centered on it. Then the bark was bent around the canoe frame and held in place with stakes driven into the ground, until the bark could be sewed to the wooden gunwales with thongs made from the roots of spruce trees. For larger canoes, sections of bark were patched and sewn together. Where birch bark was not available, that of elm, hickory, spruce, basswood, or chestnut could be used, but canoes sheathed in them got waterlogged and heavy, perhaps having to be abandoned after a brief period of use. Birch bark did not absorb water, however, and so a well-made canoe clad in it could last for ten years.[2] The bark, being "very strong and almost totally resistant to rot," was also employed by the Indians to cover their houses, and such practices were followed also by early European settlers. (Norwegians used the bark to roof their dwellings; Lapps made clothing out of it.)[3]

Birchwood itself was ideal for making thwarts, the structural crosspieces of the canoe, because it could be readily carved into the proper shape. This wood has been described as "of medium weight, fairly hard, not strong, of close grain and of uniform texture," thus making it ideal also for making shoe pegs and toothpicks.[4] Birch is preferred over poplar (or maple), which is also used, for "its softness and pliability, which afford just the amount of resistance requisite for a toothpick," and for its additional desirable quality of "retaining the forest odor and sweetness." While the name "paper birch" does not connote the quality of stiffness or soundness that is sought in a toothpick, "white birch" suggests the cleanliness that is desired. Hence, whereas the former name was favored by foresters, the latter was by Forster and other toothpick makers. Not surprisingly, the white birch has come also to be known as the "toothpick tree."[5]

White birch—along with the trimmings of other woods, such as pine, left over from the production of larger things—has also been used in the production of the variety of small wooded items manufactured by the so-called novelty mills of the Maine wood utilization industry.[6] The output of these plants demonstrated the "Yankee genius in utilizing waste material by the invention of new machinery." Toothpicks were sometimes made of pine trimmings, but mostly they were made of white birch, as were spools for winding cotton thread. At the turn of the twentieth century, items manufactured of "waste wood" would include, in addition to toothpicks, such trifles as small boxes to hold dice and dominoes, bicycle rims, children's toys, and other products used "so often and commonly that our wonder about their manufacture ceases." Great numbers of birch saplings were also cut down by what were known as "hoop pole hunters," men who typically served as area guides during summer months but in the off-season cut and shaved saplings to be made into barrel hoops. In fact, it was commonly said that "the spool factories and hoop pole hunters saved Maine's woods from being overrun by white birch saplings." The reasoning went as follows:

At one time countless millions of deer and rabbits roamed through the Maine woods, and they subsisted largely in the spring and winter on the sprouts of the white and gray birch saplings. There is no more prolific growth than the birch, and in Maine if left unrestricted the trees will soon spread everywhere and crowd all else out of existence. The destruction of the rabbits and deer destroyed nature's nice balance, and the birch trees threatened to rule everywhere. When the trees were about to monopolize all the Maine woodlands, the hoop pole hunters and the spool manufacturers discovered that the birch was the best wood for their purposes. The result has been that an enormous industry has been built up with an inexhaustible supply of raw material, and the birch saplings have been kept within certain restricted areas.[7]

Thus, before it became used for toothpicks and spools, white birch was considered a liability on wooded land. It would be declared "worthless for lumber."[8] However, soon after Charles Forster adopted it as his wood of choice, he created a "ready market" for it in Franklin County, Maine.[9] By the end of the century, although still not considered to rank among the most important of hardwoods, it was a "valuable species in the Northeast, and particularly in Maine," where the only hardwood tree cut down more frequently was the aspen—to make matches. The paper birch was known to grow "mainly on burned-over areas," and past forest fires had led to extensive stands of it in the western part of the state. In his 1909 Forest Service circular on the tree, S. T. Dana, an assistant with the service, departed from his matter-of-fact introduction to observe of the birch, "This characteristic of occupying burned-over tracts it shares with 'popple' [i.e., poplar; also spelled 'pople'], with which it is frequently associated. Both species are, in a way, the pioneers of the forest; they occupy the ground only for a short while and prepare the way for the permanent forest type natural to the locality."[10] According to another point of view, "Popple is an unlucky wood, to go in a ship or a house, because the wood of the Cross was made of it. And that's why the leaves of the popple have shivered so ever since." Whether lucky or cursed, poplar was used much less frequently than birch for toothpicks. Among Mainers, it would come to be considered "pulp."[11]

Sweet birch, which grows in Appalachian forests, is another wood that has been used for making toothpicks. Also called black birch, after its black or brown bark, this tree is a source of wintergreen oil, which can be obtained from its leaves and sap. Its twigs, when broken or crushed, give off the distinct odor of wintergreen.[12]

Still, the names of trees and their wood can be very confusing and confused. The side panel of a box of Trophy toothpicks identified its contents as "sweet birch tooth picks," but the bottom of the box suggested that they were "manufactured in the State of Maine, U.S.A. from Northern Maine White Birch by American work-men." The quotation marks accompany this statement on the box, making one wonder whether it applies to the contents or is just something said or written sometime, somewhere by a representa-tive of Hardwood Products Co., of Guilford, Maine, and Mayville, Michigan.[13] This same company has also used "Silver-Birch" as a brand name, employing another of the pseudonyms for white birch.[14]

Regardless of a wood's name or type, the cord is the unit of measurement that is commonly used to quantify an amount of cut wood, whether to be consumed in a factory or in a fireplace. Thought to be so named because the size of a woodpile was origi-nally measured with a piece of rope or string of a certain length, a cord of wood is now defined to be an orderly stack having a volume of 128 cubic feet. Thus, a cord of neatly stacked four-foot-long logs would be eight feet long, four feet high, and four feet wide. A cord of firewood, which is usually cut into sixteen-inch lengths, could be stacked three wide to make a pile of the same overall dimensions. Because loosely stacked wood contains bark and air spaces, the vol-ume of a cord does not equal the volume of wood that could be got-ten from it. It has been estimated that, on average, a cord contains less than two-thirds usable wood. Over a quarter of the pile is air, and about 10 percent is bark.[15]

A stack of finished lumber is commonly measured in board feet. One board foot equals 144 cubic inches of wood, which is the vol-ume of a one-foot-long piece of a true two-by-six. Because of the uncertainty about the air spaces, the amount of bark, and the imperfections in a stack of wood, not to mention the amount of waste in the form of sawdust, it is difficult to determine the number of board feet of finished lumber that can be cut from a cord of logs.[16] However, according to one estimate there are approximately 400 board feet in a cord. In the early twentieth century, the amount of paper birch cut annually in the Northeast was about 80,000 cords, or 32 million board feet. Of this, from 3,000 to 5,000 cords were said to be sufficient to supply the entire toothpick industry. A single cord could yield as many as nine million toothpicks, and the bulk of the wood was used for making such things as spools, shoe pegs, and dowels.[17]

The spool industry alone consumed over half the total amount of white birch cut. By the end of the nineteenth century, Maine had seventeen spool factories, with a combined output of three hundred million spools annually. These were shipped throughout the country, to be wound with fifty billion yards of thread. Great quantities of white birch, perhaps in the form of dowels, were also exported to England and Scotland to be made into spools for their thread mills. Other uses of white birch included making checkers, and filling a single order might have required the manufacture of as many as eight million of the wooden disks.[18]

Still, it was the simple toothpick that captured the attention and imagination of wood watchers, and the prodigious number of picks being produced led to such quips as, "It is feared that the enormous manufacture of wooden toothpicks is utterly destroying the forests of America; but then, the young man who spends all his salary for good clothes must have something to eat."[19] In fact, at the end of the nineteenth century it was reported that the supply of wood was "unlimited." In any case, toothpick manufacturing consumed only a small fraction of wood compared to other popular items, such as matches. As late as the mid-twentieth century, wooden match production consumed over ten times the number of white pine trees that toothpicks did of white birch.[20]

Because toothpicks are such slender items, the wood from which they are made should not contain significant imperfections. Small imperfections and blemishes could be tolerated in a product such as a spool, which would be covered with thread and labels, but because toothpicks are used in the mouth, the wood of which they are made should appear unblemished and uniform in color. While it is true that toothpicks are sold in boxes that can conceal any blemishes until a box is bought and opened, the brand that surprised its user with dark-colored or otherwise flawed toothpicks was not likely to be bought again. (Had the quality of Forster's Ideal toothpicks not lived up to their name, the slogan on the box—"Once Tried, Always Used"—would likely have been ridiculed.)[21] The toothpick is essentially a naked piece of wood, and it cannot conceal anything. Thus, the birch stock used in the toothpick industry had to be free from knots and dark heartwood (called "red heart"). It also was expected to be straight and large enough in diameter to be able to provide a sufficiently long ribbon of veneer. As a result, the best grade of birch was selected for the toothpick plants, and in the early 1910s it commanded from fifteen to twenty-five dollars a cord at the mill.[22]

The famous Forster toothpick factory that was established at Strong in the 1880s got a good deal of its wood from farmers and small jobbers, but it also bought a sizable amount from large timber operators working in the vicinity of Kingfield and Bigelow, which are located about ten to twenty miles to the north, from where it could be hauled on a narrow-gauge railroad. These operations began cutting wood in late September or October—after the growing season was over—so it would be free of sap stains.[23]

How the sap behaved had long been of prime concern to those in the toothpick industry. Cold weather keeps the sap frozen in the logs; in warm weather the "sap begins to stir and stains the wood a tobacco brown."[24] According to Benjamin Sturtevant, "in winter, the wood will keep well on hand; but in summer, it will spoil in a few days after being cut," according to a process he termed "fermentation of the sap, which makes the wood appear to be stained." Such wood was not suitable for shoe pegs, let alone toothpicks, and was "invariably thrown out as worthless, or sawed up into box stuff for boxing peg blanks." But waiting until winter to secure wood also had disadvantages, since a season of heavy snow could make the wood inaccessible.[25]

Not cutting in the summer also eliminated the problem of accelerated seasoning of the wood, and so those who depended on a good wood supply had to take their chances with the weather. Maine weather is as legendary as Mainers themselves. They have been said to be jacks-of-all-trades: "Farmers in the one month between frosts, which is called summer, hunters in the fall, lumbermen in the winter, ice loaders in the spring," referring to the crop they harvested from the frozen Kennebec and other rivers.[26] Another observer wrote, "In the woods of Maine every winter there are great camps of lumbermen engaged in felling and hauling out timber for various commercial uses. Whenever the foreman of a camp comes across a particularly fine white birch tree he orders it cut down and the main part of the trunk laid aside to be reserved for the toothpick factories."[27]

Such practices sometimes frustrated Henri Vaillancourt, the New Hampshire maker of traditional birch-bark canoes profiled by John McPhee in his book *The Survival of the Bark Canoe*. When Vaillancourt would come across a good tall, straight birch on a scouting trip, he would note its location for the time when he needed its bark. Unfortunately, sometimes he returned to find only a stump, since "a timber company had taken the tree for spindles and spools." When Vaillancourt got to a good birch first, he was pleased that he could use the wood of the tree to make a hand-carved thwart rather than have it end up as "all the chintzy two-bit things they make out of birch—clothespins, dowels, toothpicks."[28]

Wood was cut and transported to the toothpick mills principally during winter months, when sleds could be pulled over packed snow, as depicted in this photograph from about 1912.

After felling by less sensitive woodsmen, the tree trunks were trimmed of branches and cut into four-foot-long lengths, which were placed in piles eight feet long and four high, thus making a cord. Alternatively, the wood was stacked in large piles of no specific measure, a process known as "yarding." Such piles were placed where they would be accessible in the dead of winter, when hauling could be done with sleds.[29] Sleds were used because there were few roads in the local woods, but they were obviously not very effective on dry dirt or during the infamous Maine spring mud season. So their use was largely restricted to the winter months when there was snow on the ground, and preferably snow that was sufficiently compacted. A striking painting by Rockwell Kent—who was quite familiar with Maine's topography, weather, and ways—shows a large roller being pulled by a team of horses over a hill of snow in order to make a serviceable winter road.[30] Thus, weather conditions controlled when the wood supply could be replenished, and wood-hauling crews were generally not engaged in that activity from spring until the end of the year. Also, because of their dependence on the supply of wood, the early toothpick mills themselves usually did not operate year-round. According to an official of a factory at Wausau, Wisconsin, the "ideal season" for making toothpicks was believed to be between December 1 and April 15, but this was by no means a hard-and-fast rule.[31] One way to extend the manufacturing season was to begin in the fall to spray water over large stockpiles of wood, letting it freeze as the temperature dropped. As snow fell on the frozen piles throughout the winter, it was compacted, and in the early spring the piles were covered with sawdust, which kept the ice and snow from melting well into the summer, when the wood was finally used. The process had the added advantage of maintaining the "color, texture, and moisture content" of the wood, as well as keeping the ends of the logs from splitting.[32]

The maximum distance horses or oxen were expected to haul wood was about ten miles. In 1890, when there were in Dixfield two wooden spool mills in addition to the large Forster toothpick mill, this limitation and the supply of birch in the area were thought to mean that Dixfield could continue to manufacture wooden products for only about ten more years. An alternative means of transporting wood, and over longer distances, was via rail; the narrow-gauge cars of the Sandy River (later the Sandy River and Rangeley Lakes) Railroad, which served Strong, could each hold about three and a half cords.[33] Still, draft animals pulling sleds remained necessary to get the wood from the forest to the rolling stock. The introduction of the truck in the early twentieth century would greatly expand the range and flexibility of wood hauling operations.

Larger quantities of white birch were transported by rail. Here, a load of wood is carried by a car of the narrow-gauge Sandy River and Rangeley Lakes Railroad, which was formed in 1908.

The Forster toothpick plant across the Webb River at Dixfield continued in operation into the early years of the new century, but it burned down in early 1904. A new one was built to replace it and began operating about seven months later. It was soon producing one hundred cases of one hundred boxes of eighteen hundred toothpicks each day, for a total output of eighteen million, which required as many as five cords of wood daily. A lumberman could fell and yard between one and a half and two cords of birch in a day, for which he was paid as little as a dollar a cord.[34] In the boilerplate of a letter contracting for wood, the Forster factory at Dixfield required that suppliers provide white birch lumber in four-foot lengths and no less than "eight inches in diameter at the smaller end." Birch with a "sound red heart" would be accepted if the diameter of the heart was no more than about a quarter to a third that of the diameter of the lumber. In the 1911–1912 season, wood was not accepted at the factory before October 20 or after February 15.[35] Seldom was wood provided beyond late May, which gave an adequate supply to carry production into July, at the latest. When the wood ran out, the factory shut down and did not resume operation until the fall, when wood could once again be supplied in an appropriate state. (Extensive year-round toothpick production did not begin until the mid-1930s, when technical and economic conditions were favorable to such a level of activity.)[36]

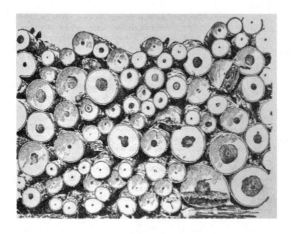

Four-foot-long bolts of white birch, four to eighteen inches in diameter, were employed in the veneer-cutting machines, but the dark heartwood was not used for toothpicks.

Only full-railroad-car lots of between seven and eight cords were accepted from large suppliers, and the wood had to be marked with a Forster lumber car tag identifying the source. Smaller quantities were brought to the factory by independent suppliers, who were issued a yard receipt from which they would later be paid. Payment for lumber was made on the fifteenth of the month for deliveries made through the last day of the previous month.[37] Payment for "first class birch delivered to the mill in the fall on wheels" was eighteen dollars a cord. Delivery of the same amount of wood by sled, which made it more difficult to move about the yard, commanded only sixteen dollars. Second-class birch brought ten dollars and poplar nine dollars a cord. During the 1910–1911 season, eleven hundred cords of birch were used by the factory in Strong and twelve hundred in Dixfield. The amount of poplar used in Strong was only about two hundred cords. (At the height of the toothpick craze, around the turn of the century, no more than five thousand cords of wood were used by the entire industry.)[38]

A great deal of wood was stockpiled at the mill site throughout the winter, as shown in this photo, taken about 1912, of the lumberyard beside the mill of the Estate of Charles Forster at Strong, Maine. The factory proper is to the left; a storage building is to the right.

As photographs of toothpick factories show, the wood was typically stored end up all around the mill yard and was left out in the open air. The toothpicks made from it retained "something of the natural sweetness of the birch and maple"; they were "soft and pliable, while retaining sufficient firmness." Kiln-dried wood was not used because it tended to yield toothpicks that were too brittle, resulting in sharp splinters and causing injury to gums and tooth enamel alike.[39] For making toothpicks, the wood was used when it was still green. In 1912, eight cords of wood were enough to supply the Strong factory for a full day's production of toothpicks.[40]

After World War I, the Estate of Charles Forster, as the patriarch's business continued to be known, would begin to purchase land in and around Dixfield and would acquire over seven thousand acres by 1929. This "well timbered land" provided the company with a "proper surplus of quality timber." As late as the 1930s, sled trains (typically consisting of four long sleds pulled by a tractor) were used to haul around the clock from Weld, which is twenty-three miles from Dixfield. For a while, tree-length timber was accepted. This was naturally more difficult to handle, but it was advantageous in that "long logs will not show sap stains when warm weather comes," the way four-foot-long bolts would. In the longer timber, the sap settled at each end of the log, which could then be cut off about eight inches from the end. The remaining portion would be "as free from stain as when cut down in the woods."[41]

This stereoscopic image of a Dixfield toothpick-manufacturing plant, taken perhaps as late as the 1930s, shows that the setting of such an industrial establishment could be considered bucolic and photogenic.

Some of those outside the toothpick industry worried more about conservation. A Kansas City high-school teacher made the newspapers in 1911 by announcing that he had calculated that thirteen million board feet of lumber was "thrown into waste paper baskets every year by people who discard partly used lead pencils." He begged everyone to economize in the use of pencils to help forest conservation. The *Boston Globe*'s editors did not take the teacher seriously, concluding that he "may be urging us next to save our discarded toothpicks and burnt matches to use for kindling wood."[42]

Not everyone was so clinical or cynical about trees. Nathaniel Hillyer Egleston, an early conservationist and author of a handbook of tree planting as well as a history of Arbor Day and its observance, wrote about the "reckless and criminal treatment of our forests in general and of our good friends the trees in particular." He believed that "nothing in nature except a man is more valuable than a tree," and decried the fact that forests had been "slaughtered" for timber and woodenware products of all kinds, including toothpicks, for which the "woods have been flayed alive."[43] Perhaps he thought that trees had volunteered to die to make the paper on which his books and articles were printed. Just as an omelet cannot be made without breaking an egg, so a toothpick cannot be made without felling a tree.

CHAPTER SIXTEEN

Trade Secrets and Closed Doors

TAKE A TOOTHPICK—any toothpick. Unlike a playing card, which the chooser can peek at before putting it back in the deck for the magician to find, a machine-made toothpick taken from a box has no obvious identifying features beyond possible imperfections. The slip of wood may be flat, round, or square with round tips, but in the best of boxes it is indistinguishable from the hundreds of boxmates with which it was packed sardine-like at the factory. Identifying a particular toothpick among its neighbors is akin to finding a needle in a haystack. Manufacturers continue to make different kinds of toothpicks because some consumers, at least, are adamant about which kind they will put in their mouth. One habitual toothpicker expressed frustration that his wife consistently bought round toothpicks when he had long had a preference for flat. Looking more closely at the two kinds, however, he did have to admit that the round style had "more class." He also acknowledged it was more durable: "The flat cannot take much abuse and usually become limp when kept in one's mouth too long, whereas the round can be used as a pacifier for a considerable time and then placed in a shirt pocket for later use." At no time in his musings did he wonder about the origin of the thing, or about how the manufacturing process might have accounted for the different behavior.[1]

Trying to divine how a toothpick was made is no mean feat. Ironically, more complicated things—like automobiles and cellular phones—might be more readily reverse-engineered than the very simple. Things made of parts, like the internal combustion engine of a car or the circuit board of a consumer electronics device, can be taken apart with wrenches and screwdrivers or probed with ohmmeters and voltmeters. Things of a whole cannot be disassembled because they have no component parts.

With a toothpick, what we see is what we've got—inside a toothpick is the same wood that we see on the outside. About the time that Charles Forster's mill was daily turning out toothpicks by the millions, the *New York Times* carried a letter about the mythical village of Squallitj Kills, which was peopled by those who did "not set much store by science, being rather of a dreamy, poetical, and speculative vein." One Squallitjan was known in the area as "Old Outsides," because of his theory "that nothing has an inside, or at least that it is impossible to prove the existence of an inside of anything." He demonstrated his thesis by taking a piece of kindling wood and observing that all that can be seen is an outside. Whittling part of it away, he revealed not an inside but a new outside. If it were an inside, he argued, you would not be able to see it. Repeating the process, he remarked, "And so on, if I had clove it up into tooth-picks and match-wood ye wouldn't see no inside, and ye couldn't find none if ye ground it all up or blowed it to flinders with dynamite."[2]

Whether or not a toothpick has an inside, it is largely the form of the outside that makes it what it is. In the best-made toothpicks, all traces of a manufacturing process are husked and polished off, leaving a smooth outer surface that bears no scratches or gouges to provide a hint of what kind of tool might have been used. In the so-called hustling process, toothpicks were sometimes tumbled for three days or more to achieve a smooth finish.[3]

If we were not aware of the history of the development of the machine-made wooden toothpick, we might seek a clue to the forming process in the overall shape of the thing itself. For example, we might think that a flat toothpick bought today most likely was made from a flat piece of wood. This belief might be supported by noting the uniform thickness of all the flat toothpicks in the box: they look as if they were punched out, cookie-cutter-like, from a uniform piece of veneer. But since flat toothpicks generally have one end wider than the other, veneer would likely be wasted if they were all cut out in the same orientation. Knowing that manufacturers abhor waste, we might surmise (correctly) that every other toothpick was punched out in one orientation and the alternating ones in the opposite direction. But questions would remain: How were they punched out? What kind of machine was used? How was it constructed? How does it operate? What kind of wood is used? How is the wood prepared for the machine? Is the wood blank reoriented after each toothpick is stamped out?

If we wanted to make toothpicks ourselves, perhaps to compete with those we had been studying, we would want to know the answers to such questions. Being given even a rough description of the machine—say, through the drawings and text of a patent—would be a tremendous help in building one, of course, and being able to buy a ready-made machine would be even better. But we still would have to learn how to prepare the wood with which to feed it and how to operate it efficiently, which would most likely involve a lot of trial and error.

Recall that when Benjamin Sturtevant, Charles Forster, and Charles Freeman set out to make toothpicks, they began with machinery designed for producing shoe pegs. They spent years developing the right combination of parts and actions—and the right process. When they achieved their end, they might easily have patented their ultimate device and process, but they chose not to do so. For all of his interest in acquiring the rights to Sturtevant's patents, for years Forster seems to have had an aversion to securing patents for equipment developed in his own mills, perhaps because Freeman and he would have had to divulge what they had labored so long to learn and protect. It would only be after their patent protection on the flat wooden toothpick expired that they assiduously began seeking protection for a new round one, even though they revealed the secret of its outsides and insides in the process. That is the price of getting a patent, of course, and for a long time Forster appears not to have been willing to pay it.

In the 1860s and 1870s, his strategy was to exploit the rights to Sturtevant's veneer and toothpick patents to the fullest. These rights prevented J. C. Brown from manufacturing toothpicks on the machine he invented and patented in 1864. Although Brown's patent, and an article describing it in *Scientific American,* may have been very helpful to Forster and Freeman, they were not about to be so open.[4] When the Forster operation was described in magazines and newspapers, the proprietor's name was seldom mentioned, and there was a paucity of information about what his machines looked like and how they worked.[5]

Before the 1880s, Forster had chosen the alternative to patenting, and that was to maintain trade secrets. He would sell toothpicks to anyone who would buy them, but if they wanted eventually to become a competitor, it would be their problem to reverse-engineer the thing, to infer the machinery, and to fine-tune the process. Although Forster never had a patent issued to himself, and Freeman did not have one until 1887, the number of patents for toothpick-making machines and related devices grew at an increasing rate. Before the 1870s there were only Sturtevant's two patents and Brown's one; by the end of the century there were more than thirty issued to a score of different inventors.

In his patent for a machine for compressing and polishing toothpicks, Freeman had acknowledged that although it had been "common to manufacture wooden tooth-picks by machinery," they remained "somewhat rough, and have not sufficiently fine points."[6] Getting the point right would continue to be a vexing problem. Machines of the kind Freeman patented "very effectually rounded and polished" the central portion of the pick, but the pointed ends were "not as well executed," according to the inventor William Dyer of Strong, who claimed to have improved upon the process by employing sand or emery paper–surfaced disks.[7] A machine patented in 1900 by Joseph Hommel of Detroit, and assigned to the National Toothpick Co. there, employed "belts to grind or abrade the end of the toothpick and produce on the end a point that is sharp and conical and is without a projecting thread or filament from the extreme point," suggesting that producing such a point was a definite problem with prior machines.[8]

However well they did or did not work, the machines continued largely to be built and maintained by the mechanics employed by the toothpick companies themselves. Such machines turned and reciprocated millions of times each day, so it is not surprising that their cutting parts would quickly dull and the machines themselves would frequently break down. According to Harry Dorr, an inventor from Providence, Rhode Island, except for the cost of the wood, the "greatest expense" in producing toothpicks was that of "keeping the knives in proper repair to produce good work." Dorr's invention, patented in 1905, consisted of employing machined blocks, which he called "templets," to which sharpened sections of shaped sheet steel were fastened. With his cutters formed from mating pairs of these templets, he was able to form flat toothpicks "rounded at one end and sharp at the other." He claimed that flat toothpicks with a properly rounded end had "never been produced before commercially," since they could not be made "in a practical manner with the old-style knives."[9]

Other inventors concerned themselves with the problem of machinery breaking. One was Charles Scamman of Deering, Maine, who at his death in 1902 would be president of the Scamman Manufacturing Co. in Portland and would be described in his obituary as "an inventor of ability, especially of appliances used in making wooden toothpicks."[10] One of his several inventions dealt with machines in which a strip of veneer being fed across the edge of a stationary knife was forced onto the knife by a reciprocating wooden block. After each impact of the block, the veneer strip had to be raised again over the knife edge, and this was done by a spring plate, which would "frequently break." He had dealt with the problem by replacing the spring action with a cam action, but this was "difficult to keep in order and required the services of a skilled mechanic." To improve the machine further, he replaced the cam system with a lever device. It took an individual with special talents to attend to the ongoing modification of the machines that made toothpicks, shoe pegs, and other small mass-produced items. According to a description of one such mechanic, "he couldn't invent a tooth-pick, but there never has been a machine which he couldn't improve upon."[11]

The machines that talented inventors and mechanics developed for the small, specialized industry were not easy to come by, but outsiders continued to try. As late as 1906, an item appeared in the "Business and Personal Wants" column of *Scientific American,* to which relevant manufacturers could write to receive the name and address of the anonymous inquirer. It read, in full, "Inquiry no. 8228.—Wanted, machinery for making wooden toothpicks."[12] Evidently there were still no readily known makers or suppliers of toothpick machinery. The industry itself continued to develop its own hardware and thereby grow and attract attention, and of the towns in which it grew, Strong attracted the most.

In 1801, the town had been incorporated into the Commonwealth of Massachusetts, of which what would become the State of Maine was then a part. Among Strong's distinguished early citizens was Philip M. Stubbs, its first lawyer, who in 1834 built the first brick house in the town. His son, Philip H. Stubbs, and grandson,

Philip D. Stubbs, were also Strong lawyers.[13] Another grandson, Robert Goff Stubbs, would eventually become a district forester in the state, but in 1912 he was a student in the course on wood technology offered at the Yale Forest School, and for it he wrote a report on the nature of the contemporary toothpick industry in Maine. Uncharacteristically, the Forster firm allowed Stubbs to observe directly its manufacturing operations at the Strong mill, which his report described in considerable, sometimes minute detail. He may have been given such uncommon access because of his family's prominent ties to Strong and to the legal profession; he was not given such ready access to any other plants and had to "rely on more or less general descriptions of them by persons who were presumably familiar with them." According to Stubbs, "Queries con-

The Strong toothpick factory of the Estate of Charles Forster, as it looked when forestry-school student Robert Goff Stubbs was given access to it in 1912. At the time of this photo, when the last of the winter snow was melting, much of the season's wood supply had already been made into toothpicks.

cerning some of the interesting but intimate aspects of the business received rather reticent and general answers at times for business reasons."[14]

But Stubbs was able to report on how toothpicks were made in Forster's Strong plant. From the mill yard, the wood was taken a half to a full cord at a time into the lower part of the factory, where it was prepared for use in the machines. The ends of the logs were squared off and the logs were cut into two-foot lengths, called "bolts." Next, a two-inch-deep hole was bored into the center of each end of a bolt so that it could be mounted on the peeling machine, which removed the bark and formed the bolt into a perfect cylinder. The peeled bolts were then placed in one of two steam boxes, each capable of holding six cords each. Steaming was necessary to thaw wood that had been sitting out in the cold Maine winter, but the process had to be done slowly and with care lest the bolts warm up too quickly and thereby season and crack.[15]

After being kept in the steaming boxes overnight, the bolts were carried by conveyor to the second floor of the factory, where the lathes were located. The Strong mill had five lathes, each operating at about 750 rpm, on which knives cut from the rotating bolts ribbons of veneer the width of a single toothpick, which were then fed into separate machines to cut off the individual picks. Even when softened by steaming, birchwood dulled the lathe knives after two to five hours, and so they had to be replaced by ones freshly sharpened in the mill's machine shop. (When cutting poplar, the knives lasted barely an hour, becoming blunt and gummed up with wood fibers, causing them to tear rather than cut the veneer.)[16]

By the early twentieth century, the Estate of Charles Forster had manufactured as many as ten different kinds of toothpicks, which were distinguished from one another by their size, their shape, or their quality. Stubbs glued actual samples of these toothpicks into his report, including a pair of orangewood ones "made by Chas. Forster when he was interested in the industry in Brazil." Probably machine-made, these are about two inches long, very slender, round, remarkably regular in shape, and pointed at one end only. They are in stark contrast to the four-inch-long imported toothpicks pasted onto the reverse of the page. Stubbs captioned these: "An imported toothpick, sold by the Cutter-Tower Co. of Boston. Branded as rosewood toothpicks and as made in Portugal.

A very light, smooth and flexible pick."[17] Though somewhat larger than the orangewood toothpicks that were being imported from Portugal a century later, the faceted rosewood picks appear to have been hand-carved by the same kinds of knife cuts.

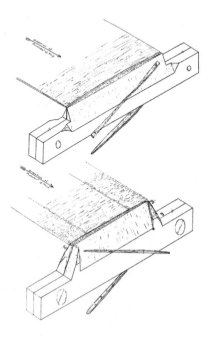

This sketch shows a veneer strip one toothpick-length wide being fed across knife edges outlining the shape of a toothpick, which was punched out by a hardwood platen. Depending on whether or not the veneer was beveled and on how the knives were shaped, the Ideal no. 1 (shown here in the bottom sketch) or the inferior no. 2 (top sketch) toothpick was produced. The basic process would change little over time.

The shape of the distinctly American toothpicks that Forster produced were artifacts of the manufacturing process. They bore, if only in their regularity, the marks of the machines that a "competent machinist and an assistant" made and repaired for both the Strong and the Dixfield plants. At the time of Stubbs's visit, Cutter-Tower was only one of several wholesalers of toothpicks made by the Estate of Charles Forster. The Forster company also sold picks directly to retailers such as F. W. Woolworth and Co., which carried them in its five-and-ten-cent stores. A case containing almost two million first-grade toothpicks was sold to Woolworth for twenty to twenty-five dollars. At the five-and-dime stores, a box of fifteen hundred Ideal toothpicks sold for ten cents, representing a 50 percent profit for the retailer. Second-grade toothpicks that were made from unbeveled veneer were sold wholesale at ten to twelve dollars a case. They came in boxes of twelve hundred and were sold two boxes for a nickel in Woolworth stores.[18]

At the time, the Forster firm considered these two distinct toothpicks to be "first class." In keeping with the tendency of manufacturers to give their products names that evoke positive images and connote superiority, the first- or high-grade toothpick made in the Forster Estate's Strong mill had been named the Ideal. In overall shape this was similar to what a century later was being called a flat toothpick, which is essentially just sliced off the end of the wood veneer as it is advanced into the machine. However, in the earlier version, each end of the flat toothpick was inclined to give a thinner point. This was achieved by beveling the edges of the veneer strip before putting it through the cutting machine. The Ideal was also known in the industry as a no. 1 toothpick. A variation was described in a 1903 stationer's catalog as being the "long-looked for wood tooth-pick": it was "shaped similar to the Imported Orange Wood, flat sharpened at one end and fine pointed at the other."[19]

The no. 2 was Forster's "second grade," with yellow birch and poplar being used in addition to white birch. It was not beveled, which meant that it could be produced more quickly. It was more truly flat when viewed from the side. The no. 2 was the thickness of the veneer throughout; it was made "pointed" at both ends solely by the shape of the cutting knife acting on the veneer. Essentially, its profile was defined by two intersecting shallow arcs, resulting in a lenticular shape, and its points were generally not as fine as those on the Ideal. Also, whereas the Ideal was polished and smoothed through the process known as hustling, the no. 2 was not.[20]

Since the shape of the no. 2 was convex, the wood between adjacent toothpicks was waste, just as is the material between round cookies cut from a sheet of dough if it is not recycled. There was more waste associated with production of the no. 2 toothpick (about 30 percent) than with making the no. 1 (about 16 percent, comprising mainly the wood lost in beveling). Such "waste" might almost

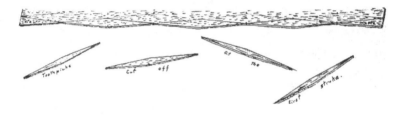

An efficient toothpick-cutting process developed at Maurice Forster's plant employed long knives with a wavy profile for cutting multiple toothpicks simultaneously from wide strips of unbeveled veneer. The wavy knife was shifted laterally the length of one-half a toothpick after each cutting stroke, thus reducing waste to a half toothpick per stroke.

be called a by-product, however, since it was burned in the power-house to generate steam. (In the early twentieth century, the mill Maurice Forster would establish in Dixfield virtually eliminated waste in cutting "low grade No. 2 toothpicks" from strips of veneer as wide as the bolt of wood was long and stacked thirty-two high. The knife, which had a wavy profile, shifted sideways half a tooth-pick length on every stroke, resulting in waste equivalent to only half a toothpick per advancement of the veneer strip.)[21]

The Forster Estate's Strong plant was powered by four 65-horsepower boilers, in which production waste was burned. The mill had its own dynamo for generating electricity to operate the machine shop and production machinery, and to light the place.[22] Even after electrical power became commonplace, toothpick and other wood novelty mills still put by-products to good use. As late as the 1980s, Forster would install steam generators in its plants at Mattawamkeag and Wilton and was burning wood waste to produce heat and electricity for the plant and steam for use in the manufacturing process. No fossil fuels were consumed, and excess electrical power was sold for use in the New England grid.[23]

By 1912, Forster had ceased making the Velvet toothpick, which had been considered a "high-grade" product. It resembled the Ideal but had a square rather than rounded broad end, and it was not husked or polished. The standard Velvet, along with a slightly longer version, was discontinued because it cost about as much to make as the Ideal, but it sold at a lower price. The other first-class toothpick made by the Estate of Charles Forster came out of its Dixfield factory. This was the World's Fair, which was round with symmetrical pointed ends. The process for making it was similar to that used for the Ideal, but with the added step of rounding and pointing the toothpick by compressing it.[24] The resulting product, which had been patented in 1891, gave Forster a new monopoly, and in the early twentieth century boxes of Worlds Fair toothpicks (the apostrophe having since been dropped) declared them to be "the Leader of All" and "the Most Popular Pick in Use."[25] The Worlds Fair and Ideal brands, their logos undergoing periodic modification and modernization, would continue to be showcased by Forster past mid-century and at least into the 1970s.[26]

Among the second-grade toothpicks that Forster had also once made was one that it designated the 1B, which had one face flat, one beveled, and the whole tapering. This rough and unpolished pick was used by florists in making wreaths. When tied to a flower stem, which was stiffened with a wire, it could be pushed deep into a moss or florists' frog base. This was one kind of toothpick for which a matchstick was not wisely substituted. On one occasion when that was done to hold dried flowers in a funeral arrangement, they caught fire in the wagon taking them to the cemetery. The incident was attributed to the bouncing action of the wagon causing a match head to rub against the wire frame, thus igniting the match and consequently the floral arrangement. In spite of its being safer than a match, it was expected that the market for the 1B toothpick would eventually be taken over by a "cheaper and better adapted article" that incorporated a wire. Such an article was being produced in New York.[27]

Half a century after Stubbs's remarkable report on toothpick manufacturing, a forest products textbook provided a comprehensive fount of information on their sources, production, and utilization. In the chapter on secondary wood products, the section on toothpicks stated that "these simple little articles now sustain a prosperous industry" and proceeded to give a one-page description of the "modern machine manufacture of flat toothpicks," which provides a verbal snapshot of the state of the art around 1960. This did not differ significantly from what Stubbs reported observing at Strong five decades earlier:

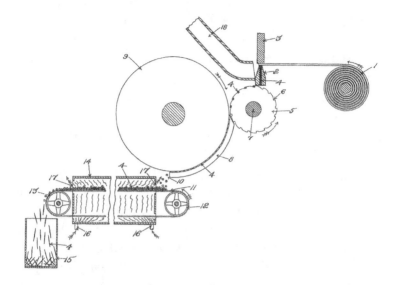

*In 1907, George Stanley and Willis Tainter patented a continu-
ous process in which flat toothpick blanks stamped from a coil of
green wood* (upper right) *were compressed into round and pointed
picks that were then passed through a drying chamber and finally
tumbled to make them smooth.*

Toothpicks are made from veneer, and in the initial stages of
processing the bolts may be debarked and then rotary-cut in a
cold condition, but the more common practice is to soften the
wood by prolonged steaming or immersion in hot water prior to
insertion in a lathe. Spurs set in milled recesses of the pressure
bar at $2\frac{7}{16}$-in. intervals score the rotating bolt just ahead of the
cutting knife so that several ribbons $\frac{1}{20}$ in. thick and $2\frac{7}{16}$ in. wide
are produced simultaneously. Good bolts may ordinarily be
reduced to 3-in. cores. Each ribbon of veneer so produced is
then fed into a pointing machine that bevels or skives the edges
to produce the thin ends required in a toothpick; the veneer is
pulled over two knives by means of feed rolls. . . .

The toothpick-cutting machine is an automatic device of
the punch-press type that cuts and forms the toothpicks. It is
operated at a speed of about 2,000 strokes per minute, each
stroke forcing two toothpicks through dies in the bedplate of
the machine. The dies are so arranged that there is little loss of

material, and approximately 154 toothpicks are obtained from each linear foot of veneer. . . .

When thoroughly dried, the toothpicks are placed in a tumbling barrel or polishing drum with a small amount of powdered chalk or shaved paraffin. The tumbling action wears off the sharp corners, and the chalk or paraffin gives the toothpick a glaze.

Polished toothpicks are placed in an automatic straightening and box-filling machine. . . . A standard packing case holds 1,152 of these small boxes, each of which contains 750 toothpicks.

A complete unit of toothpick machines consists of one veneer lathe, one pointing machine, one veneer-winding spool, six toothpick-cutting machines, one drying oven, and one straightening and box-filling machine. Such a unit requires only about 15 horsepower for its operation, and its hourly output is about 1,440,000 toothpicks. One standard cord of sound white birch wood will yield from six million to nine million flat toothpicks.[28]

The inclusion of a beveling or skiving step in the process makes it likely to have been one used at the time by the Forster Mfg. Co. A box of its Ideal toothpicks from the mid-1960s carries on its bottom this telegraphic legend: "The only Toothpick tapered thin—polished—removes food particles from between closely set teeth."[29] A flat toothpick bought today is likely to have been stamped out of veneer that has not first been beveled or skived. The result is a thin little stick with blunt ends that hardly deserves the name toothpick. It is typically very ineffective for fine picking, though it may still serve for chewing.

Although toothpicks could be cut from veneer with lightning speed, the entire process—from when the bark was removed from the logs to when the finished flat toothpicks were packed in a box—could take days, much of the time consumed in the tumbling process alone.[30] The production of round toothpicks, however, which were made in "prodigious numbers," was not described in the textbook. Rather, its authors explained that, as late as 1962, round-toothpick makers had "refrained from divulging the manufacturing procedure on the grounds that to do so would be a disclosure of trade secrets."[31] The secrets had been held for over half a century.

Whether flat or round, toothpicks had to be packed into boxes. In the beginning, that had been done by hand labor, mostly by girls and women. According to one description,

> The packing of the picks requires a large amount of practice. They must all be pointed one way before being placed in the boxes. The picks are all tangled up and pointing in all directions when the expert girls take them in hand and with deft fingers swirl and whirl them around until they are all pointing one way. They are then quickly placed in paper boxes, when they are ready for the market.[32]

One of these packers may have been the "Maine girl" who once inserted a note with her address into one of the boxes, "requesting the finder to write to her." A man from Kansas City did write, and the ensuing correspondence led to his traveling east "to see if the young lady was the sort of a woman he wanted for a wife."[33] It seems not to have been recorded whether the encounter led to marriage, but if there was one such note tucked into a toothpick box, it is likely that there were many, and other suitors may not have been far behind.

By the early twentieth century, when the high-speed machines for making many millions of toothpicks daily had reached a high level of sophistication and reliability, the need to deal with such great quantities of toothpicks shooting out from the machines and to put relatively small quantities of them into small boxes produced a bottleneck in the factory. Some of the same inventors who had devised and improved the machines for making the toothpicks began to turn their attention to developing machines to box them. In 1896, Henry Churchill of Deering, who a year earlier had invented a toothpick machine, patented a "receiving and delivering spout" for it. He pointed out that those spouts that he knew already to exist consisted of open tubes through which the picks, falling by gravity, were "apt to get crossed, whereby it becomes difficult to gather and pack them properly in boxes." His solution was to close the tube so that the toothpicks accumulated in it until there were enough of them to provide a pressure feed.[34]

The Forster plant at Strong was naturally also engaged in solving the problem of getting the toothpicks quickly and neatly into boxes—without including any unwanted bits and pieces. In 1900, when a sickly Charles Forster resided in Portland, his son, Maurice, was living in Strong and working in the plant there. Among the problems that occupied him was that of separating debris from the toothpicks. As his 1900 patent showed, he achieved his objective by blowing a current of air—perhaps using a Sturtevant fan—opposite the direction in which gravity was carrying the toothpicks, thus lifting the lighter and unwanted material toward a suction pipe.[35]

In 1907, Willis Tainter and George Stanley, the latter in charge of toothpick production at Forster's Dixfield plant, were granted patents for toothpick machines. The next year they joined Simon Tainter in patenting a "machine for boxing toothpicks." The machine was designed to take the picks coming out of the tumbler and deliver them in an "orderly arrangement" to the boxes. This was accomplished by means of a "rotary wheel conveyer having upon its periphery a series of trough shaped buckets," which

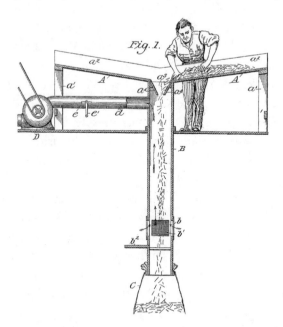

Maurice Forster employed an opposing current of air in a patented device for separating the lighter waste from a stream of toothpicks issuing from a cutting machine.

dropped the picks into a chute containing vibrating slats and a means of delivering a measured quantity into each box below.[36] In a subsequent patent, granted to Simon Tainter alone, he described an improved mechanism for depositing the toothpicks into boxes, which he pointed out had "usually heretofore been filled by hand." A still further improvement was patented in 1910. In the meantime, Charles Freeman also patented a machine for boxing toothpicks.[37] Surprisingly, none of these patents—at least upon their being issued—was assigned to the Forster business, where they can be assumed to have been developed and used.

In the eyes of the forestry student Stubbs, the mechanism of the packing machines in the Strong plant of the Estate of Charles Forster was "complex and cumbersome appearing." The operation was far from fully automated. At the bottom of a shaker assembly there was a "mechanical device that grasps the exact number of toothpicks required to fill a box and deposits them in the boxes as they pass beneath on a moving belt," but it took two girls to tend each packing machine. One assembled empty boxes and fed them into it; another took each box out the other end, folded the cover over and closed it, and placed the box in a shipping case.[38]

In the mid-1990s, while driving through Maine, the writer Sue Hubbell came across Strong and "saw a towering pile of birch logs stacked beside a mysterious-looking red factory building," which she learned was "the biggest toothpick manufactory in the United States." She was told by Forster vice president Richard Campbell ("who made an exception to the company's no-visitor policy" and gave her a tour of the plant) that it might even be the biggest in the world, though he admitted that there was "a factory in China that might be bigger." Hubbell was impressed that "the computer-driven machinery, like something out of Charlie Chaplin's silent movie *Modern Times*, with sound effects added, thunders, clunks, clatters and speeds materials on their way." She also remarked that it took "only ten humans" to oversee the production of twenty million toothpicks daily. Then she described the process:

A mechanized beak grabs a log from the woodpile and loads it onto a conveyor belt, where it is cut into lengths about half the size of a fireplace log. These smaller bolts run through a peeler and are conveyed to a steam room to absorb heat and moisture.

Soggy and hot enough to be as flexible as a wet noodle, they enter a veneer-cutting machine, from which the wood oozes like long lengths of spaghetti exactly one toothpick thick, and then go on to a slicing machine, where they are transformed into little toothpick-size blanks. On the plant's heated top floor, those blanks are tumbled and dried and then drop down by gravity into lower-level tumblers where they jounce against one another, polishing themselves, each to each, in the process. They drop into still lower machines, which point each end and feed them into welcoming boxes that are whomped shut by another machine. They are ready to be sent to the shelves of stores around the country. Excluding the day spent in the steam room, the entire process from log pile to boxed toothpick takes a couple of hours.[39]

Toward the end of the twentieth century, little had changed in the industry, except the threat of foreign competition. In one case, a three-hour taxi ride brought three Japanese businessmen from Boston to Wilton, Maine, where they showed up unannounced and asked for a tour of the toothpick factory there, but the plant manager just shook their hands and wished them a good return trip. According to a reporter, "the wood component manufacturing industry is one of the most closed-mouth trades out there."[40]

More recently, the request from an American engineer and historian of technology to tour a factory in Minnesota received the response, three weeks later, that, "regrettably, tours at the toothpick plant are not available."[41] But an attachment to the e-mail did include production figures for flat toothpicks (5.4 billion per year using 1,300 cords of wood; 21.7 million per day; 45 thousand per minute) and a description of the manufacturing process:

Birch logs are purchased in 100" lengths and cut to 24¾ inch long blocks. During the winter the blocks are heated with raw steam for approximately 8 to 10 hours to thaw them from a frozen state. The blocks are veneered at approximately 70 degrees F. on a lathe. During this process the veneered ribbon is cut into strips the length of a toothpick. At the same time the ends of these strips are beveled with a series of rotating skiving knives. The strips are wound into a roll that is fed into a chopping machine. The choppers are equipped with knives that chop the shape of a flat toothpick as the veneer ribbon passes through. The chopped flat toothpicks are conveyed through a hot air dryer that reduces the moisture from 35% to 5%. The dried toothpicks are transferred to a rotary polishing drum that removes the rough edges and polishes the surface of the tooth-picks as they tumble against each other. The polished toothpicks then travel through a "seeding" process that removes the dust and broken toothpicks. From the seeder, the toothpicks are fed to a shaking operation that orientates them length-wise. This shaking operation feeds the toothpicks through a chute that has a blade activator that allots 750 picks per stroke. This activator also drops the 750 toothpicks into a pre-formed box. These boxes are sealed, inspected, and packed into cases for shipping.[42]

The production of round toothpicks, the process for which is no longer top secret, differs from Freeman's compression method and thus results in a pick with a different profile. The process begins as above, but the steam temperature is specified to be 160°F, and the moisture content of the 0.117-inch-thick veneer is reduced to 6 per-cent by means of a hot-air dryer. Then the process continues:

The sheets coming from this dryer are graded and cut into cards. These cards are sorted for quality and fed into a molding machine which produces round dowels $\frac{1}{12}$" diameter by $2\frac{5}{8}$" in length at a rate of 12,000 per minute. The dowels are processed through a rotary polisher and fed into a pointing machine. The pointing machine consists of a series of belts and grinding stones which point the dowel at both ends. These pointed dow-els are again tumble polished to improve the smoothness and sent to a shaker arranger. It feeds a packing line that allots 250 toothpicks and inserts them into a box at the rate of one per sec-ond. These boxes are sealed, inspected and packed into cases for shipping.[43]

The number of round toothpicks produced this way was 2.5 billion per year using 2,800 cords of wood; 10.5 million per day; 15 thou-sand per minute.

Production figures have generally been easier to come by than sales figures. The former can impress the potential customer that no order is too big to fill, but the latter can give the potential competitor a target for how much business might be lured away. However, in the late 1980s, after the related clothespin business had been lost to offshore competition, a researcher for the *World Book* encyclopedia could not determine how many toothpicks were even produced annually in Maine.[44] The reason for the secrecy was explained by the then-president of Forster, Richard J. Corbin, who told a reporter, "We don't want to whet the appetite of our offshore competition." In 1993, when another reporter asked Diamond Brands for sales figures, his calls were not returned. Retailers, however, are less likely to withhold information on stock and sales if they think that it will attract new customers impressed with how many people seem to be buying toothpicks from them. We know that in 1985 the Kroger supermarket chain bought over twenty-two thousand cases of toothpicks, mostly the plain round kind.[45]

Allowing Sue Hubbell's exceptional look into how toothpicks were made may have been a tacit acknowledgment that the secret had been out for some time. Among the reasons for the silence and xenophobia had been that the industry was small and had a small profit margin; one manufacturer put the percentage in the low teens.[46] Yet, once one had the machinery, the know-how, and the markets, toothpicks almost make and sell themselves. As the above descriptions attest, except for the introduction of computers to control it, the manufacturing process has remained pretty much the same as developed by Charles Forster and Charles Freeman and their industrial progeny. But because the industry was still so tiny and its equipment relatively modest, there was little incentive for machinery companies to offer new toothpick-making machines. Hence, as Forster's establishment did in the 1860s, toothpick manufacturers generally had to continue to build their own machines in their own machine shops, sometimes at a considerable investment in development time and money. To let others learn by a plant visit what your machines looked like and how they operated was to contribute your sweat equity to their competing enterprise. But in the late twentieth century, people such as Campbell probably knew that it was a lost cause. As early as 1985, a toothpick maker in Strong could be heard saying, "Imports are killing us."[47]

CHAPTER SEVENTEEN

Other Toothpicks

EVEN BEFORE TOOTHPICKS were made by machine, the word "toothpick" pervaded the language the way the common wooden implement would come to pervade society. Yet it is not Maine but Arkansas that is nicknamed the "Toothpick State," and its residents have been known as toothpicks since the nineteenth century.[1] Their weapon of choice would appear to be the Arkansas toothpick, the name given to a kind of Bowie knife. In fact, the territorial association was firmly established by the mid- to late 1830s, and by midcentury, the terms "bowie knife and Arkansas toothpick were, for the most part, used interchangeably."[2]

An example of the weapon was once described as having a "blade some seven inches in length, broad, and gradually tapering to a strong sharp point; the edge was sharp, and the back very thick; but several inches toward the point were ground away, so that it was double-edged. It had a handle of common yellow horn, and a crossguard of silver, with a sheath of red morocco garnished in silver. On the blade, near the hilt, was engraved, in an oval, BOWIE KNIFE." A less prosaic definition of the eponymous knife required it to be "sharp enough for shaving and heavy enough to use as a hatchet. It had to be long enough to be used as a sword and wide enough to paddle a canoe."[3]

The origins of the Bowie knife are hotly debated among historians and collectors. According to one version of the story, the first Bowie knife was made from a "rasp or large file" by James Black, who was the blacksmith for the plantation of Arkansan Rezin P. Bowie.[4] The weapon was "first dipped in human blood" in 1828, during a duel that took place on a sandbar in the Mississippi River below Natchez. Rezin gave the knife to his brother and second, James Bowie, telling him that it was "more trustworthy" than a pistol and that he should keep it always with him, for it would be his friend of last resort and might save his life someday. Jim Bowie, by then a colonel, and Davy Crockett used Bowie knives in defending the Alamo. Rezin's blacksmith made no more than about "fifteen genuine bowie-knives," and among collectors these are treasured like paintings or musical instruments made by the masters. One of these has been said to have an "edge so keen that a hair may be cut with it, or one can drive it through a silver dollar without the slightest injury to the blade."[5]

Though not all Bowie knives are called Arkansas toothpicks, that did become a common term for any knife with a sharply pointed tip that was a familiar companion up and down the Mississippi and across the country. In 1836, an item about the execution of a killer in New Orleans noted, "The frequent murders there and elsewhere in the south and west generally, arise from the habitual practice of carrying arms, such as the Bowie knife, Arkansas toothpick and pocket pistol."[6]

In 1839, Tennessee passed legislation prohibiting the "sale, offer to sell, or bringing into the state for the purpose of sale, gift, or other disposition, of bowie-knives and Arkansas toothpicks" and the "wearing of these weapons concealed about the person." Philadelphia had experienced the "fighting spirit," but it had declined there by the 1840s, judging by the fact that the demand for Bowies had abated to the point that twenty-dollar knives were selling for a dollar fifty.[7] But the spirit did not die out elsewhere in the Northeast. In 1854, a newspaper correspondent filing a story from Dunkirk, New York, where he had retreated from the riots taking place in Erie, Pennsylvania, opened his report by noting that so fearful had he become for his life that he had "two revolvers in my pockets, and a fearful-looking Arkansas tooth-pick on the table before me." In Nevada, the Bowie knife came to be considered, along with a Colt revolver, as part of the "fashionable toilet."[8]

At midcentury, one did not have to go as far west as Nevada to visit what were considered "western cities." A piece of verse on the subject, which spoke of "These enterprising, fearless chaps, / So ready with percussion caps, / And knives, the bowie," rang true to an editor when he looked upon a fellow from Wisconsin, who was described as a "walking arsenal." He carried "a revolver in each pocket, bowie knives in his boots, a cow-hide in his hat, and an Arkansas tooth-pick down his back." Perhaps one of the knives was reserved for doing what its name suggested, for as could be observed still in 1877, "if the universal traveler's tale be true, a fork, or the all-useful bowie knife" served the purpose of a toothpick.[9]

Any knife with a long, slender, pointed blade came to be known as a "toothpick," the word often modified by the appropriate geographic designation, suggesting the implement's widespread use. When Alabama "passed a severe law against the carrying of secret weapons," it grouped with Bowie knives not Arkansas toothpicks but Mississippi toothpicks.[10] An adventurer on the Mexican border encountered a man who wore "a sash and belt filled with fire arms, a Kentucky 'tooth-pick' glistening in its crimson sheath among them." Campers visiting the site of the massacre of the inhabitants of an Arapahoe village some decades after the event found there "a bowie-knife of ancient pattern—a true 'Louisiana tooth-pick'—whose handle had yielded to the weather, but on whose hard blade the rust had made scarcely any impression. The edge of this knife was whetted in true Indian fashion, round on one side and flat on the other." In 1880, a visitor to Wisconsin described a "very vile and odious town, generously streaked with human gore and deeply shaded by every crime punishable by human law" and found the "troublesome burghers" to be ever ready to "plant an 'Oshkosh toothpick' in the breast of the traveler."[11] But the weapon did not everywhere take the name of the town in which it was used. Rather, sometimes it kept its name of origin. According to one story, a raucous individual, who evidently had had too much liquor, entered a San Francisco barroom and threatened to disembowel anyone who would not accept his invitation to drink with him. He emphasized his determination to buy "by drawing an eighteen-inch Arkansas tooth-pick from the back of his collar and advancing upon the table" at which some patrons were sitting and talking. They quickly scattered and escaped, and a policeman took the knife-wielding regular away. Bowie knives were among the objects removed from New Orleans prisoners in the 1840s. Other confiscated weapons included a "long stiletto" and a "Red River toothpick, with 'sharper than a serpent's tooth' on the handle."[12]

A "redneck toothpick" with a six-and-a-half-inch blade was recently offered on eBay. In another auction, a handsome knife hand-forged out of a steel railroad spike was identified as an "AR-CAN-SAW toothpick dagger."[13] A 1956 trading card from the "Davy

The practice of referring to a sharply pointed knife as a "toothpick" has been widespread since the middle of the nineteenth century. Here, a tourist visiting Alaska is advised of the fact.

Crockett, King of the Wild Frontier" set referred to the large, pointed knife shown in his hand as "Bowie's toothpick," thus after over a century of travels bringing the two legendary names together again.[14] In the early twenty-first century, just about any knife with a long pointed blade may be called a toothpick, but the term seems to have come to be applied especially to folding knives. There is hardly a day that one cannot log on to eBay and find one identified as a Texas toothpick offered at auction. By extension, anything long and narrow is liable to be referred to as a "toothpick." Thus, "spicy batter-fried slivers of onions and jalapenos" sold in fast-food restaurants are also called Texas toothpicks, as are "all natural chews for dogs" made from beef tails "smoked deep in the heart of Texas."[15]

West Virginia seems to be among the states that do not brag about a namesake knife, but natives of the mountain state have claimed that its historic industrial city on the Ohio River was the birthplace of the real toothpick. "Any place else, they'd have more accurately called it a teethpick," the reasoning goes, but "nobody from Wheeling has more than one tooth." The self-deprecating joke ignores the fact that historically toothpicks were used not only to clean the spaces between the teeth but also to pick out food from a cavity in an individual tooth. Hence, the original name of "tooth picker" was apt indeed. (A dentist—from London, not West Virginia—once promoted the idea that the toothbrush be renamed the "mouthbrush," since it is "used to clean cheeks, tongue and gums as well as the teeth.")[16]

Art historian and museum administrator Stephen Weil once described in considerable detail a fictitious National Toothpick Museum. The multidisciplinary NTM, as he frequently referred to it, appeared in many ways to be like other museums. It was "heavily endowed, well situated in a prime downtown area, installed in its own large, modern and climate-controlled building, and devoted wholly and exclusively to the collection, preservation, study, interpretation, and display of toothpicks." The museum mounted changing exhibits, sponsored lectures and symposia, and published its own quarterly journal, *History's Splendid Splinter,* in which scholarly articles on the instrument's "role in social history, patterns of forestry, and the evolving technology of toothpick manufacture" were published. The NTM director was proud of his museum, which he considered "first-rate."[17]

However, according to Weil, "virtually nobody ever visits the NTM, nor does the press frequently write about it," and he opines that "our common sense tells us that this is a ridiculous endeavor, a venture that might be acceptable enough as a hobby but which becomes grotesque and preposterous when inflated to the level of a large-scale museum." His real purpose becomes clear: to use the imagined NTM to critique developments in the museum world that had legitimized its kind by their simply "collecting, preserving and displaying" the objects that they did. When these three activities were taken as the museum's raison d'être, he concluded, "something as silly as a museum wholly devoted to toothpicks can begin to seem plausible and legitimate." What is missing from that approach, he maintained, was the coupling of objects and ideas about objects, which is what went into a real museum. In spite of Weil's desire to "conceive of museums in such a way that the NTM would be a palpable absurdity from its very inception," and the view that toothpicks themselves "are of dubious, if of any, interest," the close study of anything as both an object and as an idea is potentially intellectually rewarding and revealing about the technology and culture in which it is embedded. Perhaps not for the museum curator bent on satire, but "for the historian there are no banal things," at least according to Siegfried Giedion, who championed anonymous history.[18] If nothing else, the toothpick deserves to be understood in context, as part of the landscape and language.

The term "toothpick" is not necessarily pejorative, though it has often been used ironically and hyperbolically. In a discussion of humorous and witty sayings associated with the Civil War, the use of the term "brevet" to convey honorific promotion in rank for exceptional service was made clear by example. Thus, a private might be motivated to good conduct by the promise of becoming a "brevet colonel," a mule was called a "brevet horse," and a toothpick was a "brevet bayonet." (In French slang, the bayonet is referred to as a *cure-dent*.) Things were also demoted, thus making a horse a "mule," and a bayonet a "toothpick."[19]

In his eighteenth-century travels around the world, Captain James Cook came across land in the South Atlantic that he named the Isle of Georgia, but he did not think that the discovery would benefit anyone. According to his description, the island met the sea mostly in "perpendicular ice-cliffs." In its interior, "the wild rocks raised their lofty summits, till they were lost in the clouds, and the valleys lay covered with everlasting snow. Not a tree was to be seen, nor a shrub even big enough to make a toothpick."[20] To illustrate Aristotle's definition of hyperbole in an early-nineteenth-century context, there is the story of a decidedly poor man who boasted of his wealth: "Here, this man says he has a large house encircled with an extensive wood, when I am certain that a tortoise could walk over his house in ten minutes, and that he has not wood enough to make a toothpick."[21]

In the early years of the twentieth century, one of the most popular speakers on the lecture circuit was the Ohioan Sylvester A. Long, who would eventually serve as president of the International Lyceum Association. His agent, the Coit Lyceum Bureau, described him as "a sane idealist" who was accustomed in his performances to "start somewhere, go somewhere, and stop when he gets there," having some fun along the way. He was "slender and a trifle under six feet," a careful dresser, and "a bundle of nerves" when speaking.

In the course of a lecture, it was said, "his body disappeared and his head expanded until it filled all the stage." He delivered more than two hundred lectures a year, and his best-known was "Lightning and Toothpicks," which he gave over a thousand times, and at least once in every one of the United States. It was "a mosaic of great truths laid in a cement of humor." Since you had to pay for the truth and the humor, flyers advertising Long's speeches, including the one on lightning and toothpicks, were short on specifics: "As is the case with many books and plays, the subject is based on an incident and has very little to do with the theme. It is a practical discussion of the universal reign of law, which is made endurable by love and appropriated through habit. It is a real contribution to advanced thinking."[22]

In more self-evident relevance and allusion, the toothpick image has also been used to convey the insignificance of one thing among many like things. In a review, the music of the Russian composer Sergei Taneyev was evaluated to be "devoid of life and character and is as individual as a toothpick nestled in a box of toothpicks."[23] An episode of the television series *The Sopranos* was entitled "Another Toothpick," a phrase that was uttered when still another "insignificant" life was violently snuffed out. But toothpicks do count in politics. In the 1828 presidential race, when electoral votes were being projected, Andrew Jackson supporters observed that they had mistakenly "placed Delaware in our breeches pocket, and lost it with an old tooth pick, and a piece of waste paper."[24] Analyzing the 1904 presidential race between Theodore Roosevelt and Alton Parker, the *Buffalo Times* concluded that "the hope of the Republican party of winning the Presidency while losing New York is as slim as a toothpick." In 2004, it was said that George W. Bush's "electoral vote victory was toothpick-thin."[25]

The toothpick can also be a symbol of futility and frustration in the face of an overwhelming if not impossible task. Though baby beavers might be happy in a toothpick factory, they would have a difficult time building a dam out of the little sticks. The frustration level would have been even higher for some late-nineteenth-century freshmen at Williams College who as part of their hazing were made to "sit down in a bowl of water and row with tooth picks."[26] At the other extreme, to a small insect, the toothpick can be a lifesaver. According to one wit:

I found a cockroach struggling in a bowl of water. I took half a peanut shell for a boat. I put him into it and gave him two wooden toothpicks for oars, and left him. The next morning I visited him, and he had put a piece of white cotton thread on one of the toothpicks, and set the toothpick up on one end as a signal of distress. He had a hair on the other toothpick, and there that cockroach sat a-fishing. The cockroach, exhausted, had fallen asleep. The sight melted me to tears. . . . I took that cockroach out, gave him a spoonful of gruel and left. That animal never forgot that act of kindness, and now my house is chuck full of cockroaches.[27]

Next to knives, the things that seem most frequently to have been referred to as "toothpicks" are very narrow and pointed boats, some even being given the word as a name, or at least a nickname. Thus, everything from an inland-lake canoe to a coastal cutter might be christened the *Toothpick*. At the 1893 Chicago exposition, the world's largest searchlight was cast on everything from the Ferris wheel to the lagoon, from the top of the former of which "launches and gondolas looked like toothpicks."[28] Sometimes, boats are disassembled into toothpicks. In the harbor of Portland, Maine, the pilot who took a steamer out would generally return to port in a dory, but when the sea was too rough, he would stay aboard the liner until it arrived at Halifax, Nova Scotia. One old pilot, who had "never been away from Portland over Sunday" in his entire life, took a steamer out one time and then insisted on rowing back in a terrible sea. His plan was to get to Half Way Rock and wait out the storm, but those watching him from the rock's light thought for sure that in such a pounding sea, if it was lucky enough to be cast ashore, "a dory must surely be smashed into toothpicks." Fortunately, the keeper of the Half Way Light spotted the little boat coming through the smothering waves, and he and his helper were able to assist the pilot onto the island.[29]

All sorts of craft can be "tossed like toothpicks" in a rough sea, telephone poles can be "snapped like toothpicks" in a hurricane or tornado, wagons and trucks can be "made into toothpicks" before an oncoming locomotive, and small trees can be "crushed like toothpicks" under the force of a larger one. At the other extreme, very large trees have been called toothpicks ironically. A postcard from the 1950s pictured a fully loaded logging truck that was described as carrying "Paul Bunyan's Toothpicks." Another popular postcard from the Northwest pictured a trio of gigantic squared timbers loaded on a railroad flatcar. Depending on the sale point, the identical image could be captioned "Oregon Toothpicks" or "Washington Tooth Picks." An early-twentieth-century photograph, thought to have been taken in the Cascade Mountains, shows a gang of lumberjacks posing with an enormous log. The title of the photo is "A Washington Toothpick."[30] Anything large, long, and relatively slender when viewed up close may also be described as a toothpick when seen from afar. Thus, when a writer wanted to convey an impression of what a runway might look like to an airplane pilot aiming to land on it, he used the following image: "Imagine a toothpick pointing toward you in the middle of a long dining room table."[31]

The image of a toothpick has also been employed to signify overwhelming odds. Thus, according to a sports commentator, one football team's defense was so bad that he had "a better chance of hitting a grand slam at Yankee Stadium with a toothpick than they have of winning a Super Bowl."[32] In the 2004 presidential campaign, an opponent of John Kerry, whose tradition it was to have lunch on Election Day at the Union Oyster House (that purported cradle of toothpick use in restaurants in America) likened his effort to win the presidency to "trying to build a pyramid atop a toothpick." Of course, there is an aluminum pyramid of sorts atop the Washington Monument, which has been called "that toothpick they call a Monument."[33]

Any slender thing can be referred to as a toothpick. One veteran baseball player who had been gaining weight was remembered as once being "so skinny that a toothpick would hide him with its shadow." An umbrella that could be rolled up to the thinness of a cane was advertised as the world's "closest rolling" and sold under the name "The Toothpick." "Toothpick" is a common nickname for any tall, lanky person, or as a noun or adjective to describe someone tall and thin, as in "that toothpick Nicole Kidman."[34] It can also be used to describe an actor who does not show much emotion as being "as wooden as a toothpick." The generally characterless characteristics of a toothpick have also led to such descriptions of blandness and thinness as "the soup tastes like a tooth-pick."[35]

The toothpick has also been used as a symbol of standardization and conformity. In the late 1930s, there was a movement to require newspaper editorial workers to "conform to certain tests imposed by a membership in a national organization and that none be employed or remain employed except members of that organization." An opponent of such a practice worried about its effect on the profession of journalism: "It will have sacrificed individuality. We will have become button makers and tooth-pick whittlers."[36]

From the late eighteenth century, the use of a toothpick has been associated with idleness and relaxation. Thus, reading something in which nothing of grace or interest might be expected to be found might be done "in our easy, pick tooth, after-dinner way." Another writer, reflecting on the fast pace of progress in the early days of the railroad, envisioned steam soon being employed to set type, replacing printers who "pick up their types one by one in that lazy toothpick way."[37] Perhaps more in reference to the way it should be read than how its type was set, an Illinois newspaper was called the *Ashmore Rose's Toothpick*. A less cryptically named newspaper, though at the time one with decidedly comic intentions, was the *Arkansas Tooth-pick*, which in 1883 announced a duel between the paper's junior editor and the "senior member of the firm of sculptors" Dooley and Garret. The obvious spoof earned extended notice in the *New York Times*, but surprisingly there was no explicit mention of a Bowie knife.[38]

CHAPTER EIGHTEEN

New Uses for Old

A LTHOUGH ORIGINALLY INTENDED to be specific in its use, the
toothpick has found applications far beyond the teeth and
gums. People are by nature adaptive, creative, and inventive, capable of taking anything far beyond its stated and intended purpose.
Given a grain of sand, they will see in it a world. Given a lever, they
will move the earth. Given a toothpick, they will turn it into a universal tool.

Toothpicks have come in especially handy in prison settings,
and in a curious twist of technological fate, the quill toothpick,
which had often been cut from a handy feather pen, on occasion
found use itself as a makeshift writing instrument when there was
none other to be had. In mid-eighteenth-century London, a perfectly sane gentleman of some means was kidnapped to a madhouse,
where he was confined and denied access to pen, ink, and writing
paper. He was able to communicate his plight only after being
allowed "some Morrella cherries and a tooth-pick," with which he
wrote a note that was carried to the outside by a friendly servant.[1]

A more famous incident occurred during the French Revolution, when Madame de Lafayette was imprisoned. According to her
daughter, "With a toothpick and a little piece of Indian ink, she
wrote the life of my grandmother, on the margins of the engravings
of a volume of Buffon." The foreign minister Chateaubriand
reportedly knew of a lady, perhaps the same Madame de Lafayette,
"who wrote her memoirs with a toothpick which was dipped in
watery pine soot."[2] In another prison story, an inmate was confident
that he could plot a successful escape because he had already
deceived his jailers using "a toothpick and soot," and a scrap of
paper to compose a letter to the outside. Like Madame de Lafayette
writing in Buffon's *Natural History*, he was able to make a pen out
of a quill toothpick.[3]

The distinguished Italian historian Cesare Cantù, imprisoned by the Austrian government for his history of seventeenth-century Lombardy, during his incarceration wrote a romance "with a toothpick and soot." Even the quill's successor, the wooden toothpick, was used as a writing instrument during World War II. A confessed Nazi spy revealed that she had been instructed to dissolve special pills in a small amount of water to produce a colorless solution of invisible ink into which a toothpick could be dipped for writing secret messages.[4]

The idea of writing without a pen has produced many a fantasy. In one odd Victorian conceit, the fingers and toes talk about themselves and their talents. In its soliloquy, the little finger asserts its superiority over one of its larger brothers, "who cannot hold a pen alone." The little finger, on the same hand, "can write without a pen at all; all that I have to do is to allow my nail to grow very long, and then, when it is cut into the proper shape like a quill-pen, I can dip it into the ink and write very well."[5] Clearly, it could also be used as a toothpick, in or out of prison.

Toothpicks in a prison setting have also served as other than writing instruments. Before the Revolution, two Frenchmen who were imprisoned in the Bastille with no expectation of being released plotted an escape. The mastermind determined that their best chance to reach the outside would be to climb up through the chimney of the fireplace in their cell, and then to use fabricated and concealed ladders to get down into the moat and over the wall on the other side of it. In order to execute his plan, he had to understand the layout of the cellblock, so that he would know how to proceed. The lead prisoner had been held previously in several different cells in the Bastille, and so he knew that if anyone was in the cell above or below, he could hear them. However, he never heard noise from beneath his present cell, although he knew there was a prisoner there, and so he inferred that there was a space between the floor of his cell and the ceiling of the one below. To find out exactly what the situation was, he took advantage of the brief time the prisoners were let out of their cells to hear mass in the chapel. On one of these occasions, he instructed his cellmate to wrap his toothpick case inside his handkerchief so that, when he pulled it out of his pocket, the case would fall down the stairs. While the guard retrieved it, the determined prisoner had a chance to survey the

204 · THE TOOTHPICK

construction and confirm that beneath his cell there was indeed a significant space in which to hide ropes and other materials for the breakout. When his cellmate asked where they would get rope, he indicated that they would unravel linens and reweave them. Ladder rungs were made from their cell's wooden tabletop and crowbars from its iron legs. These were used to dig into the space beneath their cell and to dislodge the iron bars blocking the escape route through the chimney. As things were made, they were stashed away in the void beneath the cell. In time, all was ready, and they did escape.[6]

Like most everything else, a toothpick can be enlisted into service equally for matters of life and death and for matters of vanity and pastime. During World War II, a woman who had begun styling hair for the girls working in the mills in Lowell, Massachusetts, was reported to have substituted toothpicks for metal hairpins, which were unavailable.[7] It was not reported how she did this, though it may have involved a technique similar to that used for producing tiny pin curls. After the war, a woman who dressed dolls in historical costume created authentic-looking period wigs full of curls by "twisting the fake hair tightly around tooth picks."[8] This is easier to picture.

Men have also enlisted the lowly toothpick in creative ways. A poker player who was dealt a pair of deuces decided to bluff with a twenty-dollar bet. After he was raised, he himself raised, and soon there was two hundred dollars in the pot. When he drew nothing helpful, he took a toothpick out of his pocket and began to pick his teeth. He bet two hundred dollars more and continued to use the toothpick. After some time, the opposing player folded with a pair of kings. According to the bluffer, it was the toothpick that helped him win:

> If I hadn't had it to play with I would have nothing to do but sit and look my adversary in the face, and my looks would have betrayed that I was bluffing at a glance. Say, didn't you ever notice the habit some men have of having toothpicks handy when they go on the witness stand? It gives them self-possession. A man may call you a liar, and if at the same time he shifts a toothpick nonchalantly from one side of his mouth to

the other you can rely on his meaning business and backing up his assertion with right-handed argument if need be.[9]

In 1929, H. Fendrich, Inc., of Evansville, Indiana, sponsored a contest that invited smokers of its Charles Denby cigar to explain "the secret of its nation-wide popularity." Among the notable entries quoted in an advertisement announcing the winners was "The one cigar you can smoke to the toothpick stage," evidently referring to the practice of sticking a toothpick through a cigar stub too short to be held directly in the fingers.[10] It was once common-place to find in a kitchen three or four toothpicks stuck into an avocado seed or a sweet potato, making it look somewhat like a spindly-legged rocket ship about to take off from the glass of water in which it was partly submerged in hopes of its sprouting into a plant.[11]

Wooden toothpicks have long been used to amuse children, who have been encouraged to arrange them in the form of letters to learn the alphabet or to spell out words. The wooden splints have also been the props for puzzles, tricks, and other amusements.[12] Toothpicks supplemented by peas, marshmallows, gumdrops, or similar connecting devices have been used to construct everything from stick figures to skyscrapers. For older children, toothpicks and glue have been supplied for the building of structures whose strength was later tested. Constructing a toothpick bridge is a familiar high-school challenge, with the winner being the one that can sustain the most marbles or other weights being added to a small pail hung from the completed structure.

Toothpicks and other novelty wood products, such as blunt toothpicks, never-used Popsicle sticks, and modified clothespins (doll pins) intended to be painted and dressed as dolls, have been sold as materials for craft projects.[13] In 1960, the Forster corporation promoted the use of regular toothpicks in making a Christmas centerpiece. The project, which was described as being easy "as sticking pins in a pin cushion," consisted of inserting lots of toothpicks partway into Styrofoam balls, resulting in objects that resembled curled-up hedgehogs. These were then stacked up in a pile resembling a Christmas tree, and sprayed with "snow spray." The bill of materials called for eight boxes of Worlds Fair toothpicks,

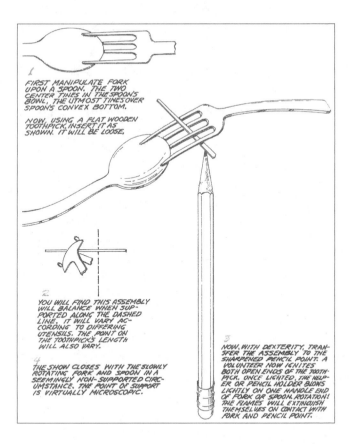

The toothpick has been a prop in many a parlor game and trick. Here, a jammed-together fork and spoon naturally orient themselves so that the center of gravity of the assembly is located at the pencil-point support on which the toothpick rests.

and Forster's employee newsletter credited the promotion with being "directly responsible for the greatest run on toothpicks in the history of the industry." The following season, the company added a toothpick wreath to its holiday promotions. Also, "in order to sell a maximum amount of toothpicks . . . and to insure a dependable source of supply for all materials" needed for the projects, Forster itself made Styrofoam balls and bases and marketed plastic snow.[14]

With or without promotions by the industry, adults have developed hobbies, if not obsessions, involving the construction out of toothpicks and glue of replicas of everything from a crucified Christ to the *Titanic* and other historic steamships. Such projects can take thousands of hours of painstaking effort and hundreds of thousands of toothpicks that machines could spew out in minutes. The Christ, which was made up of 65,000 sandwich, flat, square, and round toothpicks, took 2,500 hours of work to glue into place. The rock guitarist Wayne Kusy used 194,000 toothpicks to make a sixteen-foot-long replica of the *Lusitania*.[15]

Theodore Roosevelt is reported to have said of Americans that they "like big things," but they often have to settle for scaled-down versions of them. In the first half of the twentieth century, there was a National Pickbuilders Club from which could be obtained "plans for bridges, towers and other structures that were accurately scaled down to toothpick proportions," typically about a tenth of an inch to the foot. Around 1935, Phil Harris, a craftsman from Fullerton, California, in the cabinet- and furniture-refinishing business, joined the club and began building with a passion. His thirty-five-thousand-toothpick Eiffel Tower, which weighed only three pounds, stood ten feet high and required the use of only five tools: tweezers, razor blade, paper clips, a tiny clamp, and of course glue. After completing his tower, Harris wanted to make something that moved, so he chose to build a model of a Ferris wheel. Other projects of his involved models of great bridges, including a five-foot-long Hell Gate Bridge that could support a load of seventy-five pounds.[16]

Steven Backman, who took two and a half years to craft a thirteen-foot replica of the Golden Gate Bridge out of thirty thousand toothpicks, describes himself as an "artist working in toothpicks." New Yorker Stephen Talasnik, who grew up to be a real artist, made his first sculpture as a child of ten or eleven when he entered a contest sponsored by Elmer's glue, then a new product, and used it and toothpicks to construct a model roller coaster. Another sculptor, Michael A. Smith, of Baton Rouge, Louisiana, produced a fifteen-foot-long, 850-pound alligator by gluing together about three million toothpicks, which took him about three years. Still another artist, Stan Munro, used 1.3 million of the wooden splints to construct Toothpick City, which included replicas of the Empire State Building and other skycrapers, towers, and landmarks from around the world. His work was on exhibit at the Syracuse Museum of Science and Technology during the summer of 2006.[17]

At the other extreme, an Indian man seeking to get into the *Guinness Book of World Records* carved twenty-eight chain links from a single wooden toothpick, beating the old record of seventeen. People can use toothpicks to censor their speech, perhaps when the knife slips: "H-e-double toothpicks."[18] Nerve-wracking and frustrating as toothpick model building and carving hobbies might at times become, they no doubt beat having nothing to do. According to one verse description of a geezer who had come upon hard times, he just hung around the general store:

> *Sitting waiting for the mail—mail that never comes;*
>   *Gee, but times are hard these days—there's nothing one can do*
> *But sprawl about the store in town, a toothpick in his gums.*[19]

Educational uses of toothpicks have included teaching about the theory of evolution even when it was not approved by the state. One teacher in Kansas gave his students the task of finding and picking up out of the grass red and green toothpicks by using only plastic eating utensils, which taught the children that "red toothpicks had a smaller chance of 'survival' because they stood out more than green toothpicks." Also, children using "forks scoop up fewer toothpicks than those with spoons, an analogy to the process of natural selection."[20]

From the time they are babies, children have benefited from the many uses to which toothpicks can be put. When an early-twentieth-century advice seeker asked about outfitting a baby basket, she was provided with a long list of items, ranging from safety pins to talcum powder, which included "some absorbent cotton, a few wooden toothpicks." Another woman expecting her first child asked essentially the same question but got a more specific answer, which included a "package of absorbent cotton; package of wooden toothpicks." This advice belied the remark of a writer who appeared to be a social reporter rather than an advice columnist. She ridiculed a doctor who brought a new baby a box of toothpicks, commenting, "That's about as much as a bachelor knows about babies." However, advice columnists continued to promote a "liberal supply" of absorbent cotton and remind readers that a "box of tooth-picks must not be forgotten, as these, when wound with cotton, are used to wash the baby's nostrils and ears."[21]

Toothpicks wrapped in cotton served also for other purposes. The way to treat an ulcer on the roof of the mouth was to "take a small piece of cotton, firmly twist it on a wooden toothpick," dip it in a prescribed powder, and gently touch it to the sore. To bring relief to a baby who had a severe cold, a mother was advised to "twist a small piece of absorbent cotton firmly on a wooden toothpick, dip it in olive oil and gently insert it up each nostril several times daily." Even for the well baby, a twist of cotton on a toothpick end was to be dipped in boric acid and inserted into the nostrils just after the daily bath.[22] There was probably also advice on what to do if the cotton came off in the baby's nostril.

With all the uses to which homemade-on-demand cotton swabs were put—and the uses were clearly not restricted to babies—it is perhaps a wonder it took so long for someone to think of manufacturing the things ready made. Inventors did note the faults of sticking cotton on the end of a toothpick, however. In 1906, Harry Dorr, in between inventions relating to the cutting knives that were used in manufacturing toothpicks, patented a "toilet article" that could serve for "manicuring purposes" and as a "high-grade toothpick." In describing his invention's advantages, he pointed out that it was suitable for use on the navels of babies, its "smooth rounded end being free from splinters and peculiarly adapted to receive a wrapping of absorbent cotton."[23] However, it was not until 1923 that Leo Gerstenzang thought to make and sell little sticks already wrapped in cotton. Evidently, he was watching his wife "trying to clean their baby's ears by applying a wad of cotton to a toothpick" and got the idea to make what he first called Baby Gays but later changed to Q-tips. The now-familiar cotton swab has been called "another 'humble masterpiece' that unites materials, form and function."[24] The same might be said of the toothpick itself.

Wooden toothpicks have long been recommended for cleaning dirt from beneath adult nails, after the fingers have been soaked in warm water, perhaps having had a little borax or ammonia added. Wood was to be preferred over ivory or steel implements because those, "if used roughly or too constantly, wear a ridge on the under side of the nail from which it is difficult to dislodge foreign matter." Similar advice warned against using the "nail-file for any purpose but that of filing the nail." Rather, an "orange stick or a soft wooden toothpick" was prescribed. Even now, an orange stick is recommended for nail and cuticle care. Closer to skewer- than toothpick-size, a good specimen of this cylindrical object is made of wonderfully smooth, light, and tight-grained orangewood sharpened like a leadless pencil at one end and cut off at an angle at the other. Professional manicurists have been known to dip the stick in an antiseptic before employing it.[25]

In the hands of the clever and mischievous, toothpicks have also been used for evil. In 2003, protesters against the coffee chain "squeezed glue and forced toothpicks into the locks" of Starbucks shops in Houston and elsewhere.[26] Even if they could be extracted whole, such toothpicks would be unlikely to provide much of a lead to catch the perpetrators. But other crimes and mysteries have been solved with the help of a toothpick. In one fictional case, a witness at a murder trial accused the assistant prosecuting attorney of the crime and said he could prove it from the toothpick that the prosecutor had thrown on the floor. The toothpick was hand-whittled and matched one found at the murder scene. Both were obtained from a lame man who made and sold the toothpicks, and the newly accused was one of his regular customers. In a real criminal case, suspects might be convicted when samples of their incriminating DNA are collected from discarded toothpicks.[27]

The humble toothpick has also been used in the design of bugging devices. The San Francisco private eye and expert on electronic surveillance Hal Lipset once demonstrated before a U.S. Senate Judiciary Subcommittee hearing on eavesdropping how a "listening device could be disguised as a martini olive with a toothpick antenna." A miniature microphone was embedded in the olive's pimento, and a copper wire serving as the antenna was concealed by the toothpick. It was described as a very pry martini.[28]

CHAPTER NINETEEN

Maurice, Oscar,
and the Industry

M AURICE FORSTER had learned the toothpick business first-hand. When he left school at the age of twelve he was already "doing the work of a skilled mechanic" in the factory. Later he would be described as "uneducated," and throughout his life his signature would remain almost childlike, though his F beginning "Forster" was decidedly studied. Maurice was not without talent or determination, and eventually "he became a skilled bookkeeper and accurate accountant, a fine draftsman and a machinist of no small ability." Indeed, "his capacity for business was almost uncanny as a young boy." At twenty, he was manager of the Forster mill in Dixfield. Six years later, in 1899, he was superintendent of the toothpick factory at Strong.[1]

When his father died in 1901, Maurice was still in his twenties and was (but not by his father's will) "placed in charge of the business at Strong, with full power to give orders, purchase goods, sign checks, hire and discharge men, and handle the sales end of the business." He was managing the plant for Oscar Hersey, the sole trustee of the Estate of Charles Forster. For his services during the year 1902, Maurice was paid one hundred dollars a month from each of the mills at Strong and Dixfield, plus expenses. This compensation was in addition to the thousand-dollar annuity from his father's estate and did, in effect, more than triple the amount of money that he realized from the making of toothpicks. In his position, Maurice no doubt knew how profitable the business could be, and he could also see that he was realizing less from the family business than the green trustee for whom he worked. Hersey did need help. He may have been exposed to the industry through its legal issues, but his experience was in the law, not in the nitty-gritty of actually manufacturing toothpicks. At the time he was appointed executor and trustee, both his legal practice and the toothpick busi-

Maurice Forster's adult signature appeared to be childlike, perhaps reflecting his limited formal education. This example, affixed to a petition regarding Oscar Hersey, trustee of the Estate of Charles Forster, was signed when Maurice was in his mid-thirties.

ness were growing rapidly.[2] He had to know that experience like Maurice's was essential to the profitable running of the Estate of Charles Forster.

Through industry trusts and other cartel-like mechanisms, manufacturers in the toothpick industry had held the competition in check until the late nineteenth century, when a flood of people seemed to want to get into the business. In 1900, an advertisement in *Scientific American* told the tale concisely: "WANTED, to communicate with Tooth Pick Machinery M'facturers. Keystone Wood Co., Williamsport, Pa." Such pleas were heard like the repeated calls of tinkers: "Wanted to buy toothpick machinery."[3]

Wherever there was a supply of suitable wood, someone could see toothpicks being made. William F. Hutchinson, who was born in Maine, bought a match and toothpick factory located in Hadley, New York, which is in the Adirondacks north of Albany, and moved it to Valatie, which is near the Berkshire Hills south of Albany; shortly thereafter he died suddenly while on a business trip to Liverpool, England, where he was likely soliciting orders for his factory's wares.[4]

But the proliferation of manufacturers, not all in sympathy with the idea of a cartel, soon flooded the market with toothpicks, and inventory accumulated. As with any industry, some businesses would be able to wait out the lean times while others could not. In 1910, one-hundred-dollar shares of the Sweetwood Toothpick Company would sell at under twenty dollars apiece, and in 1915 the less robust Royal Antiseptic Tooth Pick Company of New York would undergo voluntary dissolution.[5]

In the first years of the new century, however, the future still looked bright. It was then that Maurice Forster decided that he wanted to participate directly in the growth of the industry in which he grew up, and he wanted to do so as more than a manager and a beneficiary of an estate that he did not control. He wanted to be a business owner. So, in 1902, he went into partnership with John S. Harlow, an attorney who had had experience operating a toothpick mill in Buckfield, and George Merrill, who about fifteen years earlier had established G. L. Merrill & Co., Dixfield's first spool mill. Maurice built his toothpick factory in Dixfield and named the business Forster Manufacturing Co. It succeeded by "giving about twice as many toothpicks in a box and at a lower price per case" than the competition, and by 1905 the enterprise had prospered to the point where Maurice could buy out his partners. The rapid rise and obvious success of the new Forster toothpick maker "advertised to any who might be interested that it might be a good business to engage in."[6]

The turbulent development of the industry and Maurice's participation as a competitor as well as a beneficiary of the Estate of

A photographic portrait of Maurice Forster as a mature toothpick magnate is part of the Forster Collection at the Franklin Memorial Hospital in Farmington, Maine; another print of the same portrait hangs in the administrative offices of Jarden Plastic Solutions in East Wilton, on the site where a Forster plant once operated.

Charles Forster must have made Oscar Hersey's job as trustee and his relationship with Maurice increasingly difficult, if not openly acrimonious. Hersey's own actions might well have contributed to the tension. In 1906, a trademark for wooden toothpicks—consisting of the "arbitrary word, 'VELVET' "—was registered in his own name. In his application statement, he identified himself as "testamentary trustee under the last will and testament of CHARLES FORSTER," but his applicant's declaration claimed that he believed "himself to be the owner of the trademark" rather than any "other person, firm, corporation or association." In his application for a second trademark—consisting of a design featuring two crossed toothpicks—he did not even state his association with the Forster estate and claimed himself as the owner of the trademark.[7]

At least one patent was also assigned to Hersey as trustee, rather than to the estate itself. The inventor was George P. Stanley of Dix-field, who in the 1910 census was identified as a toothpick mill supervisor, the mill being that of the Estate of Charles Forster. The invention was for a cutter to make chamfered and tapered tooth-picks with the wider end rounded, thus "giving it a smooth and waver finish, making its introduction into the spaces between the teeth much easier and lessening its liability to break off or sliver."[8] The invention was very similar to one by Harry Dorr, who had patented a cutter for toothpick machines just six months prior to Stanley's filing of the application for his. Curiously, the latter's patent was issued only five weeks after filing, an uncommonly quick turnaround in the Patent Office, especially for an invention seemingly so similar to something recently patented. It is especially curious because Dorr filed another application for a variation on his cutter—consisting of an arrangement of the knives that was designed to minimize waste—just weeks before Stanley's, and it took almost four years for that patent to be issued.[9]

According to the inventor Albert Hall of Peru, Maine, writing in 1909, "One of the principal objects of manufacturers of toothpicks, namely, the production by automatic means of a tapered, round pointed toothpick, has been only imperfectly attained." In other words, the inventions of Stanley and Dorr could stand to be improved upon. These inventors had achieved a degree of success, of course, especially in how they rounded the wider end of the pick, but Hall wished to round both ends satisfactorily. He achieved this

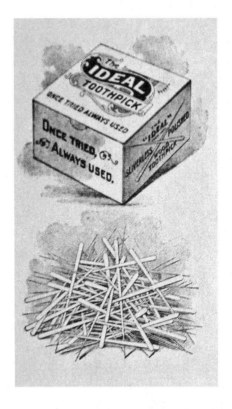

*A box of Ideal toothpicks, along with a pile
of its contents, was depicted on the cover of
an early-twentieth-century folder offering
"Tips on Tooth-Picks."*

by first cutting off the sides of the veneer strip in a pattern of alternating wide and narrow scallops. Then the toothpicks were cut out fully, at least two at a time, thus accelerating the rate of production. Hall's patent was assigned to Maurice's Forster Manufacturing Co. and gave that firm a decided advantage over the competition.[10]

The flat toothpick with two bevel-pointed ends was the one known as the Ideal, and it played a central role in the toothpick wars that were developing. The word "Ideal" had come to be used more or less generically as a descriptive term, even though it had been trademarked by Charles Forster.[11] Virtually all of the major manufacturers had their own patented knife systems efficiently to cut out flat tapered picks with one or two rounded (and beveled) ends. For example, the double knife method that William King of Phillips, Maine, invented and patented was assigned to the International Manufacturing Co. However, the technical difficulty of rounding the narrow end eventually led to it being squared off the way it is in toothpicks sold today as "flat." A machine to produce such picks was patented as early as 1917 by Edward Greenwood, also of Phillips. The fierceness of the continuing competition was evident not only in the subtle variations in the means of making an Ideal toothpick but also in the changing loyalties of the inventors, which may have been forced by circumstances. Thus, George Stanley, who had assigned his patent for making Ideals to Oscar Hersey, would assign his 1927 patent for making round toothpicks to the successor to Maurice's Forster Manufacturing.[12]

With all the intellectual-property jockeying that existed in the toothpick industry, Hersey's appropriation of trademarks and patents for himself rather than having them assigned to the Estate of Charles Forster should not have gone unnoticed by Maurice and his sisters. And the distinction should not have escaped them. After all, Charlotte had served as a witness to her father's statements in his trademark applications. And when Maurice filed applications for trademarks for his own firm, they were registered under the name of the Forster Manufacturing Co. In fact, in one document he identified himself as president of the corporation and declared explicitly that he believed that not he but "said corporation is the owner of the trade mark sought to be registered."[13] We will never know, because the matter was never put to the test, but Maurice might well have been on firm ethical and legal ground had he accused Hersey of violating his fiduciary responsibility by annexing what should have been the estate's intellectual property.

Aside from the issue of trademarks taken out in his own name, Hersey appears to have become a conscientious trustee engaged in running the toothpick business at a profit. According to his final account as executor of the estate, in the middle of 1902 the business had had assets of about $125,000.[14] The Forster children readily signed a statement approving of his accounting and asking him "not to file the said accounts in Probate Court, on account of the publicity which such accounts would make," and requesting him to "keep the same safely in his possession," with a copy kept in a safe in Strong. Perhaps these arrangements were Hersey's idea, or they may have been requested because the Forsters, or at least the women, simply wished to keep their association with toothpicks out of the public record, or they all may have wished to keep an accounting of assets private to maintain a business advantage. In any case, in their statement Charles Forster's surviving heirs (his wife had died in April, of paralysis) also acknowledged that they had each received three thousand dollars as a division of profits beyond the cash required to be kept as working capital.[15]

Hersey appears to have honored the wishes of the surviving heirs, at least at first. He kept meticulous accounts but seems not to have filed them immediately in Probate Court in Cumberland County. His first accounting as trustee, certified in October 1904, was not recorded until February 1906.[16] Two years later, the Portland courthouse suffered a fire; the Forster estate files now in the Probate Registry Office were restored through copies provided in 1910 by Hersey's attorney, George C. Wing, who practiced in Auburn. The restoration and augmentation of the files with incredibly detailed accounts appear to have been prompted by accusations and disagreements that in time had developed between Hersey and the Forster heirs and that culminated in litigation that made its way to the Maine Supreme Judicial Court.[17] This would be but the latest of a number of developments, ranging from the mundane to the extraordinary, that took place during the time that Hersey served as trustee.

During the first years under his supervision, the business had grown significantly in value and profitability. However, after the initial distribution, the Forsters saw little in the way of dividends, and their combined annuities did not come up to the commissions realized by Hersey as trustee. In addition, he was reimbursed for his expenses, which typically amounted to about twenty-five hundred dollars annually. Among the expenses were half the cost of Hersey's office and telephone, which means that he was spending, or at least charging, half his time running the Estate of Charles Forster, even as his overall practice was growing.[18]

There can be little doubt that Hersey did invest considerable time in learning about the manufacture and sale of toothpicks and in executing the business. He traveled regularly to visit wholesalers and negotiate orders. He also negotiated contracts with printers and box manufacturers for essential services and supplies, a not insignificant cost to the business. He does appear to have been earning a commission, but to the Forster heirs it seemed to come at their expense. They must have wondered what their father had been thinking when he named a trustee. The situation would have been especially annoying to Maurice, who in the meantime had shown himself perfectly capable of running a profitable toothpick business and had become fully aware of the potential benefits to be realized from it.[19]

In his defense, Hersey could point out that, by Charles Forster's specification, there was not enough cash to pay regular dividends to the heirs. Indeed, there was less than six thousand dollars available at the end of the 1905 fiscal year. However, there was also a large inventory of merchandise, empty boxes, and cases. It is in the nature of running a business to make judgments about what is the right balance to strike between cash and inventory (not to mention dividends), and observers could disagree on whether or not Hersey had struck that balance. Perhaps to assuage ill feelings, in the next fiscal year the heirs each received, in addition to their annuity, two thousand dollars. This made their share of the profits, collectively at least, about the same as Hersey's, which in the course of a year had increased by over 40 percent, to almost ten thousand dollars.[20]

In the meantime, however, the inventory of merchandise had increased from about $90,000 worth of toothpicks to almost $130,000 worth, and the amount of cash had dropped to less than $3,000. Trustee Hersey appears to have opted to appease the heirs, even at the expense of violating the terms of their father's will. That Hersey grouped all distributions to the heirs under the term "annuities," rather than breaking out what surely must have been dividends of a sort, suggests that he was aware of what he was doing.[21] In his defense as a businessman, Hersey did bring profits (and cash) up in the following two years, while paying even larger dividends, which he clearly labeled as such. But for fiscal year 1909, dividends dropped precipitously while cash swelled. Hersey evidently sensed a shift in the market, but the Forsters did not see it that way. Maurice's former partner and now his attorney, John S. Harlow, began to inspect the annual reports.[22] But the reports did not tell the whole story.

In 1906, the number of toothpicks being produced in the United States was said to be "beyond calculation—thousands of millions each year." There were notable plants operating in Massachusetts, New York, and Vermont, as well as Maine. In addition, "incredible numbers" of toothpicks were being imported, mainly from Japan, Portugal, and Italy.[23] There was enormous competition for the toothpick trade. It was in this business climate that Hersey had met with other manufacturers regarding the price on Ideal toothpicks. In 1907, he had conferred with wholesalers about meeting the price of toothpicks made by Maurice's Forster Manufacturing Co.[24]

Maurice had become a significant presence in the industry, as the actions of his competitors demonstrated. His toothpicks were packaged distinctively, with the monogram of the company proudly displayed. So, when the Cutter-Tower Co.—then ostensibly in the business of manufacturing as well as selling toothpicks—copied the packaging exactly, except for the monogram, it was sued and found liable for damages and any profits "realized from the sale of toothpicks in the imitation packages."[25] Maurice and his toothpicks were proving to be formidable adversaries, and the Estate of Charles Forster—along with the whole industry—increasingly was feeling the heat.

An advertisement for aromatic and flavored antiseptic toothpicks taken out by Cutter-Tower in the spring of 1909 provided further evidence of that company's desperation. The ad was aimed at the individual consumer, who was offered a full-sized box of the toothpicks in exchange for fifteen cents' postage and a completed coupon, outlined in toothpicks, that provided the name of the respondent's grocer and druggist. The company, which identified itself as "the largest manufacturers of toothpicks in the country," also invited correspondence "with dealers, hotel men, dentists and physicians." Just as Levi Tower had teamed up with Charles Forster to create a market for toothpicks four decades earlier, now his company was looking for ties to anyone who might buy and sell this new product. Dealers and agents, who were claimed to be "making from $3 to $10 a day," were wanted "everywhere."[26]

Another advertisement, appearing just months earlier in *Baseball Magazine,* was aimed not at the consumer but at the salesman. It told prospective representatives of the Aromatic Toothpick Co., whose address—184 Summer Street in Boston—was the same as that of Cutter-Tower, "Don't complain of 'hard times' when you

A 1910 advertisement taken out by Forster Manufacturing Co. included drawings of side and top views of a Gold Medal "flat" toothpick with tapered and rounded ends. The annotated drawings served to point out the features of the toothpick, which was about 2⅜ inches long.

can earn $30 to $75 a week taking orders for AROMATIC TOOTH-PICKS." As the advertisement explained, "These new picks, delicately perfumed and made in three flavors, sell on sight in high class restaurants, groceries and in homes." The company was willing to let salesmen carry the toothpicks as a "sideline" in exclusive territories.[27] If times were hard for salesmen, they may have been even harder for toothpick manufacturers and wholesalers.

And Maurice was relentless. In 1910, his company advertised its "flat" Gold Medal toothpicks as *"improved,"* principally by being "rounded at both ends—so that they will not pierce the tender gums nor scratch the enamel of the tooth." By contrast, the competing "ordinary toothpick" was described as "a clumsy piece of kindling wood that will splinter in the mouth or break off between the teeth." To leave nothing to interpretation, drawings showing the exact size and shape of the improved toothpick were labeled with its features. Readers who could not find a five-cent package of Gold Medal toothpicks to buy were invited to get a free sample box by providing (as perhaps they had also to Cutter-Tower) the name and address of their grocer, who would definitely be contacted by Maurice's salesmen. When it came to marketing, he was indeed his father's son, but he was not as reticent as his father when it came to associating his name with his product. The Forster Manufacturing Co. ad that appeared in the *Saturday Evening Post* was signed

"Maurice W. Forster, Pres."[28] While Maurice was putting the Forster name before the retail trade, the Estate of Charles Forster was still known mainly to wholesalers.

Early in 1911, there was a glut in the market. In a matter of weeks, the wholesale price for wooden toothpicks dropped by almost 80 percent, from thirty-eight dollars to eight dollars a case, which was less than the cost of manufacture. A new toothpick cartel was in the process of being formed, but a newspaper editorial warned that if it planned to set prices "unreasonably high," it might encounter resistance. After all, if necessary, people could "still whittle out their own toothpicks."[29]

A sample box of Maurice Forster's Gold Medal brand "improved" toothpicks was offered in a 1910 magazine advertisement. Later, full-sized boxes of similar toothpicks were "sealed" with the image of a gold medal. It was printed on a thumbnail cutout in the top flap that could be positioned on the outside of the box and glued down onto its front panel, thus securing the contents.

But the toothpick trust could not afford to let the market shrink. A promotional campaign was launched, and the newspapers carried notices that the combined output of the plants was one hundred million picks daily. In 1913, the humor magazine *Puck* saw the "inspiring power" of such a prodigious number to be "quite irresistible" to Sunday supplement editors, whom the magazine expected to report that all those toothpicks, "if placed in a straight line, would reach from Boothbay Harbor, Maine, to Walla Walla, Washington." But *Puck* was not as sharp with its arithmetic as it was cutting with its satire. Indeed, a hundred million toothpicks laid end to end would stretch over about four thousand miles, reaching well beyond Walla Walla and far into the open waters of the Pacific Ocean, where there were no customers. The flippant miscalculation by *Puck* might be seen as unintentionally emblematic of the industry's own overreaching.[30]

CHAPTER TWENTY

The Tragic Heiress

A T THE TIME that things were coming to a head between trustee Oscar Hersey and the siblings who were the beneficiaries of the Estate of Charles Forster, Charlotte, the eldest, was living in San Diego, California. She had left Maine and relocated there most likely because of Nathan Watts, a bachelor member of local society who was sometimes described as a "capitalist" engaged in real estate transactions and other financial dealings. Nathan appears to have traveled east for an extended period in 1893, at which time he may have met Charlotte, who was then living in her father's handsome house on Park Street in Portland.[1] As much as she may have liked the house that toothpicks bought, she also may have wished to distance herself from such a common business.

Charlotte's entrée into southern California society might have been given a significant leg up by her surname, and its evocation of John Forster, who played a considerable role in the history of the region.[2] He was born in England in 1815 and came to California at about the age of eighteen, settling at first in Los Angeles. He married Isadora Pico, sister of Pio Pico, who would become the last Mexican governor of California. After about ten years in Los Angeles, Forster moved down the coast to San Juan Capistrano. He acquired what had been the mission lands at that place and would live there for twenty years, coming by association to be known as Don Juan. In 1845 Don Juan Forster was granted by Mexico the twenty-six-thousand-acre Rancho de la Nación—the National Rancho, located just south of San Diego—making him one of the largest landowners in the area. (The National Rancho, consisting of forty-two square miles of land and six miles of waterfront, would eventually be developed into National City.) When Forster sold La Nación he bought the Santa Margarita Rancho from his brother-in-law. Though he would die relatively poor, Forster was "for many years a man of great wealth and lived and entertained in generous style."[3] He was someone to whom Charlotte would certainly not object to being linked.

Nathan Watts, who had a "deep interest" in local history and in preserving the recollections of its old settlers, would surely have known about John Forster, his extensive land holdings, and his erstwhile wealth.[4] Perhaps it was Nathan who suggested to Charlotte that her surname would open doors to southern California society, and that prospect may have further encouraged her to go there in the first place. The misunderstanding that would later arise about her father's being an Englishman may have stemmed from a confusion, not abated by Charlotte, of the California Forster who emigrated from England in the early nineteenth century with the New England Forsters whose roots went back to colonial times.

Charlotte ostensibly moved to San Diego "to regain her strength after a severe attack of typhoid fever." She was accompanied by a companion, who soon married a plumber; Charlotte's subsequent solitude may have intensified her dream of becoming married herself. Though she had come to be described as an invalid, her condition does not seem to have impaired her getting about. She moved up the coast to La Jolla for a while, making numerous acquaintances there. In fact, she became widely known in southern California, and her comings and goings were recorded on the society pages. Thus we know that early in the summer of 1908 she had returned to San Diego but soon went to Elsinore, a mountain resort town in Riverside County.[5] There, she could enjoy "sweat, plain and plunge baths" that were advertised as having "the best" curative properties. Elsinore was considered the "autoists' half-way house" between Los Angeles and San Diego, and so on departing the spa Charlotte could just as easily have gone either way.[6]

Leaving Elsinore without informing her physician, she went to Los Angeles, where she engaged rooms at the Westlake Hotel, which may have been a place that Charlotte knew from earlier visits to the West Coast. In those prior years, the Westlake opened the season with dancing parties that merited notice in the society pages, but in 1908 it advertised itself as a "select family hotel" with "reasonable rates."[7]

After about three weeks at the Westlake, Charlotte relocated to the more fashionable Hotel Angelus. It was so tony that it did not allow brochures for the Balloon Route Trolley Trip, then perhaps the most famous tourist excursion in the West, to be distributed in its "very ornate lobby." When Charlotte arrived, she was in the care of two trained nurses. It would later be said that she was despondent over her year-long illness and had been showing "signs of an unbalanced mind."[8] Still, it took everyone by surprise when, one Tuesday afternoon shortly after a shift change, the nurse on duty entered her room to find Charlotte preparing to leap from the window. A "fierce struggle" ensued, during which the nurse called for help, but no one responded. Charlotte then suddenly pushed the nurse away and rushed to the third-story window, from which she jumped and fell about fifteen feet onto the roof of the hotel's lobby, landing close to a railing that surrounded the skylight.[9]

This portrait of "toothpick heiress" Charlotte B. Forster appeared in a newspaper story reporting on her leap from a hotel window.

Though at first feared dead, Charlotte did survive, but her attempted suicide naturally "created a sensation in Los Angeles." In newspaper reports she was described as "demented" and her condition as "pathetic." She was also identified as the "daughter of a millionaire toothpick manufacturer of Portland, Maine," and was "said to have been quite wealthy," information that may well have been provided by Nathan Watts.[10]

In fact, at least some of Charlotte's real or imagined troubles had to do with money. The court had been petitioned on behalf of her creditors to appoint a guardian to oversee her property, which was thought not to be insignificant. Still, though her bills had totaled only $1,500, her creditors had not been able to collect from her. This may have been a consequence of a complaint filed by a bicycle rider who asserted that an automobile Charlotte was driving knocked him to the ground. The incident appears to have cost her and worsened both her financial situation and her mental state.[11]

After her suicide attempt, she was declared incompetent. In the absence of any relatives nearby, the court followed the recommendation of local friends of the family in appointing two guardians: Nathan Watts and Charlotte's personal physician. Together they were to look after the "estate and Person of Miss Forster," who, according to the petition to the court, was in possession of a "snug little fortune."[12]

About a year later, when Oscar Hersey filed suit seeking to be discharged from his trusteeship, the *Los Angeles Times* news story made no mention of Charlotte's unfortunate personal incident. Rather, it focused on the value of the toothpick business and mentioned Hersey's claim that a member of the family had "hindered him in the business by starting a rival plant."[13] But Charlotte was reported to have stated that "if an accounting was made of the business so she and the heirs would be able to define the earnings of the business, she would prefer to have Hersey remain as trustee, on account of his experience in the operation." Perhaps by paying off a seventy-five-hundred-dollar personal property mortgage for Charlotte, using cash from the Estate of Charles Forster, Hersey had bought her loyalty. Hersey's accounts show him making no such generous and extraordinary disbursement to either of her siblings. In the news story about the debate over his trusteeship, the estate was reported to be worth nearly a million dollars, though it is likely that those knowledgeable of the business climate at the time would not have valued it so highly.[14]

In the meantime, the *San Diego Sun* had carried a sensational story on its front page under the headline "RICHEST BACHELOR OF

CITY, NATHAN WATTS, TO WED A MAINE HEIRESS." The story
reported rumors that had "floated around in club and society circles" for more than a week and also gave insights into what was
known, or at least believed, about Charlotte and Nathan. Both evidently had developed reputations in San Diego that were larger
than life and better than truth.[15]

She was identified as the "millionaire daughter of New England's tooth pick king," who had come to San Diego "heralded
 as being the richest woman in the state of Maine," having been left
property valued at eight million dollars by her father, who "headed
the toothpick trust." To most newspaper reporters and readers
of the time, there was no shame in making or inheriting that kind of
money, even if it came from toothpicks. Charlotte was also said
to have had "automobiles galore and it has kept her busy trying to
spend her income." It is difficult to imagine such rumors being
started, or at least being insinuated and encouraged, by anyone but
Charlotte herself—or possibly by Nathan, whose own wealth
became inflated by association. The marriage "announcement"
recounted her "nervous breakdown" and described what had happened after she "jumped or fell" from the hotel window: "She hovered between life and death for a long time and Nathan Watts, an
old friend, for whom Miss Forster was said to have already formed
an attachment, watched over her constantly. Thus a romance
sprung up which is soon to be consummate [*sic*] in the marriage
vows and a honeymoon in Europe."[16]

If what was believed about Charlotte does not ring true, can we
take at face value what was said of Nathan? According to the paper,
even the "champion gossips, who hear everything first and are first
to believe it," could not believe this latest rumor. They were quoted
as saying, "Nathan Watts marry? Never. Nathan is proof against all
the cunning that Cupid can command. Think of the beauties who
have set their cap for San Diego's own Nathan, only to be disappointed." Still, they had noticed that Nathan was dressing differently, on one occasion appearing in "a new satin waistcoat
embroidered in pink roses" and on another "wearing a white neck-
tie and a new broad cloth suit." It could only be love, they thought,
and expected wedding bells to be ringing within a fortnight.[17]

But the principal focus of the newspaper's headline and story
was money. "Two of the largest fortunes in the city of San Diego"

were being joined. There was even speculation about the relative sizes of the fortunes, and though it was acknowledged that Nathan's might be the smaller of the two, he was not thought to be at all "shy on the shekels." He was said to have more San Diego property than he could keep track of and, like Charlotte, "would have had a hard time getting rid of his income, if he had been so disposed." And also like Charlotte, he had come by most of his wealth by inheritance.[18]

In fact, in the 1890s, Watts had been a director of the San Marcos Land Company, which developed five thousand acres north of San Diego.[19] But in the 1910 census he is identified, presumably by his own declaration, as no more than a real estate agent. The house in which he lived must have been of a good size, for he appears to have occupied only part of it as a bachelor while renting out apartments to five households. The rental income from this and other properties he may have owned would have allowed him to maintain at least the appearance of a certain lifestyle. The census record also shows his roots to have been in the Midwest, he having been born in Ohio of parents born in Indiana and Pennsylvania. It is unlikely that San Diegans knew with any more certainty the details about his background or inheritance than they did about Charlotte Forster's.

It is possible that Nathan Watts himself had learned for the first time from the newspaper story of the rumors connecting him and Charlotte. Whatever the case, the story so upset him that he went to the newspaper office and "abused" the city editor, C. A. McGrew, who responded with a blow that sent Watts to the floor. During the ensuing scuffle, Watts kicked McGrew in the face. The police were called and Watts was charged with disturbing the peace. Evidently there was some misunderstanding among Charlotte, Nathan, and the press. One point of contention may have been Charlotte's age. In the initial newspaper report of her attempted suicide, she was described as being "about 35 years old," although the accompanying likeness suggests a much older woman.[20] In fact, she was almost forty-five. But if Charlotte had misrepresented herself as well as her fortune to the thirty-six-year-old Nathan, he likely would have learned the truth when he became one of her guardians.

There is no evidence of Nathan and Charlotte ever marrying— or of her ever getting over it. When the 1910 census was taken, she was still living in San Diego, but at a different address. She correctly represented her age to the census taker and declared herself to be single. Under "occupation" on the census form was listed "own income," and to the question of whether she rented or owned her residence she responded that she rented. She shared the house in which she lived with two other single women in their forties, both from Scandinavia. One had been born in Sweden and worked as a housekeeper. The other was from Norway. Whereas for the latter's occupation it seems that the census taker had originally written "housekeeper," that word appears to have been overwritten with "companion."[21]

In that same year, Charlotte drafted a will in which she identified herself as being "of the age of forty-five years." In fact, then she was forty-seven, nearly forty-eight. Whether she misrepresented or simply rounded off her age, she was of sound enough mind to know her family situation. She mentioned her siblings, Maurice and Annie, but did not "give, bequeath and devise them, or either of them, anything," since, as she explained, their father had made provisions for them. Instead, Charlotte left everything that she owned to her "companion and friend," Bergit Olsen. Nathan Watts was not mentioned.[22]

In the ensuing years, Charlotte's life took turns both for the better and for the worse. On the plus side, her financial condition improved, presumably as a result of distributions from the Estate of Charles Forster. On the negative side, her mental state deteriorated to such an extent that in early 1917 "a friend," Roberta W. Winters, petitioned the California superior court to declare Charlotte "mentally incompetent to manage her property." Perhaps she was giving it away and was easy prey for unscrupulous types. Within a week, the court agreed with the petitioner and ordered that her money, about twenty-five thousand dollars, and her jewelry be deposited in two San Diego trust companies under the care of court-appointed trustees.[23]

Charlotte lived only about ten days under this guardianship. According to her death certificate, she died of dementia paralytica, an inflammation of the brain tissue said to be manifested early by the "development of paranoid ideas, usually notions of grandeur."[24] When her estate was appraised, her true wealth was revealed to include about thirty thousand dollars in cash, a number of rings with large stones and other jewelry, a Pierce-Arrow automobile, and real estate consisting of a house and a lot in San Diego, as well as additional lots, presumably undeveloped—but far from a million dollars' worth.[25] Her father's toothpick business had been profitable, but it had not made Charlotte a millionaire or a happy woman.

CHAPTER TWENTY-ONE

Exeunt Oscar and Maurice

Though the exact nature of the relationship between Nathan Watts and Charlotte Forster may remain speculation, there is clear evidence that the fiduciary relationship between Oscar Hersey and the Forster siblings had reached the breaking point at about the same time that Charlotte was experiencing her breakdown and public humiliation. In the fall of 1908, when she was being moved from the hospital to a sanatorium, Hersey was preparing to leave for a two-month trip to Europe, for which he charged half of his expenses to the estate.[1]

During the time Hersey was abroad, Nathan Watts was appointed one of the guardians to look after Charlotte's property, in which capacity he would have had reason to get in touch with Hersey as trustee of the estate from which she benefited. Hersey returned to Maine around Thanksgiving, and in early December he sent a telegram to the "Guardians of C. B. Forster," appropriately charging its cost of $1.41 to the estate. Later in the month, Hersey was exchanging (and recording the expense of) telegrams with Nathan Watts, who was evidently handling the fiscal aspects of the joint guardianship.[2]

Hersey and Watts must have come to some kind of understanding or agreement about Charlotte's benefits from the estate, for the exchange of telegrams appears to have ceased toward the end of January. By then Nathan would surely have discovered that Charlotte was no millionaire. Perhaps up to that time he had entertained the idea of marrying her for her money, maybe even sowing the seeds of the rumors that they would wed. Or perhaps it was Charlotte who sowed the seeds, thinking that Nathan's presumed fortune would help her out of her perceived financial predicaments. They may even have agreed not to reveal the truth about Charlotte's financial situation, thus letting the rumor mill turn.

But there was no such understanding reached between Hersey and the beneficiaries in Maine. By the spring of 1909, he had begun conferring with his counsel, George Wing, about seeking to be discharged as trustee. Hersey engaged in separate legal actions against Maurice and Annie and against Charlotte. The former two were represented by Frank W. Butler, a native of Phillips who studied law with attorneys there and in Farmington, which is near Strong and is where he practiced,[3] and the latter by Robert Treat Whitehouse, a Maine-born and Harvard-trained lawyer who practiced in Portland.[4] It seems unlikely that the Forsters were personally known to Butler prior to their retaining him, since he misspelled both their names as "Foster" in his note to the Register of Probate stating that he would represent them. The note was written on the same day that he inspected the accounts of the estate that were then on file in Probate Court.[5] That Charlotte had her own attorney reinforces the fact that either she or Nathan Watts was not in agreement with her brother and sister over the matter of Oscar Hersey.

Hersey was legally still trustee on April 1, 1910, when a "large quantity of toothpicks," valued at five thousand dollars, was in a company storehouse in Peru, Maine. The town classified the goods as personal property of the business and wished to collect $127.50 in taxes that it claimed were owed on them. The Estate refused to pay, asserting that the toothpicks were merely in transit to their final shipping destination. The Inhabitants of the Town of Peru sued the Estate of Charles Forster for the taxes, and the case was scheduled to be heard in March 1911.[6]

But in the meantime, a petition bearing Maurice's childlike signature was filed in Cumberland County Probate Court in Portland, asking that Hersey either put up additional bond or be removed from the trusteeship that controlled assets a bit in excess of a quarter million dollars. After a hearing before the probate court judge, at which Frank Butler (who can be assumed to have drafted the petition) represented Maurice, the petition was dismissed.[7] Subsequently, Maurice and his sisters brought suit against Hersey in Oxford County, where Dixfield is located and where Maurice lived, and in March 1910 the case was pending appeal before the Supreme Judicial Court for that county. The Forster children were represented by Butler and Whitehead, who, having examined Hersey's accounts, knew exactly what they were fighting for. It was not millions, but neither was it peanuts.[8]

Oscar Hersey's seventh annual accounting of the Estate of Charles Forster, which was to be his last, had been filed for the fiscal year ending June 30, 1909. Two weeks earlier, he had recorded an expense of ten dollars for "burlap, excelsior and time in packing type-writer, letter files, desk, papers and vouchers, etc. for shipment from Portland to Strong."[9] It makes little difference whether the Forster heirs were forcing him from his lucrative trusteeship or whether Hersey was trying to extricate himself from the toothpick business at a time when the bottom appeared to be falling out of it due to an oversupply and the Forster heirs were becoming increasingly mistrustful and demanding of him. From either point of view, the bottom certainly had fallen out of the relationship between the trustee and beneficiaries of the Estate of Charles Forster. Most likely as part of an out-of-court settlement, Oscar Hersey resigned his position as trustee, supposedly "to devote himself exclusively to his legal profession."[10] He got out of the toothpick business just in time.

In the still-pending matter of the town of Peru versus the Estate, the defendant was now represented by attorneys Butler and Whitehouse, and in a letter to the clerk of courts, Butler agreed to "voluntarily appear for the defendants." His letter was dated March 9, 1911, ten years to the day after Charles Forster had died and barely four months shy of when the younger Forster daughter, Annie, would die of heatstroke. The court decision, handed down the following year, was that no taxes were due. The Estate had won, and so would the attorneys.[11]

Hersey was succeeded as trustee by Whitehouse and Butler, who were listed in that order on the letterhead of the Estate of Charles Forster.[12] The new trustees evidently had a new kind of relationship with the heirs. Maurice, who had his own successful (and competing) business, was now "constantly consulted by the trustees of the Estate in the management of the trust, and his advice followed without question."[13] Whether this constituted a conflict of interest seems not to have been openly discussed, and for a while at least Maurice appears to have been able to maintain dual loyalties and Butler, who would eventually become sole trustee, to exercise remarkable disinterest. When, in 1912, Maurice applied for additional trademarks for his own Forster Manufacturing Co., it was Butler who notarized the documents.[14] Like Hersey, Butler's career would include service as a state senator. He would also become a judge. Most importantly for the toothpick story, he appears to have played a pivotal role in overseeing the assets of the Estate of Charles Forster and in shaping its future as a business.

Neither Butler nor Whitehouse seems to have had much experience in the toothpick industry when they were made trustees of the Forster estate, but they appear to have gotten savvy very quickly. Thus, when in 1912 Willis Tainter was granted a new patent relating to toothpick-packing machines, the rights were assigned to Butler and Whitehouse by name as "trustees of the Estate of Charles Forster, deceased."[15] In the meantime, Albert Hall, who most likely worked for the independent toothpick company recently established by Maurice Forster, was developing a machine and a mechanism for conveying toothpicks in "conveniently disposed piles or stacks from which they may be taken to be packed in boxes." Each of Hall's patents was assigned not to an individual or individuals but to Forster Manufacturing, as was Maurice Forster's own patent for a toothpick-packing machine.[16]

In the meantime, the toothpick industry had been evolving rapidly. In 1909, John S. Harlow founded the Dixfield Toothpick Co. and "built what was probably the best equipped plant in the business." With this new D.T. Co. factory, the town had three plants "engaged exclusively in the manufacture of toothpicks with an output of probably two hundred million toothpicks per day, and a market for all," or so it was thought by those not engaged in the selling end of the business.[17] But as we know, there were rough times ahead.

Around 1913, Maurice established a second factory for his Forster Manufacturing Co., locating it in Oakland, Maine, which is near Waterville. This was farther east than virtually all previous toothpick factories had been located, and it was no doubt the growing use of truck transportation that made it feasible. The notice prompted an editorial in the *Memphis Commercial-Appeal*, which seemed surprised that the toothpick industry had become almost synonymous with Maine: "The State of Maine at one time furnished the American Navy with spars, masts and flag poles. It is now welcoming a manufactory turning out 15,000 toothpicks a minute."[18]

All the machinery from the Dixfield plant was moved to the new Oakland one, but poor health reportedly prompted a still-young Maurice to sell his company to Butler and to Harlow, who had been operating the Forster Estate plant in Dixfield. They closed the Oakland plant in 1916, but it would be reopened after the war, when Forster Manufacturing was merged with the Berst Manufacturing Co., then of Saginaw, Michigan, and Harlow's D.T. Co., which at the time described itself as "manufacturers of high grade toothpicks," including the Oxford and Cub brands, both of which used bears in their logos.[19] The Berst company reportedly had been trying to get a foothold in Maine, and it was Ned G. Begle, president of Berst, who appears to have driven the deal and who became president of the new Berst-Forster-Dixfield Co.[20] The awkward hyphenated name, which nevertheless preserved the identities of the original companies, would be simplified in the mid-1940s to B.F.D. Co., when it would become a division of the Diamond Match Co.[21]

Meanwhile, Maurice Forster and Frank Butler had joined forces in at least two new Maine businesses. In May 1916, they teamed up with William H. Drew of Portland to incorporate the W. H. Drew Company, among whose stated purposes were "the buying, owning, leasing and selling of real estate, including timber lands," and "the building and operating of steam saw mills; the manufacturing and selling of lumber; the buying, owning, holding, and selling of the capital stock of corporations engaged in a similar business." Forster, Drew, and Butler swore to these intentions before the justice of the peace of Cumberland County, Robert T. Whitehouse. By the end of the year, the name of the corporation was changed to Forster Mills Company.[22]

Just prior to that name change, Forster and Butler joined with clerk Marion P. Dunton, of Portland, to form the Forster Corporation. Its stated purpose was also to deal in real estate, as well as to "deal in and with goods, wares, merchandise, materials and property of every class and description including mortgages, stocks, notes, bonds and personal property necessary or incidental to the business of the corporation."[23] With the establishment of these two corporations, Forster and Butler were in a position to trade assets between them and toothpick companies, including that of the

Estate of Charles Forster. They were also in a position to do long-range planning.

Unfortunately, Maurice Forster died unexpectedly in Portland early in 1924, at the age of just fifty years. From his early forties, he had been living in Portland, where of late he supposedly had "enjoyed excellent health up to a few days previous to his death," which came "immediately following an operation." His obituary, like that of his father's, was curiously written. Maurice was identified in the opening sentence as a "well known Maine business man," yet much of the second paragraph described the toothpick empire that his father had established but that had been left out of the elder Forster's obituary.[24]

It was as if the opportunity was being taken to establish the pedigree of the father's company, but in a way that did not square with the facts. According to the obituary, "many years" earlier Charles Forster had "started a toothpick factory at Strong, Me., with branches in Dixfield and other Maine towns." This "large and very successful business having been established, [it] was incorporated under the name of the Forster Corporation." Upon Charles Forster's death, which, curiously, was dated imprecisely and not very accurately in the obituary at "about twenty years ago," Maurice "became the head of the corporation, and has remained in charge until the present time." This totally ignores the fact that the father's company had been operated as the Estate of Charles Forster under the trusteeship of Oscar Hersey for almost ten years, and that during that period the son had established the competing Forster Manufacturing Co., which he had long since sold. If Maurice was "in charge" of a "Forster Corporation" at the time of his death, it was the one established in 1916 not by the long-deceased Charles Forster but by Maurice Forster and Frank Butler. If anyone was a likely candidate for authorship of the business paragraph in the obituary, it was someone who might have had something to gain by rewriting the history of the Foster toothpick business and in the process confusing the names of the father's and son's firms.[25]

At the time of his death, Maurice was evidently well integrated into men's society. His obituary closed by describing his status as a thirty-second-degree Mason and a member of the Portland Club, the Economic Club, and the Boston City Club. However, he was apparently not so well-known that his given name could not be misspelled in the announcement of funeral services for "Morris Webb Forster." Indeed, the name "Maurice" was commonly phonetically transmuted into "Morris," the way Mainers pronounce it to this day, further complicating research into his life. When would a Morris Foster be one and the same with Maurice Forster, "the last of his family"?[26]

Maurice's will was dated October 19, 1900, not three weeks after his father's and just months before his father would die. Among the witnesses to the instrument were the law partners Enoch Foster and Oscar Hersey, and it was most likely the latter who drafted the legal documents of both father and son. Maurice himself may or may not have known at the time of signing his own will the detailed terms of his father's, but Hersey may have advised him that his mother and sisters would have been taken care of by his father. In any case, Maurice's mother and sisters were bequeathed only token sums by Maurice, though since they all predeceased him, the point was moot. In effect, Maurice left everything to his wife, Lelia, whom he

This photographic portrait of Lelia Forster, ultimately the true "toothpick heiress," hangs in the Forster Memorial Library in Strong, Maine. A similar portrait, but painted in oils, is in the Weld Free Public Library.

had married almost exactly a year before signing his will.[27] It was Lelia who would become the true "toothpick heiress."

Not surprisingly, there is no mention of the toothpick business in Maurice's will, since at the time of its being drafted he had no formal interest in his father's business and had not yet established his own. The will describes in considerable detail insurance policies (complete with numbers and dates) that Maurice left to his wife, along with "all the rest, residue and remainder of my estate, wherever found or however situated," and it would be under this catchall clause that any interest in the father's toothpick business would fall.[28]

Though Maurice was said to have been a resident of Portland for eight years prior to his untimely death, his will was entered into probate not in Cumberland but in Franklin County. As executrix of the estate, Maurice's widow, Lelia, represented that he had "last dwelt" in the town of Weld in that county. The Forsters did in fact have a large house prominently situated on Lake Webb there, where they were said to have entertained lavishly. According to one recollection, their servants went out in boats to pick up guests from around the lake and bring them to the Forster dock. Wherever his last residence, Maurice's will was filed within days of his death, on a legal form in which "Cumberland" was crossed out and "Franklin" written in. The inventory of the estate was sworn to before the justice of the peace, Frank W. Butler.[29]

Lelia's initial estimate of the value of Maurice's estate was no more than about $225,000. After the appraisal, however, the value had more than doubled. The principal difference could be attributed to an "undivided surplus in the estate of Charles Forster" ($180,000) and to "200 shares of capital stock in Forster Corporation" ($234,861), contributing to the total of $467,920.[30] How such a large interest in his father's toothpick business came to Maurice, given the terms under which the Estate of Charles Forster was created, is unclear—unless, of course, it is accounted for by the role that Maurice played in running the business of the Estate of Charles Forster and in the Forster Corporation, which by then may have owned a good part of the former.

Perhaps the idea for actually incorporating the Estate of Charles Forster had been initiated before Maurice's untimely death, but papers were not filed with the State of Maine until about three weeks after it, as a result of a meeting being held at the offices of the Estate of Charles Forster establishing a corporation of that same name. From this entity may have come the stock that ended up in Maurice's estate. Among the directors of the corporation were, in addition to Frank Butler, two of Maurice's cousins, William A. Sanborn and James M. Hobbs, who now stood to share in the estate of Charles Forster.[31]

The first stated purpose of the new corporation was the "buying, owning, selling and leasing of real estate"; the making and selling of toothpicks and wooden novelties came second. Butler was identified as president; he was also listed as one of the five owners of the five shares of common stock (of one hundred dollars par value) paid in at the time of incorporation. The total amount of capital stock in the company was established as $500,000.[32] Within about a month of Maurice's death, Frank Butler, as trustee of the original Estate of Charles Forster, had filed papers to distribute the balance of the estate to the patriarch's heirs, who consisted of five nieces and nephews, including the two who were directors of and stockholders in the new corporation. The individual inheritances from the estate to these heirs amounted to less than $17,000 each, for a total "personal estate" of just under $85,000.[33] In spite of this distribution, the business continued to operate.

It may have been that the stock assigned to some of the directors of the corporation known as the Estate of Charles Forster was in exchange for services. For example, Butler's stock may have represented payment for legal services he had rendered Maurice and Annie during their period of trouble with Oscar Hersey. But surely Maurice, at least, would have been able to pay for services performed at the time. Exactly what arrangements Butler had with the Forster children is difficult to ascertain. Though Hersey's annual trustee reports remain on file in the records office of the Cumberland County Probate Court, any reports that Butler may have prepared after he took over as trustee are not part of the file, nor do they appear to be on file in the probate court of Franklin County. However his entreé was acquired, at some point Butler and a relative or two reportedly "went into partnership in the firm."[34]

Three years after Maurice's death, Lelia drafted her own will under the guidance of Frank Butler. It was as complicated as her husband's was simple. At the time of her death two years later, in

1928, of a pulmonary edema accompanied by an acute intestinal obstruction, Lelia's estate amounted to about half again as much as Maurice's, and she left generous sums to a couple dozen individuals and several institutions. In addition, she established a trust fund to benefit her mother and sister, with whom she had been living. Finally, she left to Frank Butler all the stock that she held in the Estate of Charles Forster, which amounted to $163,000, which he was to hold in trust. Dividends from the stock were to go to benefit charitable homes in Portland and Waterville, and the proceeds from matured stock were to be accumulated until there was $50,000 in a separate fund. One-half of this was to be used to construct a building in Strong that would serve as a memorial to her husband, and the other half was to be an endowment to operate and maintain it.[35]

The handsome Maurice W. Forster Memorial Building was dedicated in November 1931. It was turned over to the citizens of Strong by Frank Butler, who gave an address recounting the story of Charles Forster and the wooden toothpick industry that he had founded. It was in this speech that Butler misidentified Forster as an Englishman and his uncle's commission house in Brazil as an English business. This naturally calls into question other aspects of his history and makes one wonder why the background of the Forsters—or at least that of Charles—and the history of the toothpick business and Maurice's role in it were so inaccurately rendered. It seems unlikely that the mistakes in the newspaper article were reporting errors, since the story was identified as one that was "Special to The Journal-Chronicle," a designation normally indicating a press release that was not the work of a regular reporter.[36]

Foreign Affairs

A T THE BEGINNING of the twentieth century, American goods of all kinds were being exported around the world. The products of Yankee ingenuity and machinery were heralded as symbols of the triumph of democracy. The phenomenon was celebrated in a bit of verse entitled "The Banner of the Pork and Soap":

> *Tremble, ye monarchs on your thrones,*
> *From Caucasus to Pyrenees,*
> *Ye kings effete who hold each seat*
> *By dint of ancient pedigrees,*
> *Behold, the Yankee has arrived—*
> *Resign your jobs, abandon hope*
> *Before the splendor of his arms,*
> *The Banner of the Pork and Soap.*
>
> *Across your border see them come*
> *All jingling bolts and shingle nails,*
> *The Yankee dish-pan is their drum,*
> *Their ordinance [sic] is cotton bales.*
> *With Yankee shoes and Yankee booze,*
> *With Yankee corn and Yankee rope,*
> *They scale your walls and hail your halls*
> *Beneath the flag of Pork and Soap.*

The stanzas marched on, taunting "king and pope" alike. And embedded in the last stanza were the lines

> *The business man American*
> *Is making tooth-picks for the world.*[1]

He was indeed. Charles Forster had died with his boots on, leaving his estate to carry the banner of the flat and round. His Worlds Fair brand name had certainly signaled that it was a toothpick for all nations. Even though a box of the premium picks could cost twice what a larger box of inferior domestic or Japanese ones did, its contents must have provided the coup de grâce for the French quill toothpick, as suggested by a foreign trade opportunity advertised by the Bureau of Manufactures of the Department of Commerce and Labor in 1910:[2]

> TOOTHPICK MACHINERY.—A firm in Southern Europe desires to purchase machinery for making round, double-tapered wooden toothpicks, known in the trade as "World's Fair" toothpicks, a sample of which can be sent to interested parties. Catalogues and price lists, free on board New York, are desired, and correspondence in French, if possible.[3]

No doubt, some French-speaking southern European saw the machine-made wooden toothpick as a viable competitor to the quill, probably not unlike the way a half century earlier Charles Forster envisioned it as a successor to the hand-crafted *palitos* used in Brazil. Overproduction in the American toothpick industry in 1910 may also have prompted opportunists to try to get machinery at bargain prices. Even the Spanish and Portuguese, with their long tradition of hand-crafted toothpicks, must have sensed that it was an opportune time to mechanize. In 1913 the following foreign trade opportunity was advertised:

> MACHINERY FOR MAKING TOOTHPICKS.—A report from an American Consul states that an exclusive agency for Spain and Portugal, or for Spain alone, for the sale of machines for making toothpicks, is desired by a local business man. Correspondence must be in Spanish and prices quoted in gold pesetas. Weights and measures should be given in the metric system. References will be exchanged.[4]

Northern Europeans, too, appeared ready to get involved with American toothpick machinery, and another advertisement made it clear that not only the American toothpick but also the machines to make it in vast quantities were in demand by other potential manufacturers across the Atlantic:

> TOOTH PICK MACHINERY . . . An American Consular officer in a European country reports a mechanical engineer in his district who has had six years' experience in the United States and is at present representing certain German manufacturers of woodworking machinery. He believes that American machines, particularly in special lines, are far superior and desires to get in touch with American manufacturers of automatic tooth pick machines. . . .[5]

From the beginning of his involvement in toothpick making, Charles Forster had wanted to sell to the world, and part of his legacy was that his company did. By 1885, one-twelfth of the twelve thousand cases shipped out of Maine each year were going to Europe and Mexico and other Spanish-speaking countries as *palillos, mondadientes,* and *escarbadientes.*[6] By 1888, the plant at Strong was producing twenty million picks a day, which were "moved by pitchfork into the sun to dry like hay" before being placed into more orderly arrangements. The packaged toothpicks were being shipped (in cases of one hundred boxes of two thousand toothpicks each) via the Sandy River and Maine Central railroads to Portland, from where they were "distributed to every part of the United States and Europe." In 1907, toothpicks were being exported to Australia, Canada, England, France, and Germany, as well as to Mexico.[7] Yankee ingenuity was flooding foreign countries with its products.

But exported picks did not always reach their end users, as in the case of the 1914 "toothpick famine" in Mexico City. All but one store there had exhausted their supplies. However, the owner of that single store refused to sell the toothpicks, electing rather to distribute the essential items gratis. Even though each patron who asked was given only six picks at a time, the stock was not expected to last more than a few days. Restaurants had run out also. Some had resorted to providing diners with straws as a substitute, but these were not considered satisfactory. How could such a situation arise when there were idle factories north of the border?

The whole of Mexico had become accustomed to being supplied exclusively with wooden toothpicks made in New England and imported through jobbers in New York. The "very low price" of the American toothpicks in Mexico drove out all competitors. Even with cheap labor, Japan could not compete in the Mexican market with the products of efficient American machines. The French quill, which had been "scarce and dear at all times," was no longer available. Throughout the famine, there were some stocks of American toothpicks at the Veracruz customhouse, but no one was willing to pay for them because exchange rates were so high. It was a Mexican standoff.[8]

As long as exchange rates were agreeable, trade with other countries usually went better. In 1923, what was believed to have been the first foreign order of any kind in America received via radio was filled by the Forster factory in Strong. An entire carload of toothpicks was shipped to Japan.[9] However, according to a 1925 dispatch from Strong, it was South America that was the "largest market," and so Charles Forster's original goal of exporting his product to the Southern Hemisphere had indeed been fulfilled.[10]

In fact, the distribution of American toothpicks had become truly worldwide. One box design was quadrilingual, bearing on successive sides translations of the English-language brand name, Universal Perfected Tooth Pick, in French, German, and Italian, and declaring its contents to be, respectively, *cure-dents*, *Zahnstocher*, and *stuzzicadenti*.[11] Whatever called, the humble splinter was universally sold and used. In about 1930, the Forster company received in the mail an empty toothpick box and an accompanying letter, which read,

> Herewith I am sending you one of your boxes of toothpicks, minus the picks, which has been all the way around the globe. We bought this box in Chefoo, Shantung, China, where we have lived for the past twenty years. It is only one of the many boxes we have used. We returned to the United States traveling by way of Siberia and Europe, so that this box has circumnavigated the globe.[12]

Another indication of the extent to which a familiarity with Forster toothpicks had penetrated the world market is suggested in James Joyce's early-twentieth-century novel of growing up in Ireland, *A Portrait of the Artist as a Young Man*. In a scene that takes place near the library, Cranly, one of protagonist Stephen Dedalus's university friends, is described as "leaning against a pillar, . . . picking his teeth with a sharpened match." Later, Cranly reappears, "picking his teeth with care" until he "dislodged a fig seed from his teeth on the point of his rude toothpick and gazed at it intently." Even then, he continues to work on his mouth, "sucking at a crevice in his teeth." Joyce, a master of free association, embeds this latter scene in a discussion of the name Forster. One of the students parrying over trivia describes among others the Forster family that descended from the Flemish king Baldwin the First. Another student asks, "Where did you pick up all that history?" The intertwining of the name Forster with Cranly's toothpick work—he was "rooting again deliberately at his gleaming uncovered teeth"—and the pun on the word "pick" seem too apt to be coincidental, especially for Joyce.[13]

In addition to developing a growing reputation abroad and a thriving export trade, toothpick manufacturers such as Charles Forster had looked to diversify. After all, only so many toothpicks would ultimately be needed to supply the world. But there were other things that could be made of wood in similar ways, and the toothpick manufacturers had discovered them in the late nineteenth century. Then, at least one observer returning from a summer trip to Maine remarked that "the whirring machinery, propelled by primitive water power, seemed solely intent upon chewing up all the lumber that was fed to it, and in return for this kindness it spit out at the other end the greatest quantity of little sticks, splints, and square and circular pieces of wood." The remarkable machinery was installed in "queer novelty mills" that produced so many of the indispensable items of everyday life from wood good for little else: "The timber used in the mills is generally considered almost worthless, and the factories annually try to find new novelties that can be made out of the lumber rejected by the ordinary timber man. It requires a special sort of machinery to make the various novelties, but after the machinery is once built, the products can be turned out by the millions at a mere nominal cost."[14]

Among the "new novelties" that such factories found in the early twentieth century was the Popsicle stick. The Popsicle was invented accidentally in 1905 by Frank Epperson, who was eleven years old at the time. He had left his fruit-flavored soda outside in cold weather, and the "pop" froze to the stirring stick that had been left in it. The ice on a stick proved to be tasty, and young Frank eponymously named it the "Epsicle." In 1923, a grown-up Frank Epperson applied for a patent for a "frozen ice on a stick," which he called the "Epsicle ice pop," but which his children renamed the "Popsicle."[15] Naturally, the mass-produced Popsicle needed mass-produced sticks, and so a new wood novelty item was developed.

The Popsicle stick is essentially just a thicker, longer, and wider flat toothpick with blunt rather than pointed ends; it can also be viewed as a narrower version of the wooden tongue depressor, once a very familiar item in a doctor's office. Such relatively thin, flat devices could be made in much the same way some toothpicks were, but the machine at the B.F.D. Oakland plant that packaged them into cellophane-wrapped bundles was "of secret design."[16] Modified toothpick-making machinery could also produce flat wooden spoons, sometimes called "woons," which served for eating ice cream out of Dixie cups, and cocktail sticks "shaped somewhat like an arrow."[17] Other products of the Maine novelty mills that fell into the category of "veneer goods" were flat items formed of very thin slices of wood (as many as forty could be gotten out of an inch-thick piece of wood), and used for small boxes for butter, berries, and the like. Not all wood products are flat, of course. Unlike Popsicle sticks, wooden lollipop sticks were traditionally round, requiring a manufacturing process akin to that used to make dowels. In this case, slats of a thickness appropriate to the desired diameter were passed through rounding machines not unlike the kind used to shape round pencils.[18]

At midcentury, the importation of wood toothpicks into the United States was "insignificant," but most that were imported were from Japan and Portugal. Quill toothpicks continued to come mainly from France, but as a luxury item. In 1967, Hammacher Schlemmer was selling a box of 250 "hand-sharpened; individually sterile-packed" quill toothpicks for $3.75, or one and a half cents per pick.[19] In 1977, a box of fifty Le Negri goose-quill toothpicks featured in the Apothecary Catalogue of Caswell-Massey, which called itself the "oldest chemists and perfumers in America," cost $4.50, or nine cents a pick. Granted that the price may have been what the traffic would bear in a catalog that offered such hard-to-find oral-hygiene needs as a "smokers toothbrush" handmade out of boar bristles and the "world's original red toothpaste," the cost of quill toothpicks was nevertheless generally relatively high at a time when wooden toothpicks could be had for a fraction of a cent.[20]

Not all foreign-made toothpicks were expensive, however. In the late 1960s, Forster Manufacturing was contacted by Japanese representatives proposing to make Forster toothpicks and other woodenware products in Japan for less than they could be made in the United States. Soon, other countries, both in the Far East and in Europe—including Korea, Taiwan, Belgium, and Denmark—were also bidding to make Forster products abroad. The company turned down those offers, assuring its employees that it would continue to maintain the policy of having its Ideal and Worlds Fair lines of products "made by American workmen."[21] But even if Forster woodenware continued to be manufactured exclusively at home, the company felt threatened by the growing competition from imported goods. An advertising campaign consisting of ads headed "WARNING!" spoke directly to newspaper readers: "You May Be Getting Cheated When You Buy Imported Toothpicks." Consumers were asked to look before buying to see if the toothpicks were smooth, if the flat ones were thin on both ends, if the round ones were "tapered long enough to be useable," and if they were packed by machine. "NOT ALL IMPORTED TOOTHPICKS MEET AMERICAN STANDARDS," readers were advised.[22]

Even as late as the latter 1970s, it was stated that there was "no import problem." But in 1979 the domestic wood utilization industry was beginning to feel threatened by the possibility of duty-free imports from Brazil, Chile, and Honduras, which made such items as ice-cream sticks and toothpicks. The Carter administration saw lowering trade barriers as a means of helping third-world countries better their position in the world, but the vigilant toothpick industry still appeared to be protected from imports. When Honduras sought duty-free status for its toothpicks, it did not receive it. Brazil's request did not even get a hearing. By the mid-1980s, however, the situation had changed, and a "flood of cheap toothpicks from the Far East and South America" began taking its toll on the American producers.[23]

The export-import trade has always been a double-edged sword. During the Great Depression, American toothpick producers acknowledged that their selling below cost domestically was an unfair business practice that would be considered a violation of the Code of Fair Competition, to which all major players in the domestic toothpick industry had agreed. It was the first of more than two dozen acts and practices forbidden under the code: "(a) Selling below cost; except for export."[24] And they eventually did take advantage of the exception. In the early 1990s, Forster was accused of dumping flat wooden toothpicks onto the Canadian market. The flat were the most popular type in Canada, and Forster was selling them to its Canadian importer, Cowling & Braithwaite Co., Ltd., at half the price charged in the United States. Since Cowling & Braithwaite had been the distributor of flat toothpicks made by Canada's last domestic producer, Keenan Industries Ltd., of Owen Sound, Ontario, the dumping action was threatening the very existence of toothpick making in Canada. In fact, Keenan, which made Kaybee brand toothpicks, closed its plant in 1996.[25]

To some, toothpicks had become a commodity, and they were considered "one of the few manufactured products that are the same the world over." But to Mainers, toothpicks produced in their state were special. "They're tough, tasteless and attractive—and you won't find their ends feathering or breaking off in your teeth like those inferior ones Japan tried to peddle in the United States from timber it had imported from Siberia."[26] That Maine's toothpicks were tasteless was a supreme compliment. Unless a deliberately flavored one was desired, a toothpick should not impart the sometimes unpleasant taste of the wood to the mouth.

In the summer of 2004, in Maine, I bought a box of toothpicks made in China but imported into Canada, and which had no doubt found their way into the United States in the wake of the North American Free Trade Agreement. The top of the box of eight hundred round toothpicks (*cure-dents ronds*) had a small cutout so that the shafts of the picks could be seen through the cellophane. They looked uniformly round, but it is of course how the points were formed that would be the true test, and the points were not visible through the window. At home, I unwrapped the box and was pleased to find that the points appeared to be as well made as the shafts. But the taste is the acid test. I tried one in my mouth, and found that it had a distasteful flavor and was uncommonly soft.

After I had chewed the pick for only a brief time, its end began to split away from the body, which is not good.[27]

While some imported toothpicks are indeed inferior to home-grown ones, some do have a special cachet. In the late 1960s the Old Homestead, which claimed to be "New York's oldest and finest steak house," advertised complimentary "imported quill tooth-picks." Similarly, the hand-carved toothpicks imported from Portugal have long been a fashionable alternative to machine-made ones. In the early twentieth century they were called "fancy," and a century later they still were considered special.[28] Not nearly so uniform or familiar-looking as standard American toothpicks, the exotic orangewood kind are sold as a sophisticated alternative to the frilly-topped ones for use with hors d'oeuvres.[29]

The Japanese word for toothpick is *tsumayoji* (or *yoji* for short), which combines the ideas for fingernail (*tsuma*), willow (*yo*), and branch (*ji*) to specify the thing that can be used in place of one's nail to pick the teeth. While women's toothpicks were indeed made from the branches of the tree, men's were traditionally cut from the harder wood of the trunk. Other trees used for Japanese toothpicks have included balsam. Another common material was bamboo, from which it would seem that everything necessary for a household could be made.[30] According to Japanese tradition, toothpicks were made from different woods during different seasons: "willow for spring, cherry for summer, and chestnut for fall and winter."[31]

In the early twentieth century, Japan was right behind Portugal in exporting the greatest number of toothpicks to America. Some users criticized the "rough, unsanitary slivers made in Japan," but others praised ones handmade from "fine reeds." They were frequently enclosed in hand-painted cases made of strips of wood and were said to be "delicate and thin as tissue paper, and nevertheless strong and pliable" and suitable for being carried in a vest pocket.[32]

In 1900, Japanese toothpick makers were paid just over two cents a day for their efforts. (Portuguese toothpick carvers were paid as little as three cents.) To put the implications of this in perspective, it was pointed out that "a thousand toothpicks may be bought in Japan for as much as it costs only to pack and box 5,000 of American make." But the Japanese were also thought to practice a form of just-in-time manufacturing: "The village toothpick-cutter splits his tiny splints with micrometrical accuracy, but regulates his output with equal exactness, so that he shall not have one more than may be required to provide for the next day's need in rice and pickles."[33] In contrast, in mid-1908 the Estate of Charles Forster had over $120,000 worth of toothpicks in inventory, which must have represented in excess of five thousand cases in warehouses around the country.[34]

The Japanese have been known to be avid toothpick users. According to an attendee at a banquet that took place in Tokyo prior to World War II, "after the meal was finished every member in the audience began a very thorough toothpicking operation which continued uninterruptedly for several minutes."[35] Traditional "Japanese toothpicks," sometimes referred to as *kokeshi*-style, after the wood-turning process used to make dolls and other craftwork, are widely available and somewhat fashionable in the early twenty-first century. Indeed, they can be found in a wide range of quality, being made in China and Korea as well as in Japan. These toothpicks are—as is expected of a real or even a faux turned-wood product— round, but they are pointed at one end only. The shaft near the blunt end is decorated with a pattern of incised rings or other forms of reduced section suggestive of the finial of a lamp or the turnings of a baluster. (To the Japanese it is remarkable that "both sides of American toothpicks are sharp."[36] Of course, the earliest toothpicks that Charles Forster made, albeit flat, were pointed on both ends. In the early twentieth century, a brand such as Little Pirate highlighted on its boxes the "two points" of the toothpicks inside. But American makers never mentioned that to use the second point of a toothpick the fingers would have to touch the first one after it had become soiled.)

In fact, the function of the unpointed end of a Japanese toothpick is not purely decorative. Typically, there is a series of grooves encircling the toothpick, and the laws of mechanics dictate that when bent it will break at the groove more remote from the blunt end. The broken toothpick not only signals that it has been used, but the shorter, broken-off piece provides a small clean part that may "serve as a rest for the pointed end after use," so that what had been in the diner's mouth does not touch the common table. In this regard, the severed toothpick end plays much the same role as does a rest for a dinner knife or pair of chopsticks.[37] The custom of

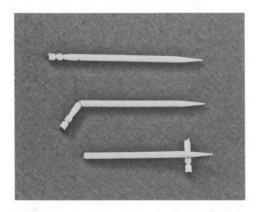

Traditional Japanese toothpicks are pointed at one end only. The grooves encircling the unpointed end not only are decorative but also enable that end of the pick to be broken off to indicate that it has been used and to provide a rest to keep the soiled point from touching the table.

breaking the toothpick evokes that associated with the tufted *fusa-yoji*, which was supposed to be "bent in half prior to disposal" in order to "prevent a variety of evil consequences."[38] The French were not so superstitious or fastidious. When the toothpick was introduced into France in the late sixteenth century by the Spanish minister Antonio Pérez, it became fashionable and was often served stuck into dessert fruits. After being used, the toothpicks were not broken in half but were "either thrown under the table or put behind the ear for decoration."[39]

Not all Japanese toothpicks are round, but virtually all of them are single-pointed and usually demonstrate some creative craftsmanship in their making and packaging. In the course of writing this book, I came into possession of a set of toothpicks in a decorative hand-crafted wooden case about the size of a deck of playing cards. It appears to have been designed to sit on a table or be passed around after a meal. The toothpicks inside are individually wrapped in black-and-white patterned paper that is glued around its edges to declare the contents clean and unused.

Just a bit of color can be glimpsed through the top of the wrapper, the only part that is not completely enclosed. One edge of the wrapping paper is thoughtfully cut at an angle to facilitate the opening of the small package. Peeling away that edge reveals a toothpick whose unpointed end is rolled in a piece of tissue paper that can be unfurled like a small banner. The top edge of the banner is colored, which is what was visible. The banner proper is printed with a saying reminiscent of something found in a Chinese fortune cookie, and presumably one can have chosen one's lucky color in selecting a toothpick from the box. (The blue-topped banner that I unfurled reads, "Nothing is lost for asking.") With the fortune removed, the toothpick is fully revealed as having been cut from a piece of twig, perhaps from a willow branch, for the brown bark stands in contrast to the white wood of the rest of the pick. Its point was formed by cutting away the bark from the lower half of the splinter, and its point was further refined by being beveled. The result is a modest toothpick of uncommon organic beauty.[40]

Bamboo, though subject to becoming mildewed when damp, is among the materials used for exotic toothpicks, and it lends itself to

These sandwich picks made of bamboo make artistic use of the different colors and textures of the plant's inner wood and outer skin, a portion of the latter being knotted to provide a decorative finish.

forming ones with striking aesthetic qualities.[41] I recently ate in a museum café that served its sandwiches pierced with toothpicks of an appropriately artistic design.[42] (This is an example of a device being identified by its function even before its form is fully revealed.) The picks were made of bamboo, and their design—like that of the Japanese fortune toothpicks—took full advantage of the inside-outside qualities of the material to give the toothpick a subtle two-tone appearance, the pale green outer skin and white inner wood nicely complementing each other throughout the design. The tip was neatly sharpened to a symmetrically angled point (like that of a shoe peg), but the most distinguishing feature of the sandwich pick was its unpointed end. Although the shaft of the pick was about four inches long, it clearly had been shaped from a piece of bamboo about two and a half inches longer. This length of inside wood had been stripped away, leaving that amount of pliable green skin to be tied into a decorative knot. The effect is striking, and the sandwich skewers are more appropriate for a fine museum café (or even the institution's collection) than common wooden sandwich picks tipped with permanent waves of colored cellophane.

Those kinds of toothpicks are also frequently used as food picks and labeled as such. I recently acquired a package of bamboo food picks (*pique-aliments en bambou*) made in Taiwan for the American and Canadian trade. The picks look very sharp and handsome in the plastic bag, and although they are a bit more slender than a typical toothpick, at two and a half inches long they certainly appear to be suitable for use as one. But when I put one in my mouth, the smaller diameter made it feel uncommonly thin. When I chewed on it, the hardness of the bamboo was disorienting. When I used it between my teeth, the sharp, hard point seemed dangerous. When I got the end soaked with saliva, it softened and separated to such a degree that it was ineffective as a toothpick. Even with toothpicks, first impressions can be deceiving.[43]

Though the essential proportions of the ordinary toothpick are nearly universal, customs surrounding its presence and use have varied as much as have the shapes and sizes of the picks themselves. The dress of a dandy in mid-nineteenth-century China included a toothpick, though probably not a wooden one, "hung at his button, with a string of valuable pearls."[44] Japanese of the late nineteenth century were described as carrying their toothpicks "in their back hair, and always [using] them after eating anything." (The sixteenth-century French Huguenot admiral Gaspard de Châtillon, comte de Coligny, carried his toothpick "variously in his beard or behind his ear.")[45]

People of all times and places seem to have had opinions about how to carry and what constitutes a properly designed toothpick. In 1965 in Sweden, a wood carver named Henning Eklund tried some American-made gum-massaging toothpicks made of balsa, which had been recommended for his use. Being intimately familiar with the nature of woods and very discriminating about wood products, he was disappointed in the American imports and unhappy that they were being sold in Sweden, which is known for its fine wood-work. Eklund believed he could make better toothpicks and sought the cooperation of a dentist. He was put in contact with Bo Krasse, who was a member of the Faculty of Odontology at the University of Lund. Krasse and his colleague Hilding Björn organized a panel to test different toothpicks, and those made of linden were judged best. Subsequently, Eklund formed a business to make and market toothpicks endorsed by Krasse and Björn under the brand name TePe, which is suggestive of both the English *toothpick* and the Swedish equivalent, *tandpetare*.[46]

TePe "interdental sticks" (*tandstickor, bâtonnets, Zahnhölzer, tandenstokers*) are similar in design to the Johnson & Johnson Stim-U-Dents that so disappointed Eklund in the first place. That is, they are short, more or less triangular in cross section, and doubly tapered to a rounded point. Similarly shaped and packaged tooth-picks have been made and sold in Australia as "medicated gum massagers" under the name Interdens.[47] Australia is also the source of birchwood toothpicks impregnated with tea tree oil. These have been sold as Tea Tree Australian Chewing Sticks.[48] There seems to be no end of the variations on the simple splinter that have been developed around the world. However, variations on a simple thing can make their use less than obvious. Thus, packs of TePe dental sticks impregnated with fluoride come with instructions: "Before use, moisten the dental stick in the mouth, for improved pliability, durability and fluoride effect. Use with the flat side towards the gums."[49]

Like Stim-U-Dents and Interdens, TePe sticks come attached to one another at the base, with the assemblage rather resembling an

Individual interdental stimulators can be broken off from comb-like forms that are often sold in matchbook-like packets that serve to keep them clean and protected in the pocket. It was American-made balsawood Stim-U-Dent picks that prompted the development of the Swedish TePe brand, which are made of linden and birch.

accessory for a barber's hair clipper or a section of a comb, with the unseparated toothpicks appropriately resembling the comb's "teeth." The group of toothpicks usually comes in a matchbook-like package that keeps them clean and protected until one is broken off to be used. This type of pick is intended to fit more naturally into the interdental spaces adjacent to the gums, which tend to be somewhat triangular, and is meant to stimulate the gums as well as serve as a plaque remover. The TePe sticks come in three sizes—very slim, slim, and wide—with the slenderer ones being made of birch, a wood harder than linden. As we can choose among different hardnesses of toothbrush bristles to suit our tastes, so TePe users can choose among the different types of sticks.

Coming attached as they do, interdental sticks are naturally made by a different process than are individual toothpicks. When Bo Krasse encouraged me to visit the TePe factory in Malmö to see for myself, I took him up on it. The factory is located very near the Swedish approach to the spectacular Øresund Crossing, the sixteen-kilometer-long fixed link—comprising a series of bridges, including a world-class cable-stayed structure, and a tunnel—that makes it possible to drive across and under the water to Denmark, one destination on a recent tour my wife and I made of Scandinavia. We were escorted around the factory by Bertil Eklund, the son of the founder and now head of the company. He showed us first how the wood was received as five-centimeter-thick longitudinal slices of tree trunk, its bark clearly visible along their long edges. The timber was interleaved with spacers to help it to dry out before being processed. We were told that only about 10 percent of the wood we observed would end up as toothpicks; the great amount of waste would be burned to generate heat.

To make toothpicks, the slabs of wood are first cut into pieces of lumber five centimeters square in cross section, and then any imperfections—such as knots—are chopped off, leaving short pieces of flawless wood that can be fed into machines that shape the combs of toothpicks automatically. The forming process is not easy to witness, since the cutters that do the work are largely hidden behind plastic shields designed not only for safety but also to capture the large amount of sawdust that results. While the machine room is loud (approximately one hundred decibels, making earplugs mandatory), it is also remarkably clean because of the effectiveness of the dust-removing equipment. The finished combs are beautifully smooth and graceful pieces of woodwork. But the wooden toothpicks, TePe's first product, now constitute less than 10 percent of the company's output, it having developed lines of toothbrushes and other oral hygiene products. Such diversification is desirable, even perhaps necessary, in the modern global economy, no matter where the manufacturing takes place.

CHAPTER TWENTY-THREE

Old Gold and Good Wood

EARLY TOOTHPICKS of precious metals were associated with wealth and status, but after the use of wooden ones became so widespread in the late nineteenth century, owning a gold or silver toothpick was no longer an exclusive marker of class. In 1886, the Brooklyn Jewelry Manufacturing Company offered a "solid rolled gold toothpick sent free to any Lady or Gentleman" who would send in the names of fifteen people to whom the jeweler could send its "Catalogue and Book of Engravings."[1]

During the century or so that the name Forster was synonymous with wooden toothpicks and the firm supplied the world with a simple, smooth, polished, and "antiseptic" wooden alternative to metal and quill, the expensive and fancy toothpick lay increasingly forgotten in drawers and boxes full of family trinkets. On occasion they were coaxed out of hiding: "Intent on raising a political war chest" for the 1936 election campaign, the Southern Women's National Democratic Organization began a drive "for old gold, silver or any other precious metal." At a kickoff event at the Waldorf-Astoria, the contributions included "broken watch chains, wrist watches, cake knives, discarded tooth fillings, gold bridges, wedding rings," and "a silver tooth pick."[2] Some of the contributions may have been the remnants of lost loves or bitter divorces, for generally such things were too sentimental or curious to be given away, thrown out, or sold. The pseudonymity of eBay has changed all that, of course, and it has provided an alternative to donating or throwing out seemingly anything. All sorts of toothpicks of another era can now be viewed in digital images and bid for on the Internet.

A nostalgia craze begun toward the end of the twentieth century revived the nonwooden toothpick. Restoration Hardware, the homewares chain that "has capitalized on the classic taste of highly educated baby boomers" with large disposable incomes, has offered its customers a silver-plated toothpick. There has also arisen a market for new kinds of toothpicks to grace a table of hors d'oeuvres or to complement a stark martini glass, as a visit to a fashionable wine shop or kitchen accessories store will reveal. According to one observer, "trendy hosts opt for handmade stainless-steel picks," ornamented with jewels.[3]

Ostentatious accessories are not unique to recent decades, of course. In the early twentieth century, *Puck* ridiculed modernist bad taste in the form of "ivory toothpicks with diamond settings."[4] Among toothpicks of questionable taste recently offered on eBay have been a "silver toothpick with twisted handle and turquoise top" and a three-inch long gold "sword toothpick with black leather sheath."[5] Years earlier Dr. Barnes' Toothpick, made of a more flexible material, was protected by a sheath that was a replica of an old I. W. Harper whiskey bottle. The connection between spirits and picking the teeth was made in a flyer headed "Speaking of Hygiene," which came with the toothpick. I. W. Harper whiskey was described as the "perfection of purity and maturity" and the Dr. Barnes' Toothpick as the "acme of cleanliness and effectiveness." Furthermore, when "used in moderation, Harper is a hygienic aid to the digestive organs, stimulating the secretion of the gastric juices and causing a healthful animation of all the faculties." The toothpick had the "endorsement of all the progressive members of the dental profession," and "dropping the whole thing in boiling water is a perfect sterilization."[6]

In the mid-twentieth century, one reporter remarked on the different types of toothpicks that were available and on their aesthetic values: "The more ornate kind—those of colored plastic and of quills—are to be found mostly in showy hotels, night clubs and restaurants. Some wooden ones are there, too, wrapped in paper containers advertising the places. There are also wood picks with a peppermint flavor."[7] It should not jar sensibilities to speak of the aesthetics of obviously designed, if not ornate, toothpicks made of gold, silver, ivory, and other dear materials. Criticizing a diamond-studded gold toothpick as garish or admiring a fancy one for the cleverness or whimsicality of its design might be considered as appropriate as admiring a fine piece of jewelry and appreciating the creativity of its artist. However, it is equally meaningful to speak of the aesthetics of a wooden toothpick and to compare its lines to those of another. Hand-crafted Portuguese orangewood toothpicks can evoke a sense of delight in their smoothness, uniformity of color, and long slender facets that are artifacts of the knife. By comparison, even the best machine-made toothpicks might seem to be rough, dark, and featureless. However, the making of them was once declared to have become "almost a fine art," and when considered on their own terms, their manufacture and their history can yield useful insights.[8]

Among the features that one might use to judge the lines and body of a machine-made wooden toothpick are its overall shape, its shaft, its points, its color, and its texture. Naturally, the effectiveness of the machinery in producing a given toothpick would be a criterion in judging an individual pick, but any judgment of the type should be made on the basis of as perfectly executed an example as can be achieved. Perhaps the aesthetic acme of the mass-produced wooden toothpick was reached in the late nineteenth century. The rolling process of compressing a splint of white birch gave it a rounded shaft that tapered smoothly into gracefully pointed ends. The toothpick was of a piece, with no tactilely awkward or visually distracting transitions between shaft and points. It was impossible to say where the one started or the other ended. The line of its taper was organic. This was the toothpick known as the Worlds Fair.

By contrast, a typical double-pointed round toothpick made today looks like it is composed of three parts: the shaft and the two ends. The rigidly cylindrical shaft has no personality, and its ends look as if they were pointed with a pencil sharpener. There is no grace to the abrupt transition between the shaft and ends. As clever a functional design as the so-called square/round tip toothpick is, it is even more of an aesthetic abomination, looking very much like a square pencil sharpened at both ends. The flat toothpick is just that—flat. It seems almost two-dimensional, and any tight rounding of its wider end seems out of all proportion to its body. Its narrower point is not even a true point.

The American Stim-U-Dent and the Swedish TePe dental stimulators are generally made and packaged not as individual toothpicks, we will recall, but as comblike groups.[9] The assemblage has a very pleasing appearance in its own right, with the repetition of its doubly curving lines presenting a pattern of individual grace making up an equally graceful ensemble. However, when one of the toothpicks is broken off from its neighboring group, both suffer aesthetically from the rough separation surfaces left behind. The single toothpick looks squat and bulky for its size, and the remaining group no longer appears to have the right proportions.

The single-pointed Japanese toothpick, with its round slender shaft terminating in a series of encircling scoops and grooves, has long had a presence in the West. It is a striking design, made to seem all the more so when its functionality is understood and appreciated, but the whole pick has no aesthetic unity. The severe point is too far away stylistically from the chopped-off and ornately decorated end, whose detail evokes more the turnings of Western furniture than the simple and minimalist lines of Japanese art. When broken after use, according to custom, the toothpick body and the separated head on which its point is designed to rest seem unrelated to each other. The idea is nice, but the execution looks ragged and clumsy.

Most plastic toothpicks have virtually no aesthetic component. Their shapes tend to be less appealing than wood splinters, and their colors are as subtle as those of paint balls. The texture and taste of the plastic are foreign to the mouth, and the point it carries is hard on the teeth and gums. However, one plastic toothpick has gained wide acceptance. The short white plastic pick with the holding end formed like the head of an old railroad spike has become an expected part of the Swiss army knife, but its concept is hardly of recent design. In 1819, a London newspaper reported on a "beautiful specimen of the art of cutlery" made in Sheffield by a Joseph Rogers & Sons craftsman who completed it after "28 days of close application." The "elegant knife in miniature" was only five-eighths of an inch long and weighed a mere quarter of an ounce, but among its "30 instruments" was a silver toothpick.[10] It certainly was not designed to reach the teeth in the back of the mouth.

In 1890, Arthur Corey of Council Grove, Kansas, patented a toothpick made of vulcanized rubber, whose characteristics he described as including "safety, integrity, adaptability, elasticity, strength, durability, cleanliness, inexhaustive supply, and cheapness." Rubber toothpicks were claimed to be "vastly superior to those made of metal, wood, or quills."[11] Among the materials that Corey left out of his list of inferiors was another relatively new one: celluloid. It did not escape the attention of New York City inventor

James Hills that it could compete with rubber, if not with quills, for a toothpick made of the plastic would have "elasticity, tenacity, durability, convenience in shape, variability in thickness to serve in removing particles wedged between teeth having little space between them." Furthermore, a celluloid toothpick would be transparent, which was said to be "desirable, for the reason that it may be used without being seen." In one embodiment, this toothpick was a thin, flat piece of plastic with one serrated edge. It was thin enough to be drawn between teeth that were close together, and it could be doubled over to serve for working between more widely spaced teeth. (A different inventor pointed out that the relatively broad flat surface of a celluloid toothpick could hold advertisements. Another one found more advertising space by impressing the corners of a business-card-size sheet of celluloid into the shape of multiple toothpicks that could be broken off as needed.[12]

Aluminum was still a relatively new industrial material in 1905 when Robert Freeman of Nashville proposed making a toothpick out of it. In his patent, he pointed out the failure of common materials to "provide a toothpick which will obviate the objections inherent in all kinds of toothpicks heretofore proposed or made." After a litany of faults of quill, wood, and metal toothpicks, he claimed for aluminum the positive traits of "softness, pliability, malleability, strength, and imperviousness to the chemical action of secretions of the mouth and food elements."[13]

The problem with making toothpicks out of hard, strong, and durable materials was that such qualities could lead to damage of the teeth and gums and would be dangerous if the toothpick were accidentally swallowed. This concern led Albert Baird of Colorado to invent a "composition toothpick," which consisted of isinglass and chrome alum dissolved in gelatin. When the mixture was molded and dried into, say, the form of an arrow, the resultant toothpick would "retain its shape and rigidity a substantial length of time so that the work which it is intended to perform can be completed." However, if it were swallowed, such a toothpick "will not stick in the throat and will readily dissolve in the stomach."[14]

Another approach to allaying the hard and unyielding surfaces of certain toothpicks was to make them out of a material that would soften with use and conform to irregular interdental spaces. A toothpick that resembled a miniature canoe and was made of an

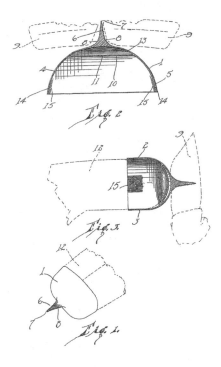

*One inventor solved the problem of using a
toothpick to get at hard-to-reach places in
the mouth by devising a pointed rubber
prosthesis-like device that could be fitted
onto the tip of the tongue.*

appropriate wood was patented in 1935 by Francis Grant of Detroit.
The idea of fashioning the toothpick in a "dugout" form was to
allow its sides to be more flexible and thereby more conformable.
Another embodiment of the invention is suggestive of today's interdental sticks, but with a shallow groove along the straight edge. The
inventor's choice of material was balsa, which he described as "quite
soft and compressible," and the toothpick made of it, "when moist-
ened by saliva from the mouth, will soften and readily conform to
the contour of the teeth." Grant's patent was assigned to Stim-U-
Dents, Inc., then an independent corporation of Michigan.[15]

Stim-U-Dents and similar toothpicks have come to be widely
admired, but the short, stubby sticks are not nearly as effective as

floss for dislodging something annoying from the tongue side of teeth. That problem had earlier been focused on by the California inventor Russell Lunday, who used "thin rubber or other pliable material" to form a toothpick attached to a cup-shaped base that could be "secured to the tongue of the user" by a slight vacuum. Once in place, the pointed prosthesis-like device could be used to remove "particles of food from the inner surfaces of the teeth," something he believed "impossible" to do with an ordinary tooth-pick.[16] Presumably, Lunday also solved the problem of how to pick the teeth without opening the mouth, at least after the tongue-pick was installed.

The Norwegian inventor Rolf Barman believed he had solved the problem of unsightly toothpicks by inventing a wooden one of such exquisite form that it looked more like a Brancusi sculpture than an interdental implement.[17] But no matter how artistically designed, a more or less rigid wooden toothpick could hardly be expected to fit into every nook and cranny in the mouth. The perennial design problem was nicely stated by Edward Barnett, who described in clinical and technical terms the geometry that the toothpick had to negotiate and the constraints that countless inventors had tried to satisfy:

These areas between adjacent contacting teeth, i.e., the interdental spaces and the interproximal tunnels, are actually like a passageway with a somewhat triangular cross-sectional shape. The base of the triangle is the gum or gingival tissue; the sides of the triangle are the proximal surfaces or side walls of the contacting teeth; and the apex of the triangle is the incisal or occlusal contact area of the two adjacent teeth.

Quite often the openings to these tunnels and spaces are blocked by slightly swollen or edematous gum tissue. Therefore, in order to enter the spaces or tunnels, the cleaning instrument must be sufficiently resistant to bending perpendicular to its longitudinal axis to enable it to depress or displace the gum tissue blocking the entrance or exit to the tunnels or spaces. Furthermore, the posterior interproximal tunnels are often quite tortuous, i.e., the path of the passageway is circuitous. Therefore, the instrument must be sufficiently bendable to fol-

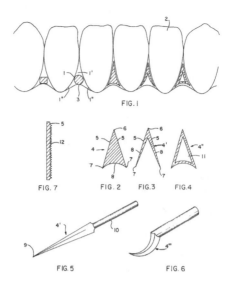

Toothpicks have been made in many different cross-sectional shapes in order to better fit into the gaps between teeth, which many inventors have recognized are more triangular than square or round.

low this tortuous tunnel as it contacts the hard surfaces of the teeth and firm healthy gingival tissues. It must also have sufficient strength to dislodge food debris and loosely adherent calcular material from the walls of the tunnel or space. It must also intimately conform to the walls of the sides of the tunnels and spaces and must have sufficient abrasiveness to remove the dental plaque without injuring the tooth or gum tissues. Additionally, it must be able to fit into the usually narrow space between the anterior teeth.[18]

What an engineering problem! After stating it, Barnett spent considerable time explaining why Stim-U-Dents and other devices did not provide a satisfactory solution, thus leaving room for his improvement. His solution took the form of a tapered triangular pick made of an elastic and deformable material surrounding a core that was stiff in the long direction but bendable transversely. It was fitted with a larger base to provide a fingerhold.[19] Even the most elementary of design problems can be stated in the most unintelligible of terms and be solved by the most ungainly of solutions.

It should thus be no wonder that people with something stuck stubbornly between their teeth often take matters into their own hands and enlist into use as a toothpick something designed for another purpose entirely. George Franklin Grant, the son of former slaves, graduated from Harvard Dental School in 1870 and later taught at that institution. He was also an avid golfer who wished to devise some means of "obviating the use of the usual conical mounds of sand or similar material formed by the fingers of the player on which the ball is supported when driving off," which he did. In 1899 he was granted the first patent for a golf tee, the pointed end of which looks suspiciously like the business end of a toothpick.[20] Perhaps it should come as no surprise that the first wooden golf tees were produced in Norway, Maine, in toothpick-making country.[21]

Like so many other things in the mid-twentieth century, golf tees came to be produced also out of plastic. One company made the connection between the tee and the teeth by marketing the Pick'n'Tee, which was described as a combination golf tee and cocktail pick. Since a standard-sized golf tee would generally be considered too short to be used to spear hors d'oeuvres, a Pick'n'Tee consisted of two golf tees connected together point to cup, thus making the assembly about the right length to pick up a meatball or a cube of cheese. When it had served that purpose, the double-tee toothpick could be broken in two and be used as golf tees.[22]

It is certainly not difficult to imagine even a single golf tee being used in a pinch to pick the teeth, but not everyone is a golfer. One confirmed bachelor attributed his discovery that a hairpin was the perfect tool for cleaning out a clogged pipe to his finally getting married at age fifty-nine. In fact, he found his wife's hairpins to have dozens of unintended uses, including as toothpicks. But, unless straightened out, ordinary hairpins, like matchsticks, are generally too short to be used easily between the inside back teeth. When a bicycle repairman was having difficulty getting at a piece of bread crust stuck near one of his molars, a boy who had been observing him suggested that long "bike spokes come wonderful handy for that job."[23]

When I have found myself without a proper toothpick, I have on occasion resorted to using the corner of a business card. I have also folded and refolded an ordinary sheet of paper to give it a sufficiently stiffened point. But these ad hoc tools generally work only on the outside crevices between teeth. For more general picking I have found that a mechanical pencil, one of which is almost always in my pocket, is very effective. Though I do not extend the 0.5 mm lead very far out, lest it break off beneath the gum line, it is usually sufficiently pointed (certainly no less so than a modern flat wooden toothpick) to dislodge whatever is annoying me—on either the cheek or tongue side of my teeth. Of course, it is difficult to conceal the fact that I am using a pencil as a toothpick, and so I try to do it in private or at least when no one appears to be looking. My use of the pencil is not unique, for Charles Dickens wrote in *The Old Curiosity Shop* of Mr. Slum "using his pencil as a toothpick," though evidently not very discreetly.[24]

Boxed and Unboxed

I N THE NINETEENTH CENTURY, wooden containers were used to package and ship a great variety of things. Elihu Beach Estes ran a hardware store in Port Kent, New York, which is located on the western shore of Lake Champlain. Much of his stock would have come in kegs and barrels. Estes filled empty barrels with a local black sand that he had discovered was very effective for use as "blotting sand," which was sprinkled across documents written in ink to aid in the drying process. He transported barrels of the sand to such cities as Boston, New York, and Philadelphia, where he sold them. The sand was also found useful for producing smalt, a pigment used by sign makers. In time, Estes came to make wooden shakers from which blotting sand was dispensed. The shakers, which were handsome enough to sit on a writing desk, could be described as small round wooden boxes with holes in the snug-fitting top.[1]

Webster Estes, one of Elihu's sons, promoted the idea of making a variety of treen, which is a plural noun designating small domestic wooden objects, such as pillboxes. (Other treen, or treeware, included wooden spoons, bowls, and plates.)[2] By the 1890s, E. B. Estes & Sons was "the preeminent pillbox supplier in the world" and "the most extensive manufacturers of turned-wooden boxes" of all kinds. The company's boxes, which were used to package a wide variety of "powders, wafers, lozenges, tablets, liquids, drugs and chemicals," were advertised as "artistically designed, truly and smoothly turned from selected White Birch, and neatly and finely finished."[3] Among the "specialty wood boxes" made by Estes around the turn of the century was a two-inch-diameter one that at three and three-quarters inches tall was large enough to hold a good number of Estes' Imported Orange-Wood Tooth Picks, which the sharply printed label claimed to be "the finest picks in the world." When the handsome box, a little masterpiece of woodworking art, was emptied of the Portuguese toothpicks, for which Estes & Sons were the "sole agents for the United States," the company would likely have been the source of a replacement supply, no doubt sold in less durable packaging.[4] The box, which would have been a proud addition to many a kitchen table, could also have been refilled with domestic toothpicks, of course, though, being shorter than the Portuguese, they would have looked lost and maybe insignificant in the wooden container.

This E. B. Estes & Sons turned wooden box dating from around 1900 is a handsome piece of workmanship. The box originally contained orangewood toothpicks imported from Portugal.

Headquartered in New York City, the Estes firm had factories in Maine, Massachusetts, New York, New Hampshire, Pennsylvania, and Vermont at the end of the nineteenth century. Later, plants were added in North Carolina and Tennessee. A staggeringly wide variety of treen and larger things were manufactured, and among the company's wooden products were baseball bats, dumbbells,

Indian clubs, tenpins, and bowling balls. They also made shooks, which were kits of a sort ready to be fabricated into boxes, as well as handles for everything from furniture to tools, plus dowels, clothespins, and, naturally, toothpicks.[5]

In the 1920s, the treen business declined with the increasing use of glass containers for pharmaceuticals, and Estes divested itself of its woodworking factories and contracted its business to one focused on smalt used in making color signs. The leaner enterprise was renamed the Clifford W. Estes Co., which by the turn of the twenty-first century had evolved into "the absolute leader in the coloring and coating of sand and gravel" and "the undisputed leader in aquarium gravel," thus in a way bringing the family business back to its roots on the shore of Lake Champlain.[6] At the same time, orangewood toothpicks were still being imported from Portugal by the E. B. Estes Division, in packages describing the picks as "strong, smooth, pliable" as well as "hygienic" and "durable," which were many of the same words used on the box of a century earlier.[7]

The common paper packaging of the orangewood toothpicks being imported from Portugal in the early twenty-first century may

Orangewood toothpicks hand-carved in Portugal are still imported into and sold in the United States in distinctive paper packaging.

not be very much different from what Estes was bringing to this country in the nineteenth. Though the individual picks are nearly uniform in color and length, they vary noticeably in their pointedness, their thickness, and their faceting—all artifacts of the cut of the knife. Nevertheless, about thirty picks are gathered in a piece of paper the color and texture of newsprint. The paper is not much greater in size than a chewing gum wrapper or a cigarette paper, which means that there is not much room to roll the thirty toothpicks into a cylindrical pack and crimp the ends. Indeed, a smaller number of picks rolled into a similar package may be mistaken for a Spanish cigarette. The *Washington Post* once reported that a practical joker offered such a package of toothpicks to his companion. After he had some difficulty lighting the "cigarette," his amused friend informed him that "toothpicks were not made to smoke."[8]

Generally too thick to be mistaken for a cigarette today, ten packs of thirty Estes imported orangewood toothpicks each are gathered and held together with a very thin rubber band. The resulting bundle is wrapped in a whiter piece of paper printed with a label in orange ink, giving information about the contents but not exactly how many toothpicks are contained therein. Whereas the individual component bundles are held together strictly by folds and tucks, the outer wrapper is kept snug by a glued lap joint. Its folded-over ends are closed with glued stickers imprinted, in orange ink, to identify the contents as Estes' Orange Wood Toothpicks. The entire package of three hundred or so toothpicks, though significantly larger than a 250-count box of Forster round toothpicks (now likely imported from China), weighs considerably less, thanks to the low density of the wood. The importing company is a descendant of the firm of E. B. Estes & Sons, which did business with the Estate of Charles Forster in the early twentieth century.[9]

The three-inch-long Estes orangewoods, which in their present form have reportedly been imported into the United States since the 1960s, look almost identical to those that Charles Forster found in Brazil a century earlier. When similar picks were sold by the Cutter-Tower Co. in the early twentieth century, they were advertised as Portuguese rosewood toothpicks, though it seems unlikely that they were that.[10] In their more recent appearances in America, they have been called SteakPiks by those who "never liked the after-dinner chore of manipulating those tiny toothpicks." They have been given a masculine image by being called "a meat-eater's best friend," "tooth chisels," and "Terminator toothpicks." But they also have a feminine side, having been termed "magic toothpicks" by quilters, who have used them as forms to achieve exquisite finishing details.[11]

Charles Forster's machine-made toothpicks were packed in paper or pasteboard boxes, which had to be designed and fabricated. A separately printed label was affixed to early boxes, giving the potential customer essential information about their contents and possibly also a caveat about imitators.[12] Eventually, information about the contents was printed directly on the box stock, thus making for more colorful and brand-specific packaging.

Securing a supply of the boxes meant incurring no small expense. In the early twentieth century, toothpicks manufactured by the Estate of Charles Forster were packed in boxes manufactured in New York by the Nevins Church Press, and the Holt Brothers print shop in Dixfield was responsible for printing boxes that carried distinctive brand names and artwork for vendors throughout the country.[13] The boxes were sent flat and unassembled to the Forster mills, to be formed, filled, and shipped to their destination. In 1945, Forster Manufacturing Co. acquired an interest in the Farmington printing firm of Knowlton & McLeary and five years later owned it outright.[14]

Once a box of toothpicks has reached its ultimate destination, a new problem of microdistribution arose. In a private home, where its contents could last for years, a paper or pasteboard box could easily become rather soiled and worn before the toothpicks were used up. This might especially be the case if the box was left out on the kitchen table, where greasy fingers could reach in for a toothpick to work on the residue of barbecued ribs stuck between the teeth. For that matter, the toothpicks remaining in the box would themselves have become soiled before too long. Such a condition would have been intolerable in a public or commercial setting.

Creative minds naturally got to work on the problem. New York inventor George Buch seemed not so much interested in keeping the toothpicks sanitary as making the package serve double duty. His 1895 contrivance consisted of a squarish box so constructed that when opened its sides could be folded down in a manner that "adapts it to be used not merely as an ordinary paper box for containing the toothpicks when offered for sale, but also permits of its additional or secondary use as a convenient holder or receptacle for the picks when these are exposed on the table or sideboard ready for use."[15] Oscar Weik, of Wausau, Wisconsin, also developed a package that could double as a holder, but in addition he attacked the problem of trying to keep the toothpicks from being unduly handled. His idea was to pack the toothpicks in a round container whose bottom was shaped like a cone. This geometry, supplemented by the force of gravity, allowed the picks to arrange themselves naturally in such a way that a hole remained in the center of the annular supply, "into which a finger may be inserted for ready extraction of a single article without handling of the others." Of course, inventors typically overstate their accomplishment, and though Weik's handiwork was an improvement, it was virtually impossible for a finger to touch only the toothpick it was extracting.[16]

One way to keep toothpicks clean for personal use is to carry them around in smaller-than-box-sized quantities. In Graham Greene's 1978 novel, *The Human Factor*, two of his characters have an exchange prompted by the use of a Waterpik, from which issues a pressurized jet of water that might be thought of as a very clean toothpick:

> "Amusing little gadget, that of yours. Fashionable, too. I suppose it really is better than an ordinary toothbrush?"
>
> "The water gets between the teeth," Daintry said. "My dentist recommended it."
>
> "I always carry a toothpick for that," Percival said. He took a little red Cartier case out of his pocket. "Pretty, isn't it? Eighteen karat. My father used it before me."[17]

Not everyone inherits a Cartier toothpick case, of course, and there are less extravagant alternatives. A small wooden tube closed by a mating wooden top—a poor and distant relative of the Estes turned box—has been sold as a souvenir of Maine. The one I purchased at the gift shop of the Desert of Maine tourist attraction near Freeport is labeled as a "toothpick holder"; if sold in a needlework store, it might just as appropriately be labeled a "needle holder."[18] But one need not travel far to buy such a purpose-made case. A clever acquaintance with an inventive mind carries "a sup-

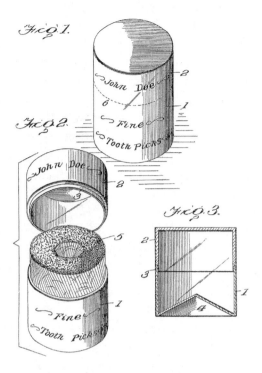

Toothpicks packed in a round cardboard container with a conically shaped tin bottom would naturally arrange themselves in such a way that a space was left in the center to allow a finger to gain easy access to the contents.

ply of toothpicks in a container that originally held spare leads for mechanical pencils."[19] Though dozens of the now-familiar very thin polymer leads can fit easily into the small diamond-shaped plastic case in which they are sold, with plenty of room to spare, only a half dozen or so flat toothpicks can fit comfortably into the one from which I emptied my remaining supply of Pentel 0.5 mm HB leads. Though the leads come in 60 mm lengths, the plastic case that they are sold in is about 10 mm taller. This leaves enough room inside to fit four standard-length round toothpicks, which measure about 3⅝ inches or 67 mm long, with sufficient space around them so that one can easily be shaken out when wanted. My acquaintance did not say whether he cleaned the lead container of any residual graphite dust before employing it to carry toothpicks. It could, of course, be done with a little bit of tissue wrapped around the point of a toothpick.

The personal toothpick case dates from ancient times; the modern open toothpick holder is intended to contain a small supply for use at a table that multiple people share. Charles Forster encountered such holders in Brazil, where they were traditional, but they were not common in America until the late nineteenth century. Once introduced in the 1880s, however, they proliferated in styles, sizes, and degrees of seriousness. Toothpick holders constituted "part of the table set in the Victorian period, and they were sometimes a part of a condiment set that might also contain cruets, shakers, and/or a covered mustard" bowl. Such holders "continued in vogue until about 1910." One advertised in the early twentieth century was gold-lined and inscribed with the phrase, "Take your pick." In 1908, it was offered on sale at half its regular price of ninety-eight cents. A German holder has been made with three handles spaced 120 degrees apart, making it easy for anyone around a table to pick it up and pass it.[20] The variety of designs seems endless.

Toothpick holders have been made out of wood, metal, glass, and china, among other materials. They have taken the form of animals or people standing beside or carrying baskets or other receptacles into which the toothpicks could be placed. They have been elegant; they have been whimsical. When it was still fashionable, "china and glassware sets always included a toothpick holder." But that practice declined after Emily Post declared that "it wasn't polite to pick your teeth at the table."[21] Today, there are voluminous illustrated catalogs of the huge numbers of domestic toothpick holders from times past that serious collectors and dealers use to identify and price their finds and wares.[22]

Toothpick holders and similar appurtenances, being naturally small items, have long been made to serve as convenient souvenirs, tchotchkes, and gifts. In the early twentieth century, a "personally-conducted tourist" could be proud of his "plaid toothpick box, bearing a picture of Scott's monument, from Edinburgh." Toothpick holders, typically about the size of a large shot glass and variously decorated in formal and fanciful patterns, could be pretty or plain. In a Christmas story, a little girl remarks poetically, "My prettiest purchase was filled with toothpicks, my plainest with chocolate creams."[23] A century later, such treasures (mostly without the contents or the poetry) might have been offered in antique malls or on eBay, along with a miscellany of other toothpick-related items, including holders from towns and cities around the globe. Indeed, more anonymous toothpick holders also proliferated, and models could be had in silver and porcelain, glass and wood. Among avid collectors, such holders are known simply as "toothpicks," and their pursuit is a well-established hobby. In 2005, the National Toothpick Holder Collectors' Society, which dates from 1973, had a membership of about five hundred. Its *Toothpick Bulletin,* which is published monthly, seldom mentions what the holders were designed to hold and seldom shows them in any other state than empty.[24]

Toothpick holders in the form of animals have been legion. Many examples of static holders were cast in metal with holes for the toothpicks arranged and oriented so that when inserted they were fanned out to allow one to be taken easily without touching the others. These work very well at tables set for hors d'oeuvres. Porcupines and hedgehogs are natural candidates to be fitted all over with toothpicks, and turkeys and pheasants have also been employed, with the toothpicks fanning out as if they were tail feathers. Such holders continue to be used with elaborate buffet arrangements. One silver-plated nickel example by the French designer Christofle and offered in the early twenty-first century by Bergdorf Goodman on New York's Fifth Avenue has been self-described as "very Bergdorf." It took the form of an elephant with a monkey sitting atop it, somewhat in the manner of the Hindu deity Indra; the monkey holds up an open umbrella, through the holes in which toothpicks could be arranged. Bergdorf's ad in the *New York Times* showed it holding single-pointed Japanese toothpicks.[25]

Even before dealing with the problem of passing around a supply or dispensing a single sanitary toothpick, there was another difficulty to deal with. This was the all-too-common experience of opening a box of toothpicks only to have them spill all over the floor, as in the famous toothpick scene in the movie *Rain Man.* When a waitress drops a whole box of toothpicks, resulting in a confused, pickup sticks–like pile of them on the floor, Raymond Babbitt, the autistic savant played by Dustin Hoffman, looks down at the seeming randomness and announces immediately, "Eighty-two, eighty-two, eighty-two . . . toothpicks . . . two hundred forty-six total." His brother Charlie (played by Tom Cruise), who as a child pronounced Raymond "Rain Man," gloats when Ray's number does not agree with the quantity printed on the box (250), but the waitress tells him that the difference was accounted for because four toothpicks had already been taken out of it.[26] The dramatic effect must have been lost on toothpick makers, who would know that "each box count is accurate within five toothpicks" only, though one Forster president once suggested that each box contained at least 250 and insisted that "a few extra are included in case some inferior picks sneak in."[27]

Most of us are not able to count toothpicks as fast as Raymond did, but we all certainly spill them. In the early 1960s, this problem concerned Forster's president, Theodore Hodgkins, who admitted that toothpicks were among those long and slender items that had "historically provided a difficult problem in packaging where it is desired to employ a folding box." Anticipating the *Rain Man* scene, he recognized that once a "box was opened, the entire toothpick contents were uncovered, with the inevitable result that unless the box were handled with extreme caution, the contents would be accidentally spilled." He also understood that because such boxes were glued closed to keep their contents secure during transit and display, the purchaser, "in opening the box initially, frequently will tear the bendable margin or tongue on the cover" so badly that the box could not be properly reclosed.[28]

To address these problems, Hodgkins invented and patented in 1966 a box incorporating a small hole in one side through which only a tiny fraction of the contents could be dispensed at a time. The dispensing hole was covered by a flap reading "LIFT UP" that was raised and the perforated seal broken by the purchaser, and since all other flaps on the box could remain closed and sealed, the contents would not spill or be exposed to an "unsanitary, uncovered condition" over the time it would take to use up the whole box.[29] A contemporary box of Ideal toothpicks (bearing the statement that it was then "the only toothpick tapered thin," the way the original Forster ones were) carries the legend "hygienically packed" and is identified as being a "patented dispenser box." The opened box of Ideal toothpicks that I bought on eBay still has its "LIFT UP" panel undisturbed, suggesting that its onetime owner did not find its new dispensing feature desirable or obvious. The box of 750 flat tooth-picks, which is contemporaneous with the Hodgkins patent and which sold for twelve cents at the time, was opened in a conventional way. At some point, its flap was held closed with a piece of cellophane tape.[30]

Toothpicks have continued to be packaged in sanitary dispensing boxes that work on a principle similar to Hodgkins's, some having a scored hole covered by a small flap that reads "OPEN HERE." When I once looked at the toothpick display in a Wal-Mart store, I found that the only box of Forster's left had been torn open at that very place, suggesting to me that some shopper who desired a toothpick but did not want to buy a whole boxful took the imperative instructions literally and broke into the box then and there.[31] The box had been opened so carelessly and left so obviously in disarray that no one could have missed the fact that it had been tampered with. (I did not try to find out if the borrowed toothpick had been returned to the damaged box.)

All kinds of packages displayed on store shelves are subject to tampering, and not always because of petty thievery. The Tylenol scare of 1982 resulted when seven people died after someone inserted cyanide into packages of the over-the-counter pain reliever, whose capsules had been opened and reassembled without leaving a trace of what had occurred. The incident led to the end of separable capsules in such medicines and to safety packaging that has become so familiar today as to be expected on drug and food products. The idea that soiled or perhaps poisoned toothpicks could be inserted into public dispensers was anticipated in the early twentieth century by the Swedish inventor Jonas Sahlin. When the body of his patented toothpick stand was depressed, a single toothpick was pushed out of the hole in its top. To prevent the "re-insertion into the stand of a tooth-pick having been used," the opening allowed for toothpicks to pass in only one direction—out.[32]

The undesirable effects of spilling and soiling toothpicks was addressed anew in the 1970s by the inventor Reinaldo Rela Zattoni, of Brazil, who devised a squarish "safety package." Looking otherwise like a regular box, and owing much to Hodgkins's design, Zattoni's also needed never to be fully opened. Instead, the toothpicks could be shaken out one at a time through a small hole in one corner. Another Brazilian inventor, Hermogenes Rella, may have anticipated that tampering might occur to Rela Zattoni's toothpick "safety box," and so improved upon it. In order to ensure that toothpicks in the box were kept "in good hygienic condition" and were dispensed "only through contact of the human hand of the user," Rella made a more securely locking flap.[33]

Concern over the cleanliness of toothpicks had long been recognized in the toothpick business. When machines replaced hand packing, the boxes were quick to carry legends reading "hygienically packed by machinery."[34] Carroll H. Brackley, president and manager of the Strong Wood Turning Corp., established in 1926, once pointed out to a reporter that "toothpicks made at his plant are never touched by hands during the entire manufacturing procedure," which was no doubt a good thing since a human hand could do little but slow down the process. In the mid-1990s, Strong Wood Products, Inc. was producing two billion toothpicks a year, compared to Forster's twenty billion. Like Forster, other Strong mills manufactured a wide variety of wood products throughout the years, even—as part of the World War II effort to conserve metals for military use—a wooden lipstick tube.[35]

CHAPTER TWENTY-FIVE

Woodpeckers and Other Dispensers

T HINGS BEGET THINGS. As fundamentally simple as the toothpick is, with an apparently singular purpose, inventors and
users alike have caused it to evolve and morph into seemingly countless variations on its theme and have supplemented the elemental
object with accessories and accoutrements and an infrastructure
that can virtually smother a single toothpick with attention. Early
on, hotels and restaurants faced the problem of needing to supply a
large number of unrelated and not necessarily fastidious customers
with clean toothpicks. Keeping the items fresh could be a challenge. Once, a "frontiersman" leaving a South Side Chicago restaurant asked the cashier if he could take one of the toothpicks from
the box on the counter. When he was told yes, he took one and used
it aggressively right there. After finishing with it, he returned it to
the box, remarking to the cashier, "I don't see how yer can 'ford ter
keep them things. I notice so many folks take 'em off an' never bring
'em back!" As soon as the conscientious frontiersman went out the
door, thinking he had done the honorable thing, "the cashier emptied the box under the counter, with disgust marked on his every
feature."[1]

Perhaps in part to avoid such a waste of good toothpicks, most
restaurants did not put the whole box out all at once. Rather, a small
amount of toothpicks was placed in a more appropriate and attractive holder—perhaps a small bowl, glass, or saucer—on the counter.
Still, what customer was capable of taking one without touching
some of the others? The Denver inventor Harold Prommel, in his
1915 patent for a "sanitary toothpick-dispenser," stated the problem succinctly:

Heretofore, as is well known, it is customary in restaurants,
hotels and in public eating houses generally, to have a quantity

of tooth picks exposed in an open receptacle so that it is impossible for the person supplying himself with one or more picks, to avoid touching others which still remain in the receptacle. It is evident that under this practice, it is impossible that the tooth picks shall be perfectly clean and free from germs. Furthermore, where tooth picks are kept in an exposed receptacle, they become soiled by reason of the settling of dust thereon, as upon all other articles.[2]

Prommel's solution was embodied in a dispenser in which the toothpicks were held in an enclosed hopper and one or two were dispensed into an inclined tray by the action of a lever, not unlike the way drinking straws are at many a fast-food restaurant. The device looked like a miniature slot machine.

Another problem that was at least partially solved by the design of clever dispensers—intended to be filled with toothpicks from a box holding a much larger supply—was that of restaurant customers "carrying away a pocketful of tooth picks."[3] Even though they did not cost very much at all individually, cumulatively toothpicks could be a not insignificant expense for a small business.

Vending-machine-like devices made to dispense one at a time items such as postal cards and cigarettes were patented in the late nineteenth century in the United States and several European countries. A very basic design, intended for complimentary use in delivering matches in cigar stores and thus requiring no coin, consisted simply of a notched slider that the customer could pull out from a box on the store counter. Another coinless dispenser for "matches or similar articles" was operated by a crank located on its side. It looked somewhat like an upside-down pencil sharpener and operated on the basic principle that subsequent toothpick dispensers would employ. Others, generally requiring a coin to activate the mechanism, were patented expressly as cigarette dispensers, designed to keep the cigarettes fresh as well as to deliver them one at a time. A later type of cigarette dispenser, dating from the 1930s, was designed specifically for use in the automobile. With the touch of a lever tray, a single cigarette would drop into place in the tray, with one end in proximity to a lighter. The lighted cigarette could be taken out of the tray with a minimum of distraction to the driver.[4]

Given the technology developing in such devices, it was relatively easy to design reliable dispensers for larger items such as cigarettes and pencils, where small variations in size could be accommodated by the mechanism. Toothpicks, however, because of their small diameter, were likely to clog and jam in the rotating parts. Thus, there was plenty of room for improvement—and for further patents. The problem of jamming could be especially acute with cheaper toothpicks, boxes of which could include relatively large variations in size, especially when broken ones were present. There was also variation in size from brand to brand. During World War II, when there was an attempt to standardize the products of industry generally, the National Bureau of Standards proposed to fix the length of the flat toothpick at $2\frac{3}{8}$ inches and the round at $2\frac{1}{2}$ inches. While this would be of some benefit to dispenser designers, the unstandardized diameter would continue to frustrate them.[5]

For better-made toothpicks that were rather uniformly round and of a consistent diameter, the same principles that delivered cigarettes could reliably deliver toothpicks one at a time. In the early 1950s, Alfred Schupp of San Antonio, Texas, was issued a patent for a device that required no coin to deliver a single cylindrical object, such as a match, pencil, cigarette, or toothpick from a closed hopper. The simple dispenser resembled a miniature vending machine, and it was operated by turning a knob on the side of the case. This action rotated a cylinder with an appropriately sized notch just large enough to receive a single item from the supply; continued rotation carried the object to the front of the dispenser, where it could be picked up for use. Variations on Schupp's dispenser became common features at the checkout counters of greasy-spoon diners and better restaurants alike. In the 1950s, toothpick manufacturers, including Forster, introduced sanitary dispensers to promote use of their product.[6] From the point of view of a restaurant, dispensers had the added advantage of making it less likely that individual customers would take more than a single toothpick at a time, thus getting more out of the toothpick budget. Another once familiar dispenser operated like the automobile cigarette dispenser. Rather than having to turn a crank or a knob, all a person had to do was push down on the lever tray, activating the mechanism to drop a fresh toothpick onto it.

Many much more fanciful designs for dispensing a single toothpick have been devised. In 1895, the inventor Heinrich Staub of Potsdam, Germany, was issued a U.S. patent for a "receptacle for holding toothpicks or other similar articles which are liable to be soiled, injured, or wasted by handling." His invention consisted of a cylindrical container mounted on an ornate pedestal. By depressing a plunger on the top, a single toothpick was dropped into a miniature boot standing at the base of the device. It was designed such that the toothpick leaned outward, so that it could be taken up easily.[7]

In 1911, Louis Tangen, a subject of Norway residing in Duluth, Minnesota, patented a "server for toothpicks" that was mounted on the back of a stylized metal turtle. Like many such devices that would follow from it, Tangen's operated by pushing a plunger into an enclosed compartment full of toothpicks. This action worked a "hook or latch" that retrieved a single toothpick from the store of them inside the turtle's shell. While one object of the invention was to "prevent the wholesale appropriation" of toothpicks by the customer, its main purpose was to "prevent unsanitary handling of toothpicks by a great many individuals successively attempting to take one or two from a larger stock." The inventor's concept included a small pair of tongs intended to be used to refill the reservoir from a box, "whereby handling of the toothpicks may be further avoided."[8]

Some dispensers are more dynamic; they have hidden mechanisms that activate moving animals to deliver one toothpick at a time. Among the most frequently employed servants have been birds, whose beaks were designed to retrieve and offer to the operator a single pick from an abundant supply. A dispenser patented in 1913 by the California inventor Abednego Hughes took the form of a stork, whose beak was a "resilient bifurcated stem," which is patent-talk for something like a flexible two-pronged fork. A plunger caused the body of the stork to pivot about its legs and dip its beak into a receptacle full of toothpicks. In order to have it retrieve one in its bill, the stem/fork had to be attached sideways, somewhat like the cockeyed stuck landing gear of the JetBlue airliner that had to make an emergency landing at the Los Angeles airport in 2005. As the stork's cocked bill was inserted into the supply of toothpicks, one became wedged between the tines, and when the plunger was released the stork sprang back to a vertical position carrying the toothpick with it. The inventor Hughes noted that although his patent drawings showed a stork, "some other shape could be imparted to the pivotally held member."[9] In fact, many others have been.

One of the most popular was in the form of a cast-metal assemblage representing a woodpecker sitting on the end of a hollow log. A spring kept the bird upright until it was pushed down to insert its beak into the log, which contained a supply of toothpicks. When released, the woodpecker returned to its upright position holding a single toothpick in its beak. The toothpick was in fact held between

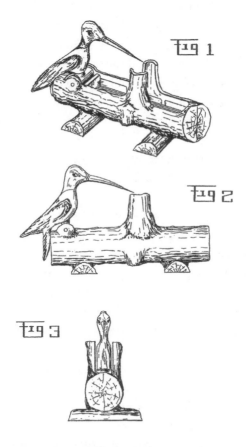

Many fanciful toothpick dispensers have been devised and patented, including this one in which the beak of a bird can be pushed into the hollow log containing toothpicks to retrieve one for the taking.

a pair of barely noticeable pins sticking out of the beak. The log was often cast in a bronzy brown metal and the bird painted the red-headed woodpecker colors of white, black, and red. The clever device was also produced in monochromatic plastic versions, some of which were marketed as the Woody Woodpecker Sanitary Toothpick Dispenser.[10]

Derivative designs featuring animated servants have multiplied. Like the crank- or knob-operated devices, these too owed a lot to the prior art embodied in cigarette dispensers and had the stated object of providing something that was "novel in appearance, entertaining in effect," and capable of simulating the action of a "bird, animal, or human figure." In particular, these mechanisms were "readily operated with one hand."[11] In one more recent variation, when the head of a stylized crow is pushed down, a tray holding a single toothpick slides out of the base of the dispenser. As the crow's head continues to be pushed down onto the toothpick, the wooden splinter wedges itself into the slightly opened plastic beak so that when the head is released the bird springs back up to present the toothpick to the operator. One variation on the device has a figure representing the butler Jeeves reaching down to grab a toothpick, which he holds out to be taken from his outstretched hand.[12]

Not all dispensers are so fanciful. Glass and plastic dispensers of various designs that operate on the principle of shaking out a single (clean) toothpick at a time have long been used. Some of them look like saltshakers, but with a larger hole or holes in the top. One patented design for a tabletop holder combines somewhat the idea of a shaker with that of a rotary dispenser.[13] As simple and convenient as it may seem, the shaker kind of toothpick dispenser can get clogged easily. With more than one toothpick point simultaneously wedging into the hole that is only large enough to let the body of one pass, the result is a miniature logjam. Another type of dispenser remains sitting on the table while its top is pushed down to project a single toothpick out the top. Inside the dispenser is a conical bottom with a recess at its apex just large enough to admit a single toothpick. When that toothpick is in place, it is vertical, and so pressing down on the device causes it to stick out the central hole in the top. When that toothpick is taken out, another falls into place from the supply inside, to be dispensed the next time. Though this device may also be subject to jamming, it does arguably have the advantage of a simpler operation.[14]

An alternative approach to keeping toothpicks clean until used is to enclose them in paper. When polite Edwardian society considered it provincial to regard toothpick use as impolite, some of the "world's most fashionable restaurants—the Ritz, the Savoy, Paillard's, and so on—set toothpicks on every table, each toothpick sealed in a little paper envelope." At a proper Japanese dinner, a guest could find on the table "a long envelope decorated in colors, in a little pocket of which on one side are toothpicks and inside are two fresh pine sticks," or chopsticks.[15] Japanese toothpicks could also be found wrapped all by themselves.

In most American restaurants today, we tend to find ordinary wooden toothpicks wrapped individually in white paper or clear plastic. The technology to do such a thing owes a great deal to the soda straw industry, which dates from the invention of the paper drinking straw by Marvin S. Stone in 1886. (Before then, drinking tubes used in America were natural straws imported from Europe, because no suitable domestic ones could be manufactured at a competitive price when the required hand labor was taken into account.) Like toothpicks, machine-made straws had an "infinitesimal" unit cost and most of them were not purchased directly by their ultimate users, who were largely patrons of restaurants and soda shops. Because of the technical similarities involved in wrapping the two products (not to mention the fact that many stores that bought large quantities of one also bought the other), a company such as the Hygeia Antiseptic Toothpick Company also manufactured wrapped straws.[16]

Wrapping such things as straws and toothpicks has long been accomplished without using glue or paste, and we are all familiar with the crimped edges of the small packages.[17] The paper has often been indistinguishable from that used to roll cigarettes, which in the early days of individually wrapped items led to disputes between toothpick manufacturers and customs officials over what duty was owed on their importation.[18] The wrappers can also be imprinted with an advertisement, of course, but even unwrapped toothpicks can carry the name of a restaurant, hotel, or other establishment or business. One machine for imprinting was patented in 1905 by Howard Barlow of Providence, Rhode Island. His apparatus pro-

duced flat toothpicks with a broader-than-usual center, thus allowing ample space for a name or message. Barlow's patent argued that
the resulting toothpicks provided a "very neat and novel means of
advertising."[19]

But Barlow's idea was not completely novel. In the summer of
1900, Boston councilman George R. Miller was looking ahead to
running for alderman the following year and began "distributing a
small leather toothpick holder filled with toothpicks, which his con-
stituents doubtless will highly appreciate." According to the *Boston
Daily Globe,* the holders were inscribed on one side with "Ninth
aldermanic district, wards 16, 20 and 24; vote for George R. Miller,
1901." The other side of the holder supported his plea: "Experience counts; George R. Miller, two years in the common council,
1899–1900; for alderman, 1901." In the unlikely event that the
users of the toothpicks might forget the name of the politician to
whom they owed their thanks or the office into which they were to
vote him, the toothpicks themselves were inscribed "Alderman,
George R. Miller." Some of the other candidates for the nomination threatened "to make the toothpick an issue in the campaign"
and debated the toothpick's "merits and demerits from the esthetic
as well as political point of view."[20]

Toothpick holders and cases have been used for other than
political advertising, and questions of aesthetics and taste seem
always to be an issue. In the tradition of corny slogans imprinted on
advertising giveaways of all kinds, one pocket toothpick holder was
emblazoned with "Pick a Pontiac from Munson Pontiac" of Marion, Indiana. In Pennsylvania, it was once possible to obtain quill-
like toothpicks made of celluloid that were imprinted with the
slogan "Pick Your Teeth Then Pick Your Vocation at the Carlisle
Commercial College."[21]

But most toothpicks themselves remain unsullied with advertising; that is reserved for their wrapper. Many restaurants today seem
to have done away with the Schupp-type toothpick dispenser and
replaced it with a holder full of wrapped toothpicks, perhaps thinking it a more sanitary solution. However, as anyone who has tried to
take a single transparently wrapped toothpick from such a supply
has no doubt discovered, it is very difficult to remove only one
without pulling out some additional ones. This leads to customers

taking more than one, spilling some onto the counter or floor, or returning the extras to the bowl or saucer. One means of restoring order to the counter was patented in 1971 by Allen Harriman of Lewiston, Maine, who assigned rights to Forster. His idea was to load a dispenser with a "continuous web" of wrapped toothpicks, somewhat like a roll of paper towels, and as one wrapped toothpick was removed from the dispenser it would pull the next one into place. According to a 1977 classified advertisement seeking distributors and exporters for wrapped mint-flavored toothpicks, wrapped picks had become a public necessity.[22]

It is not only toothpicks distributed in public places that are subject to unsanitary conditions. Inventors have also worried about toothpicks carried loose in the pocket or purse, where "there exists a sanitary problem" and, furthermore, where the tip can be broken off.[23] This was seen as a problem as early in the history of mass-produced wooden toothpicks as 1883, when Gustavus Schimmel of Detroit patented a "tooth-pick package" that consisted of a simple piece of fabric to which "any of the ordinary toothpicks" could be "caused to adhere," not unlike the way the moveable wooden slats do on a roll-top desk. With the toothpicks in place, looking as if they were only partly cut out from a strip of veneer, the piece of fabric could be rolled up and put in the pocket.[24]

While split or sharpened wooden matches had long been used for toothpicks, cardboard book matches did not hold up to the task, though a corner of the matchbook cover itself will do in a pinch. The idea of packaging toothpicks like matches has been an enticing idea. In 1936, the Long Island, New York, inventor Joseph Doll patented the idea of coating cardboard with a material that would harden sufficiently to serve as a toothpick. The "book of tooth-picks" could be fashioned in one piece and folded to be made into the book, thus keeping the picks relatively protected and clean until one was torn off for use.[25] A similar concept, but with the item made out of plastic, was patented in 1956 by Robert Briggs of Massachusetts. Arguing that toothpicks of molded plastic would be too brittle and those of extruded plastic would be too flexible, he proposed giving the latter a "longitudinal crimp" to stiffen them sufficiently. This caused them to look like miniature dugout canoes.[26]

In a still further variation on the book of toothpicks, William

Cameron of Connecticut incorporated a variety of tip shapes into the book, each being differently suited to the "location one desires to gain access to for the removal of foreign matter from the teeth."[27] Patrons of Planet Hollywood have been able to pick up a matchbook of short plastic toothpicks, albeit all being of one shape, with the imprinted cover reminding them of their visit to the theme restaurant.[28] In toothpick books, the plastic items are either formed integral to the book cover or stapled into a cardboard one the way matches are.

Another way of keeping toothpicks clean in the pocket was conceived of by three Japanese inventors, who patented a portable toothpick case. It is flat like a credit card and appears to operate on a principle somewhat related to that of a Pez dispenser. One inventor solved the problem in a different way. He patented a reusable "antiseptic toothpick" that is encased in a hollow cylinder filled with antiseptic liquid through which the toothpick is telescoped out for use and in to be cleansed for the next use.[29]

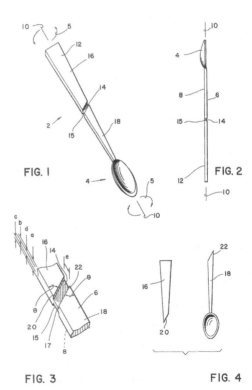

By scoring the handle of a plastic eating utensil, one inventor provided a means whereby a toothpick could be broken off after eating. The remaining part of the spoon's handle could serve as a second toothpick.

Since a toothpick is needed most just after eating, combining one with an eating utensil must have made perfect sense to the inventor John Feaster of Orange, California. His idea was to score deeply into one end of a Popsicle-stick-like implement the shape of a toothpick that could easily be punched out and separated, leaving a two-pronged fork of the kind that has been customarily provided with orders of french fries bought at the beach or at a fair. His patent, granted in 1959, suggests that the implement be made of wood.[30] George Adolfson, of Largo, Florida, incorporated a toothpick into a plastic beverage stirrer in the form of a small spoon. The device was formed with a deep score diagonally across the handle, so that a pointed corner would result when the end was broken off. Because the bowl end of the broken stirrer would also have a pointed corner, it could serve as an additional toothpick. Hence the inventor's description in his 1979 patent of a stirrer that was "capable of being broken into two toothpicks." Another variation on the idea incorporates a more conventionally shaped toothpick into the handle of a plastic fork. The Brooklyn inventor Mark Sanders has designed a set of plastic eating utensils with hook-shaped toothpicks incorporated into the handle. Although such implements have not been readily available in America, they have been reported to be used in Korea.[31]

Sometimes a toothpick is needed after eating food that does not require that any utensils be used. Snack foods such as peanuts, pretzels, and especially popcorn are certainly in this category. Perhaps out of her own frustration with such a situation, Lisa Bell of Covington, Kentucky, designed popcorn packaging that incorporates a toothpick. For more formal dinners and special occasions, when you want to "add a touch of Oriental beauty to your table top," there are colorful "hand folded origami holders" each containing Japanese toothpicks. It is also possible to buy personalized toothpick packs for use as wedding favors. These have been advertised on eBay with the luring invitation to "be the first" to have them for all the guests at the rehearsal dinner and reception.[32] The ingenuity of inventors and marketers seems to know no bounds of imagination or decorum.

CHAPTER TWENTY-SIX

Talking Round a Toothpick

CONSIDERABLE AMBIGUITY about toothpick use remained in fin de siècle America and would persist well into the twentieth century. Entire classes of society were characterized by their toothpick use, as they were by their use of other products and practices of hygiene—and by their embrace of politics. In the early twentieth century, Lady Gregory and William Butler Yeats defended their Abbey Theatre production of John Millington Synge's *The Playboy of the Western World*, in which Irish peasants were depicted as "less than holy." Young religious revolutionaries rioted against the play's "mocking ironies and wild paganism," which attracted opposing demonstrators from Trinity College, led by Lady Gregory's nephew. Of the ruckus she wrote to Yeats, "It is the old battle between those who use a toothbrush and those who don't."[1] She may just as well have been speaking of a toothpick—and how it was used.

A 1901 essay on "democratic gentility" was prompted by an incident in which "an eminent naval officer opposed the promotion of warrant officers on the ground that they lacked social qualifications." He was called a "snob," a "coward," and a "conceited ass" by a U.S. senator "all aglow with the spirit" of democracy: "Everybody knows that manners, family habits, clothes and like irrelevancies down to the smallest details of toothpick and napkin management are the chief bonds or barriers between men and between nations; that snobbery in one form or another is eternal and omnipotent, and bigger than humanity itself."[2]

Contemporary evidence for the truth of such a statement was not hard to find. A passenger on a steamship cruising the Great Lakes spent time sizing up his fellow travelers. His eyes fell on a dark-eyed, dark-haired girl whom he recognized "at once as a type of the rustic Canadienne." She was sitting in a chair, when he observed her draw up another chair, on which she put her feet, then leaned back and lit a cigarette. Soon, another woman, an "austere lady," came in from the main saloon, and she also appeared to have a cigarette in the corner of her mouth. But in fact it was a wooden toothpick. *Ladies' Home Journal* certainly did not approve of such behavior, dictating that "properly trained" people did not "use a toothpick at the table, or in the presence of another person." *Methodist Review* advised the properly trained "up-to-date young man" that whereas "a toothpick should be indulged in only in that spot to which Scripture enjoins us to retire when we are about to pray, a meerschaum pipe is a perfectly well-bred article for public wear, and one which enables him to fulfill agreeably that law of his being which suggests that he should always be putting something in his mouth."[3]

The use of toothpicks by women came in for special criticism. One octogenarian, writing about life as he remembered it to have been in the nineteenth century, considered toothpick use to be "of recent origin" and had strong opinions about it: "In Boston on any day between twelve and two o'clock, nearly every third woman met in the vicinity of Winter and West streets, has a tooth pick between her lips. This practice is made more vulgar when at table the hand is held over the mouth, for thus its vulgarity is acknowledged by those who persist in it."[4]

In restaurants at the turn of the century, toothpicks were ubiquitous, but how they were presented varied greatly, as did how the presentation impressed diners. A young couple moving from Boston to the West found the food in a Chicago hotel "most satisfactory," but the service and setting were wanting, since "they always insisted upon serving toothpicks with the finger-bowls, no matter what your age or previous condition may have been."[5]

In 1907, *Ladies' Home Journal* provided some guidance for the business girl. The aspiring clerk was told to greet and address customers by name, approving of "Yes, Mrs. A," but not the "tabooed" form, "Yes, lady." According to the magazine, " 'Lady' goes hand in hand with chewing toothpicks in public. Let us drop both." But not everyone did. In an etiquette column, a writer described a beloved woman who had few faults, but among them was "using a toothpick without the slightest effort at concealment, and making the most dreadful sounds in otherwise cleansing the teeth."[6] On another occasion, the magazine carried a warning to girls who were careless in their dress and manners when they thought they could not be observed by anyone who knew them. The advice giver reported seeing a "woman of international reputation riding in the street car with a toothpick in her mouth, which she used diligently from time to time." Readers were reminded, "Too careful attention cannot be given to the toilette of the teeth, the nose and ears before leaving your own room."[7]

Whatever was put in the mouth, there were those who insisted that it was wrong to fish out its residue with a toothpick. One writer found poetry in legumes, and a metaphor for more prosaic things:

> Try a peanut. Ah! you smile and say that they are vulgar? Well! that depends entirely upon circumstances. Toothpicks and finger-nail files are often fine works of art, but when you use them as you stand in the queue, waiting to struggle to your seat to see Bernhardt, or at the street corner watching for the next trolley, there are some ill-natured persons who will say that the dainty things are vulgar. It is merely a mistake in logic, putting the blame on the instrument instead of on the agent. Peanuts are good food.[8]

Sometimes neither the agent nor the instrument is to blame, but the times. As far back as the early 1880s, styles of footwear with sharply pointed toes were in vogue, and they were called "toothpick shoes," or simply "toothpicks."[9] But they came in and went out of style as quickly as they have in more recent times. In 1884, contemporary observers referred to the "modern toothpick shoes so much affected by the dudes" and speculated that people of an earlier time would have thought they were made for a "curious kind of human beings." Just after the turn of the century, hints of the "old-fashioned" toes were reappearing in women's shoes but were "not looked forward to with any pleasurable sensations." At about the same time, the latest in men's shoes were also returning to a point, "though far from the old tooth-pick toes."[10] Even the Transcendentalists were enlisted in the debate: "If Thoreau tried to put his feet into a pair of 'toothpick' or French toed shoes to find that they hurt or pinched, he would tell the whole French nation that he did not like them. If there were not other shoes he would make some for himself." Years after the fad passed in America, toothpick shoes imported from Mexico and Europe were described as having a comical look about them. In 1926, an advertisement for a new book, *The Elegant Eighties,* asked, "Do you REMEMBER? The Tooth Pick Shoe?" The style was described as one of many "hilarious recollections" of life forty years earlier.[11] The style returned in America in the early twenty-first century, though generally without reference to the toothpick. In Liverpool, England, a boot with an "extreme point for extreme people" was being sold as a "winkle picker," a style and name that first became modish in Britain in the 1950s.[12]

Fashionable people have always gone beyond fashion. In 1905, the humor magazine *Puck* reported that at Mrs. Astor's ball favors for the six hundred guests "included diamond-studded knives and forks, hand-painted napkins, silver toothpicks, and soda-mint tablets." The rest of the world had to make do with lesser favors. On another occasion, *Puck* noted that when timber was scarce the imagined Richleighs were said to be very lavish entertainers when at the close of an evening banquet "each guest was presented with a solid wooden toothpick." There were always those who had not fancy silver souvenirs in their pockets but plain wooden toothpicks in their mouths. At about the time that Mrs. Astor was throwing her ball, *Puck* also noted that a Bishop Potter was defending a person's right to worship in shirtsleeves, and the natural extension of such informality was to allow him also to "say his prayers with a toothpick in his mouth if it pleases him to do so."[13]

Some people do not feel comfortable without having something in their mouth. An early-twentieth-century book drummer, "one who travels about drumming up and making sales of literature," was described as "talking round a cigar that had apparently grown fast to his mouth." In his zeal for selling, he made the distinction between a novel as merchandise and one as literature. The former was something to stock in a hundred copies and the latter in one. At some time during his spiel, the cigar had been supplanted, and the drummer began "talking round a toothpick."[14]

A literary press agent of the period confessed to using newspaper stories to gain publicity for books—mostly nonfiction merchandise. There were many cooperative authors, and the more controversy their books stirred up, the more books they sold. One writer, after a story appeared about the topic of his latest book, asked "all his friends to write letters to the newspapers about it." The tactic, which was reminiscent of Charles Forster's scheme for getting retailers to stock toothpicks, was quite effective in generating a fuss and thus interest in a book that might otherwise have been lost among better ones. And the topic did not much matter, according to the agent. A fabricated "red hot" discussion might be started "with the merest trifle—the use of toothpicks in good society." This was likely to attract considerable attention, for just about everyone used the common toothpick at home, and a shopping list might easily include, among the milk and eggs, "one box tooth picks."[15] It would be a box of wooden toothpicks, of course, and very possibly left open on the table at home. One rural versifier confessed to loving to dine among "the ultra swell," but he was disappointed when waiters would not give him crackers for his purée or consummé "served in thinnest Haviland." He lamented the situation and sought his simple pleasures elsewhere:

> Let me go back to Ruralville
> Where simple viands are in place,
> Where toothpicks are in fashion still
> And menus show of French no trace.[16]

The toothpick seemed not to be a threatened species in the more domestic restaurants. In a story that appeared in *McBride's Magazine,* about "a movie actress who did not have a Mary Pickford income," the starving young woman was sitting in a cafeteria, debating with herself about whether she could afford to spend a nickel. She watched the line of people passing the cashier's counter, "helping themselves to tooth-picks as they went, and throwing down nickels for packages of spearmint-gum." She finally relented, and consumed so much food that she found herself forty cents short of the tab. She sat at her table until the cafeteria was nearly deserted and, when she saw that the cashier was away from her station, walked past the counter and "made a motion as if to lay down the check and the money." In doing so, she was nervous, of course, and her eyes appeared to play tricks on her, magnifying her act; "as she withdrew her arm the convex lump of toothpicks swam before her eyes" and seemed to shout, "Thief, thief!" She ran out the door, dropping the check as she rushed to escape. As happens in romances, a young man of her acquaintance saw her running away, picked up the check, and paid her bill. In typical *McBride's* fashion, the couple did, of course, eventually get together and presumably lived happily ever after.[17]

The fast-food restaurant may seem to have been a latter-twentieth-century invention, but the concept of a quick meal was nothing new. Early in that century, there was a proliferation of "quick-lunch rooms," which were thought then to be a "peculiarly distinctive American institution." At the time, the quick-lunch room was said to be "the creature and the symbol of that hustle and hurry which has enabled the American business man to do ten times as much work in a decade as can be done by all his foreign competitors together." But it was also decried as the "great mart in which food adulterators turn into gold the honest hunger of innocent men" and the source of "dyspeptic death." In short, it had become as fashionable then as it later became again to bad mouth these establishments. Still, they had their defenders and frequenters, both of whom thought they provided a good value. Although they were evidently individually responsive to local tastes, collectively the quick-lunch rooms shared common traits: "Every one of them claims to serve unsurpassed coffee, every one makes pie the chief article of diet, and every one furnishes toothpicks ad libitum."[18]

Before World War I, the public use of toothpicks offered in American cafeterias was coming under increasing fire from those more likely to eat in French restaurants. One letter to the editor of the *New York Times,* while approving generally of an editorial on the "use for napkin rings," lamented the absence of a postscript on "the future of the toothpick." The writer would like to have seen the wooden kind and the "quill inclosed in the dainty antiseptic package" both "shelved with the Saturday night bath and spelling matches."[19] In fact, the future of the toothpick would be much like its past, and class distinctions would determine it as much as it defined them.

Politicians and sycophants have long come to Washington from a wide variety of geographic regions and socioeconomic backgrounds—and they have often been cheek to jowl. During the Spanish-American War, Richmond P. Hobson, an assistant naval constructor, had become the "Hero of Santiago de Cuba," having survived a suicide mission designed to detonate and sink the explosives-laden coal-supply vessel *Merrimac* and block the entrance to the harbor, thus corralling the Spanish fleet. The *New York Times* editorialized on heroism and pride and quoted from a letter that his mother wrote proclaiming that "it took no Santiago to proclaim him a hero" to her.[20] When President McKinley referred to Mrs. Hobson informally as "Mother Hobson," it raised the ire of the social commentator Harry Thurston Peck. He deplored a head of state "departing from a high standard of linguistic propriety and descending to anything that is cheap and common" and compared the base locution to a low-class habit:

> If a President in the midst of the inauguration ceremonials should cock his hat over his left ear and chew a wooden toothpick, he would deserve, as he would certainly receive, the severest sort of censure not only from his political opponents but from his mortified and disappointed friends as well. Now when Mr. McKinley spoke in public of an estimable lady as "Mother Hobson," he was verbally and figuratively cocking his hat over his left ear and chewing a wooden toothpick, and the effect of it was most unpleasant.[21]

In spite of what Peck thought of his president's verbal manners, McKinley was a careful dresser. During his first administration, he was described as "one of the best-groomed men who have ever been in the White House," yet it was not uncommon to find someone seeking a position in the government to be "picking his teeth with a quill" while he waited in line to plead his case to the president. But it was an improvement over the "days when Senators and Congressmen and Cabinet officers chewed wooden toothpicks in public and spat tobacco-juice with equal facility at every point of the compass." When running for office, McKinley was said to have suggested there be "a toothpick plank in his platform, but nothing came of it."[22]

William Jennings Bryan, the Nebraska politician and newspaper editor who lost to McKinley in the presidential races of 1896 and 1900, was himself an inveterate toothpick user. So was his unsuccessful competitor for the 1896 Democratic nomination, Richard P. Bland, a congressman from Missouri who threw his support behind Bryan. On one occasion during the campaign, the populist Bryan was traveling with his family in a rail coach, having refused the railroad's offer of a special train. When the train stopped in East St. Louis before crossing the Eads Bridge from Illinois into Missouri, Bland joined the party. At the first stop across the river, Bryan and Bland stood together on the railroad car platform, and Bryan graciously told the assembled crowd that his rival had been "really much more entitled to the nomination" but that circumstances had decided differently. According to a report of the incident, "Mr. Bland stood at Mr. Bryan's elbow reflectively chewing a big quill toothpick."[23]

The first race between Bryan and McKinley was hotly contested and prompted a lot of election bets. Naturally, there were winners and there were losers, one of whom had to "roll a peanut a mile with a toothpick."[24] The 1900 presidential race between Bryan and the incumbent McKinley was just as hotly contested, and even more creative bets were made. One of them was described as "the most peculiar bet ever put on record." It was reported on even before the winner was known:

> In the case of Bryan's election the McKinley advocate is to march to the foot of a steep hill which rises for more than a mile near the city limits. He is to be escorted by the Republican campaign drum corps, by his successful rival, and by as many of the townspeople as may care to attend. Arrived at the bottom of the hill the task before the losing politician is to roll a peanut from the base to the top of the hill with a toothpick. He is obligated not to touch the peanut with anything but the toothpick and must stay in the beaten road during the entire journey. In case he does not get the peanut to the hilltop in a single day's work he must sleep on the ground under guard and begin again in the morning.
>
> The same penalty will be paid by the Democrat if Bryan is defeated, and it is noticed that the Bryan advocate has recently been taking golf lessons and showing in other ways his fear that the work will fall upon him.[25]

And it did. Another lost bet involving toothpicks and peanuts required a Bostonian to use a toothpick to roll a peanut "from the state house on Beacon St. entirely around the common." The event was scheduled for the Thursday after election day, rain or shine, with the roller allowed to be rubbed down every twenty minutes. A caddy was to carry extra toothpicks, and extra peanuts were to be on hand "in case the original one may be lost in a sewer."[26] But putting the loser on the ground with a single peanut and a single toothpick associated him with the small and insignificant, adding insult to injury.

After his second unsuccessful presidential campaign, Bryan began editing the *Commoner*, a weekly newspaper. Though he himself was known as "the Commoner," the quill toothpick that he, like Bland, used might have set him somewhat apart from the wooden-toothpick crowd. Bryan's habit of chewing the quill became a minor issue when he was preparing to make his third run for the presidency. In the summer of 1906, upon returning to America from a long foreign trip, the American Anti-Toothpick Society, formed by a group of women aboard his ship, let it be known that Bryan's habit was more distasteful than most, for he used the same toothpick well beyond its usefulness. According to a report in the *New York Tribune*, "he gets a good goosequill toothpick and he cultivates an affinity for that particular quill. He keeps up a partnership with the toothpick until it no longer possesses resiliency." The report continued by noting that to Bryan "a good quill toothpick is as dear as bimetallism and lasts almost as long."[27]

"Bimetallism" was the term introduced to designate the monetary standard of employing a predetermined ratio of two metals, such as gold and silver, so that each could constitute legal tender without the different coins changing their relative value. It was obviously a practice that, once begun, should have had a long life. However, in 1873 the U.S. Congress had demonetized silver, which angered western silver-mining interests and debtors, who would have liked to have had the option of paying off their debts with silver rather than gold. A compromise of sorts was reached in 1878 with the passage of the Bland-Allison Act, which obligated the U.S. Treasury to purchase at market value each month between $2 million and $4 million worth of silver from western mines and mint it into silver dollars that would be legal tender. President Rutherford B. Hayes vetoed the bill, but Congress overrode the action. The Hayes administration complied by purchasing and minting only the minimum required by law. When prosperity returned in the 1880s, the issue subsided and the nation continued in a monetary state of "limping bimetallism."[28]

Proponents of a more liberal policy toward the lesser metal came to be referred to as "silver knaves," and Bland and Bryan were counted among their number. The issue became prominent again in the 1890s, when the country found itself in a serious depression. The "free silver" advocates were principally Democrats, and Bland led the fight in the House of Representatives. His lieutenant on the floor was Bryan, who had become the representative from Nebraska by advocating tariff reform and the free coinage of silver.[29] It was at that party's 1896 convention that Bryan gave his famous "Cross of Gold" speech, in which he contrasted the "idle holders of idle capital" and the "struggling masses, who produce the wealth and pay the taxes of the country," concluding his peroration with this admonition: "You shall not press down upon the brow of labor this crown of thorns, you shall not crucify mankind upon a cross of gold."[30] It was largely this speech that gave the nomination to Bryan over Bland.

Years later, when he returned from that long foreign trip, bimetallism was thus still a perfectly apt metaphor to use for Bryan's enduring toothpick habit. On his arrival in New York it was said of him:

He found the right kind of a quill toothpick in St. Petersburg last winter and carried it all the way along the route. He chewed on it when he shook hands with the Mikado in Japan, stuck it back into his vest pocket after hobnobbing with all the satraps of the Orient, and kept it in service until he reached the Battery.

On the way over on the Princess Irene the old quill toothpick became a social and political issue. It had become frayed, splintered and unlovely. Mr. Bryan became gracious, as usual, with the women on the voyage home. The old quill also was presented in polite society, but was cut dead.

On its last appearance it looked like a linen dish rag. It had arms and legs, which were spread out like the drying pelt of a raccoon on a woodshed door. Moreover, it had lengthened. It had become so elongated that once it was mistaken for a book agent's sample volume and at another time it was mistaken for a dried fish from Japan.

Mr. Bryan, however, did not dissolve the partnership. Like other "old saws" it seemed to be indispensable. The climax was reached when Mr. Bryan "drew" the quill after lunch on Wednesday. It was mistaken by one of the crew for an old whaleman's harpoon which had been stowed away in the hold. Mr. Bryan was asked not to remove the machinery from the ship.[31]

According to the women who had founded aboard the ship the American Anti-Toothpick Society, such behavior should have prevented Bryan from ever becoming president, for Article 1, Section 1 of their organizing document "disqualifies politicians from citizenship when they use a quill toothpick more than one week." When Bryan was sent a copy of the statement, his only comment was, "I stick to my friends. Our sisters can't vote, anyway." Whether the 1908 election turned on bimetallism or the toothpick issue, Bryan lost his third bid for the White House, this time to William Howard Taft. The rotund politician's popularity had led to the quip "Everything looks like Taft." *Puck* rejoined, "No, not everything. For instance, there are flag-poles, tooth-picks, . . ."[32]

At the 1912 Democratic convention, Bryan played a significant role in getting the presidential nomination for Woodrow Wilson. In appreciation, President Wilson appointed the pacifist Bryan secretary of state, a position he held until 1915, when he resigned in protest of administration policy in the wake of the sinking of the *Lusitania*. (The potential effect of his pacifism on foreign policy had been alluded to years earlier when the *Los Angeles Times* editorialized, "If Bryan were to become President the Roosevelt big stick would look like a toothpick.") Today, Bryan may be best remembered for opposing Clarence Darrow in the so-called Scopes monkey trial.[33]

By the 1920s, unhealthy, unsanitary, and uncouth toothpick practices were considered by some at least to be largely a thing of the past, but this did not spell the end of toothpick-related concerns for America. One doctor, who wrote a newspaper column called "How to Keep Well," would become convinced that poor teeth were bringing the nation "considerable ill health, incapacity, and discomfort" and looked for someone to "put the toothpick into good society." As the second administration of Woodrow Wilson was drawing to a close, the doctor had given up on Mrs. Wilson being the one to do so. However, with the inveterate toothpicker Warren G. Harding the president-elect, the doctor wondered if Mrs. Harding would "be sure enough of her position to set a toothpicking example for the nation."[34]

Whether for hygiene or pose, in the 1920s the better-off everywhere continued to use toothpicks of silver and gold, if lost-and-found notices are any indication. One P.S. offered a fifty-dollar reward, "no questions asked," for the return of his monogrammed watch, chain, medallion with big garnet, and gold toothpick, which he evidently lost on the Upper West Side of New York.[35] At the same time, American dentists "pronounced against the use of the toothpick," presumably of any material. By the 1930s, toothpicks made of precious metals, at least, were becoming scarcer. An advertisement for imported silk neckties boasted of the wide variety available, capable of reflecting moods "from jitters to jubilation," and stated that "duplicate designs are as rare as gold toothpicks."[36]

The cultural differences between European and American toothpick practices continued to shock even some sophisticates. In 1928, Gerard Lambert entered his three-masted schooner, *Atlantic,* which held the sailing-vessel record for crossing the ocean, in a race between New York and Spain sponsored by King Alfonso XIII. Lambert and his family were invited by the king to the palace in Santander, where they had several formal dinners. At these, the well-traveled Lambert, who was in his forties, witnessed something he had "never seen before or since. Each guest had by his plate a gold toothpick. This was not a decoration. It was used vigorously and with evident delight," confirming that Spanish society had been immune to anti-toothpick movements.[37]

In the late 1950s, retractable gold toothpicks were still offered as gifts for gentlemen. Some seemed to carry them into the 1970s. A "gold pocket-watch fob with sharks-tooth & gold tooth-pick" lost on the Upper East Side carried a fifty-dollar reward. However, at the same time gold toothpicks were advertised as "the unusual" gift "for the man who has everything." The picks came in a leather case and could be monogrammed or fitted with an inset diamond. In whatever style, they did "solve a delicate problem in a brilliant way!" albeit one that was millennia old.[38]

Less costly toothpicks served for many unrelated purposes. The American diary of a British novelist recalled playing a literary game called Tittleteetum, in which players were to guess a poet's name by the arrangement of toothpicks representing famous lines of poetry. When he had played it as a youngster, the toothpicks were "borrowed from the servants' quarters." Once, his father caught him using a pick for its intended purpose and was dismayed, saying he would rather he had had "six drinks of Scotch than use one toothpick." A "former Kentucky feudist" who had been convicted of killing a federal agent was described as remaining stoic when the death sentence was imposed upon him. As he listened to the judge, the killer's "inevitable toothpick hardly wavered in his hand."[39] But sometimes the toothpick was on the bench. Florida circuit judge Truman Futch was known as "the Whittler." The nickname seemed apt, for during a preliminary hearing in a racially charged case in which three men faced the death penalty for raping a woman the judge "whittled several sticks of wood down to toothpicks."[40]

In the late 1960s, an observer of toothpick use was offended by what he saw some young women doing on the elevator in a modern Midwest office building during lunchtime.

All of them had toothpicks and were industriously engaged in plumbing the interspaces of their molar structure. Well it was a sight to sicken even the strongest man . . . it was ghastly. These tiny bits of wood had transformed them from lovely young ladies into plebeian savages. I've got a fairly strong stomach, but it was almost too much for me. Why hadn't their parents taught them that picking the teeth in public is as socially unacceptable as running around in the nude?[41]

Amy Vanderbilt would no doubt have sympathized with him. At midcentury, she drew a sharp distinction between European and American practices: "In Europe if a bit of food catches in one's teeth at dinner it is quite proper to remove it adroitly with a toothpick, using a table napkin as a screen. In America, however, one suffers. If you can't dislodge the offending bit with your tongue (and even such maneuvers must be unnoticed by the assemblage), leave it there until you can remove it in privacy."[42] The idea of asking for a toothpick in America seemed to be out of the question, but twenty years later, the maven of manners refined her advice. Perhaps to assuage suffering Americans, she added that "many cultured Europeans" also suffered food stuck stubbornly in the teeth. However, under a section titled "Food in the Teeth," she allowed:

> If food gets wedged in the teeth and can't be dislodged with the tongue, don't use a fingernail but on leaving the table ask for a toothpick. Dislodge the offending bit in private. If this is a frequent problem it is wise to carry your own emergency equipment. There are little triangular toothpicks in packets like matches. They are called Stim-U-Dents and are available in drugstores. They are handy to have.[43]

Letitia Baldridge, who revised and expanded on Amy Vanderbilt's advice, added, "If you have nothing sharp like a toothpick with you, often the rinsing of your mouth vigorously with water will correct the situation." Baldridge also let it be known that "one thoughtful hostess who throws large barbecues all summer keeps pretty little jars of toothpicks in all the bathrooms, for use by her guest after their corn feasts."[44]

However, the idea of chewing on a toothpick has for some time now remained something not approved of by those who considered themselves of the better classes. When a new appointment to the Eagleton Institute of Politics at Rutgers University was announced in 1974, he was described as displaying a "total intellectual involvement with political science" while having "earned a reputation as a maverick," which stemmed in part from his "outlandish appearance." This included "chewing toothpicks as an antidote to smoking."[45] In the latter decades of the twentieth century, the toothpick was well established as a prop for negative characters. A UCLA professor of social psychology associated with the school's Afro-American studies program wanted to avoid clichés in a film he was producing for television. He was disappointed in a "hoodlum scene" in which he found "blacks cast in negative stereotypes, complete with 'head rags, toothpicks and mean demeanors.' "[46]

Not everyone objected to using the toothpick to enhance the believability of a character. The actor Taurean Blacque began dangling a toothpick from his mouth when his singing coach told him that he had to give up smoking cigarettes, and he used it to his advantage when he played a junkie. One drama reviewer admired his "confidently drinking from a can of beer without taking a toothpick out of his mouth." Blacque adopted the toothpick as his "trademark," saying people hardly recognized him without one in his mouth. He confessed to keeping toothpicks "in all my suit pockets."[47]

Baseball players have had a long association with toothpicks, from a tall, lanky pitcher described as an "elongated, attenuated toothpick twirler" to a shortstop who "played every baseball game with a toothpick in his mouth, batting, fielding and sliding head first to break up double plays." The pitcher was Bob Groom, who played for the Washington Senators (among other teams) early in the twentieth century, and the shortstop was U. L. Washington, of the Kansas City Royals in the late 1970s and early 1980s. Washington said he was "just carrying on a tradition," because his father always used a toothpick, and the habit relaxed him. However, he stopped the practice when mothers expressed concern that their children might emulate him. In place of a toothpick, he began chewing gum and sunflower seeds.[48] Ted Williams, the legendary Boston Red Sox outfielder and the last player on any team to bat over .400 for a season, was known as the "Splendid Splinter," itself a long-used sobriquet for a toothpick—and for a bat.

Aside from sports writers, journalists and critics generally seem to look down their noses at toothpick use. In a review of the documentary film *Roger and Me,* about General Motors and unemployment, the filmmaker Michael Moore was portrayed as displaying a "down-home wonder," which was presumably reinforced by his having "a toothpick stuck in the corner of his mouth, wearing a down jacket, jeans and the sort of cap that should have the name of a feedlot on it." Elsewhere, a buyer at a 4-H livestock auction in Minnesota was characterized as having "a toothpick in his mouth and bifocals on his nose." An Arkansas truck driver was described entering a rest stop "with a red water jug in one hand and a toothpick in his teeth."[49] It is not only homey rural and working-class folk in America that toothpick use characterized, for "carrying a toothpick almost permanently in the mouth" may be considered "jaunty and macho in several Mediterranean societies." In America, the equivalent may be classified as "cool." But everything is a matter of degree, as illustrated by the practice of "having a single toothpick dangling out of one corner of your mouth. This is cool.

But stick a second toothpick in the opposite corner and you've gone all the way round to uncool."[50]

A late-twentieth-century book on table manners declared, "Toothpicks, fairly successfully banished in England and America, have never been entirely rejected from the European Continent; it would be interesting to know just who uses them today, when, and what the strictures are."[51] Certainly they are employed or improvised by more people than would care to admit to the behavior; they are used whenever and wherever toothpicking can be done out of sight, or at least out of mind; and they are generally accepted within the bounds of what currently passes for etiquette. Charlotte Ford, writing for the twenty-first century, advises in snappier language what can be done if a piece of basil from the pasta gets caught between the teeth on a first date: "Never use a toothpick at the table, which is akin to using a fork to clean your nails. Toothpicks are a strictly private matter. Take a sip, not a swig, from your water glass. If you feel this has not dislodged the pesky perpetrator, cover your mouth with your napkin and attend to the matter. Discretion is key, as it would be in poor taste to be overtly witnessed in this manner."[52]

A somewhat opposing view was presented by Professor Pompilius McGrath, also known as the Professor of Etiquette, who in an eponymous book for young readers appeared on the New York scene in the early 1990s with a new lecture, "The Toothpick: This Misunderstood Little Friend," which was purportedly delivered to a standing-room-only audience at Madison Square Garden. The lecture began with a recitation of "no-nos of human behavior," which concluded with "using the nail of your little finger to remove food from between your teeth." After thunderous applause, the professor announced a half-hour intermission, during which the audience would "be served some steak and a toothpick." They were told that it was imperative that they eat, since their having done so would be part of the lecture, but they were also told not to use their toothpick.[53]

When the curtain went up after intermission, the professor was onstage finishing his steak. He asked the audience if they were comfortable after eating theirs, or if they were bothered by pieces of meat stuck between their teeth. He said that he did not have any annoying leftovers between his teeth, because after his meal he had used a toothpick, which he held up for all to see. (It was a quill!) He had applied his "misunderstood little friend" so discreetly that none of them had noticed. He then pointed to illustrations of the dos and don'ts of toothpick use. In the first, a man was sticking a toothpick into his wide open mouth, thus calling attention to its use. In the second, a woman was covering up her toothpick with her other hand, "broadcasting" what she was doing. In the third image, a man, "completely casual, with the utmost nonchalance," was employing a toothpick in a way that no one noticed, according to the professor. This is what he had done before their eyes.[54]

The audience members were then invited to take up their toothpicks, with the reminder that it would require some practice to get it right. The lights went out and everyone proceeded to work the toothpicks, as evidenced by the faint sounds of clicks and blips that filled the theater. When the lights went on again, the professor asked how everyone felt now, and they responded with "thunderous applause of appreciation." With that the lecture proper came to an end, but the professor went on to present "an erudite albeit condensed history of the toothpick," after which he was carried about the Garden's aisles in triumph. On the following day, "New York's most prestigious newspaper," the *Etiquette Times*, reported the lecture on the first page. This was "to the relief of all its readers, who for years had sacrificed their comfort to the mistaken belief that the toothpick meant bad manners."[55]

A fictitious professor of etiquette may have been able to use a toothpick without anyone noticing, but few others are able to pull that off. According to one observer, "Few men or women, unless they are trained, professional magicians, can use a toothpick discreetly." Even the once polite gesture of raising a newspaper or napkin to screen the activity had long fallen out of favor.[56]

Some advice columnists seemed hesitant to endorse or decry the toothpick, perhaps fearful of offending one or the other segment of their readership. A woman whose boyfriend had "recently developed the habit of keeping a toothpick in his mouth constantly"— chewing on it, swiveling it around, and sucking on it—asked Ann Landers only to confirm that it was a dangerous thing to do. Ann complied but did not discuss the propriety of the habit. Ohio Toothpick Lover, another reader, described how a toothpick changed her life, after fifteen years of smoking. Instead of using cigarettes, she now chewed toothpicks, having one in her mouth constantly. Ann congratulated her on kicking the nicotine habit but made no comment on the new one.[57]

Etiquette experts have generally been more direct. Miss Manners, writing in 2001 on toothpick use, noted that although currently considered "vulgar," the practice "has not always been socially condemned," and that there have been "times and places when the visible use of the toothpick has rallied formidable support." As evidence, she offered the fact that "artifacts exist that attest to periodic support of the toothpick by the rich and the powerful." Among the things she offered in evidence are "gold toothpicks, and small silver boxes with lids, intended to hold toothpicks on the formal dinner table." Yet she admitted, "At the moment, toothpicks are considered disgusting, but finger bowls are considered snazzy to a suspicious degree; both are intended for cleaning up while still at the table." She found "such inconsistency" in the condemnation of the toothpick difficult to understand, considering that society "has not managed to come up with a better solution than surreptitious swipes of the tongue and short absences from the dinner table."[58]

The Fatal Martini

TOOTHPICKS SOAKED IN poison have been given to unsuspecting victims since ancient times.[1] But in the late nineteenth century, amidst the strident social debate over toothpick use and etiquette, there was a heightened concern about the physical dangers associated with the wooden slivers that had become ubiquitous. The newspapers of the time increasingly carried reports of unfortunate accidents, and the medical journals described bizarre and tragic instances of toothpicks gone astray.

One such case occurred in the summer of 1884. A prominent Chicagoan, while visiting a resort hotel on Lake Michigan, encountered one of his former wives strolling along the piazza. What happened next was described as a "toothpick tragedy":

> He had not seen the wife in question for several years, but in a momentary absence of mind he forgot that she was another's and advanced to kiss her. He did not notice that she had her toothpick in her mouth, and in her embarrassment—for her latest husband was close at hand—she forgot to remove the toothpick. The result was that the unfortunate leading citizen's eye came in contact with the sharp point of the toothpick, which penetrated to his brain, killing him on the spot.[2]

That the sharp point was sticking out of her mouth makes it almost certain that the inadvertent weapon was a double-pointed wooden toothpick, but other kinds could also inflict injury, if not death. In the early nineteenth century, when metal, quill, and improvised toothpicks were still commonly used, a London man was so vigorously applying a quill pick to dislodge food from a tooth cavity that he punctured the roof of his mouth, which led to his demise. Almost a century later an old man who stepped on a toothpick that had been dropped onto the rug in his bedroom suffered a gash that ultimately led to his death, and a Connecticut man who had merely pricked his thumb with a toothpick developed blood poisoning that led to the amputation of his arm. Before that, a boy from a family given to hemorrhage almost bled to death after he inflicted but a "slight wound" to his tongue.[3]

It would seem that everyone should have known that tooth-picks can be dangerous, if not deadly. In 1889, at the height of the toothpick-chewing debate, one physician attributed a "very large number of human ills" to the practice. The mastication of the little stick, he explained, resulted in small pieces of wood being swallowed. He knew of one man who actually ate toothpicks deliberately and had sought out the doctor to obtain some help for his lacerated stomach. However ingested, the small particles of wood can lodge in the stomach or intestinal wall and cause fatal problems.[4] At the same time, another doctor was warning that the fine splinters resulting from chewing toothpicks can also be inhaled into the lungs, where they can produce serious irritation that can lead to consumption. The doctor recommended that those who must use a toothpick for hygienic purposes use a quill. As for chewing wooden toothpicks, he was adamant: "It's an awful habit, and I'd like to see people stop it."[5]

When one nineteenth-century dentist was asked about such warnings and advice, he admitted to an ethical conflict. On one hand, he acknowledged that in his years of practice he had become aware of the many deleterious effects of toothpick use, but on the other hand he confessed that his business was fed by toothpick chewing: "It destroys the gums and the teeth, its chewed fibers lacerate the bronchial passages, get into the throat and head and induce catarrh, bad breath and general disfigurement of the mouth. If I were a philanthropist instead of a dentist I should advise everybody to avoid using toothpicks."[6]

Almost a century later, there appeared an article in the *Journal of the American Medical Association* reporting that in recent years toothpick-related injuries had occurred at the rate of over eight thousand yearly. Children under five years of age hurt their eyeballs and ears most frequently, but children between five and fourteen years old had the greatest overall injury rate. The survey found that youngsters rarely died from their toothpick wounds; that fate occurred most often in adults.[7]

According to the medical journal researcher, "Although they are long, slender, hard, sharp and indigestible, they are rarely considered objects of potential injury and death." Another researcher warned that although most ingested toothpicks "pass harmlessly through the digestive tract," those that do not can be fatal.[8] A. J. "Ben" Haug, president of the Forster Manufacturing Co. at the time the report was released, issued a dissenting view of the dangers afforded by toothpicks. They had to be sharp, he insisted, to accomplish the purpose for which they are specifically made. He likened making toothpicks with blunt ends to dulling scissors.[9]

But like scissors, whether sharp or dull, toothpicks can be used for more than their intended purpose. In the late nineteenth century, a man doing time in the penitentiary for a shooting wielded an iron toothpick to cut his own throat rather than face an appeal. He did not die of his injury, but he later succeeded in killing himself by tearing his handkerchief apart, forming the parts into balls, and stuffing them down his throat. Toothpicks used where they were not intended have also been the source of considerable injury and discomfort. With the decline in the employment of spoon-ended ear picks came an increase in improvisation to clean out the aural cavity, and toothpicks have been among the implements called into ad hoc service. Injuries to the eardrum were commonly attributed to such unwise insertions of a sharp object.[10] Other parts of the body have also been invaded with foreign objects, including the urethra. In the case of one woman in the mid-nineteenth century who was having difficulty urinating, the silver toothpick she inserted slipped back into the bladder, from which it had to be extracted.[11] In a Victorian analysis of more than a hundred cases of self-introduced foreign bodies (including toothpicks) that made their way into the male bladder, almost one-third of the "patients acknowledged that the objects had been introduced to promote sexual excitement."[12]

But even when toothpicks were used for their intended purpose, they have caused injury and pain. Dentists have long decried the use of hard-pointed metallic toothpicks and have generally recommended the more pliable quill over the stiffer wooden "abomination," especially when of the cheaper kind. But even the quill was opposed by some dentists. One, who saw in London "a box of pretty-looking 'diaphanous' toothpicks imported from France," was "sorely grieved . . . to think that so much ornamentation should be bestowed upon such injurious articles."[13]

It has been said that even in ancient times toothpicks made of gold and silver were known to injure the gums. Thus, metal picks were shunned in favor of softer ones, such as those made of porcupine quills. But, perhaps for lack of a quill, sharp metallic objects like pins continued to be employed to pick ears and teeth, even though such practices were thought to result in the development of cancers. In the early nineteenth century it was suspected that the brass and quicksilver used in making the pins was to blame.[14] Still, many people have continued to utilize slender pointed articles of all kinds for other than their intended purpose. On one occasion, while visiting an apothecary shop, a "burly farmer from the neighborhood" picked up a vaccine point, which was lying among a supply of them on the counter, mistaking it for a toothpick. Unfortunately, he pricked his gums and thereby administered the vaccine to himself. His mouth subsequently swelled with a virus.[15] Even the use of a softer object as a toothpick can lead to infection. Actinomycosis, which is referred to as "lumpy jaw" when it afflicts cattle, has been known to have been introduced into humans when an infected spear of hay or straw is put into the mouth.[16]

Some things put into service as toothpicks are not much different from things used as food, and so it is understandable that parts of them may be inadvertently ingested by those who are presumed not to know better. Thus, it is perhaps from the habit of keeping a piece of grass in his mouth that the unsophisticated country boy came to be called a "hayseed." Toothpicks of all kinds have continued to be the vehicle for jokes and derision aimed at those whom presumed sophisticates looked down upon. As an example: "Two Turks were at a French banquet. Toward the conclusion of the feast one Frenchman selected a toothpick from a tray of those useful implements lying near him, and politely passed the receptacle on to his neighbor, who, however, peremptorily declined his offer, exclaiming, 'No, thank you, I have already eaten two of the accursed things, and I want no more.' "[17]

Doctors and dentists have long had a love-hate relationship with the toothpick. At least one dentist—for who else would have been likely to have published "Ode to a Toothpick" in *The Dental Review*, which was "devoted to the advancement of dentistry"—put his feelings into verse. The semi-anonymous F.D.T. recognized the hazards associated with the toothpick, but on balance was quite unashamedly adulatory:

> *A little splint of wood are you,*
> *Tough, but often very brittle too,*
> *Sharpened both your ends should be*
> *Though oft the point we fail to see.*
>
> *A comfort, true, you are to many,*
> *In fact methinks I know not any*
> *Who would refuse your kind assistance*
> *To remove some slight resistance*
> *From twixt the teeth; yet shame on thee*
> *Who make us call Mephistofole,*
> *As you sometime split or break in two,*
> *When between the teeth we urge you through,*
> *And leave a part of you to dwell*
> *Beside the grub we love so well.*
>
> *Yet we love thee little stick,*
> *And after dining, our teeth pick,*
> *And many a thought with you bring out;*
> *A sage conclusion, a dream about*
> *Sometime ago, a winning smile*
> *Of a pretty face, us to beguile,*
> *And oft we do with reason bless*
> *The points you urge on business,*
> *So, I swear true, through thin and thick,*
> *And will befriend thee, my toothpick.*[18]

Just as a stray bristle from a toothbrush can get under the gums and cause inflammation, so can a splinter or the broken end of a toothpick do the same damage. But the problem of chasing down an elusive food particle has long been a postprandial obsession. Among a list of early American "miseries of human life" was the occurrence of a "fishbone, or other substance, stuck between your two hindmost teeth; then, in your endeavours to remove it with a toothpick, only wedging it tighter than ever."[19] In attempting to extricate the bone or another elusive piece of debris, it could become necessary "to plough up the gums and sow splinters of wood with badly made and unpolished rough wooden toothpicks." Such plowing and sowing could go deep. One dentist reported a man who came into his office with a very sore and loose tooth that showed no signs of filling or decay in it or its vicinity. However, upon examination, the dentist found "about half way up the root" something that proved to be "a portion of a *wooden tooth-pick,* about a quarter of an inch in length, which had been broken off in picking the teeth." Another dentist, who had been "toothpicking deep" between his own molars, broke off a long piece that became "buried wedged tight." He tried to get the piece out himself, but in the end had to seek the help of a professional colleague.[20]

A dentist getting hoisted by his own petard may be amusing, but anyone swallowing even part of a toothpick is no laughing matter. In the 1880s, the thirty-five-year-old "champion colored pugilist" George Godfrey insisted that he was not too old to fight, asserting that he didn't drink and ate three square meals a day, four when he was hungry. He added that his digestion was perfect. In fact, he bragged that he could "digest a wooden toothpick." But less fit individuals were warned by *Puck,* "It's all well enough to chew a wooden tooth-pick in front of Delmonico's; but when you swallow the tooth-pick, it ceases to be funny." It can also be difficult to diagnose. One gentleman who visited his doctor with discomfort in his chest was advised by the physician that he had consumption. However, after a few days, he had a fit of coughing that ejected "about half a wooden toothpick," after which he began to recover.[21]

In the 1940 season of the New York World's Fair, a visitor from Brooklyn, "while looking at a case of fabulously valuable gems in the House of Jewels, accidentally swallowed a toothpick." He was taken, successively, to a first-aid station at the fair, then to nearby Flushing Hospital, and finally to Queens General. An attendant at the latter hospital was reported to have said that the patient "was discharged in satisfactory condition and that he left saying he intended to return to the Fair." It is not known whether he actually did, or whether he suffered any subsequent ill effects. One can only hope that the apparent absence of a follow-up story indicated that he did not.[22]

Not everyone is so lucky. Although toothpicks can "pass through the gastrointestinal tract uneventfully and are eliminated in the feces," that is not always the case. One woman house worker was reported to have "kept a toothpick in her mouth much of the time, and she would often chew on it and sometimes would swallow the pieces." She had to stop working when she developed "a rather mysterious stomach problem," which "turned out to be several large abdominal abscesses caused by pieces of toothpicks she had swallowed over the years," creating intestinal punctures. She became critically ill and recovered only after multiple surgeries. Another woman was found to have a six-inch-diameter "pancake-shaped abscess" filled with pus that contained "a fragment of wood that was about a third of the size of an ordinary toothpick."[23]

Sometimes a whole toothpick is the culprit. A Minnesota doctor, reporting in 1896 on a death due to a perforated colon, described finding the point of a toothpick that he characterized as Japanese. It had been "found to be bent by a 'greenstick' fracture, so that the thick end formed a very acute angle with the pointed end." Today, the same condition might be blamed on a toothpick made in China to look like a product of Japan. But the consequences of swallowing any kind of toothpick can be just as deadly. During a four-year period ending in 1982, three deaths were attributable to toothpicks, with one being the result of "perforation of the bowel by a toothpick."[24] An autopsy performed on a woman who had been found unconscious by her husband on a Friday night showed her death to have "resulted from suffocation and peritonitis following perforation of intestines" by a toothpick.[25]

Just as certain district attorneys are said to be able to get a grand jury to indict a ham sandwich, so some ambulance-chasing lawyers can build a suit around a toothpick. During the 1960s, the Sands Hotel and Casino in Las Vegas made available to its customers toothpicks in little red boxes of the kind that more commonly held wooden matches. However, the practice was eventually "scrapped due to lawsuits involving people wanting to sue over choking" on the picks.[26] It is harder to establish the provenance of the generic toothpicks that are made available in restaurants today.

A California man who had abdominal pains on and off for over six months finally went to the emergency room after he began experiencing fever, chills, and rectal bleeding. He was put into intensive care and given massive blood transfusions, but his condition deteriorated. Exploratory surgery revealed that a wooden toothpick had lodged in his intestine and punctured his colon and a key artery, allowing eleven liters of blood to accumulate in his abdominal cavity and gut. The man did not survive. After his death the family did remember that he had mentioned swallowing a toothpick some six months earlier. He had gone to an emergency room then, but nothing was found at the time. He had endured the painful condition of carrying around "a lethal time bomb" until it was too late.[27]

Not all such incidents are fatal, depending on the path the toothpick takes within the body. A Spanish man survived having a toothpick inside him for about fifteen months before he was diagnosed with something unusual in his liver, which proved to be an abscess containing a toothpick, which he survived. (The cause of a pain can go unnoticed for a significantly larger object than a fragment of a toothpick. When a Colorado man complaining of a toothache visited a dentist, X-rays revealed a four-inch nail embedded in his skull. The origin proved to be in an incident with a nail gun, which had backfired about six days earlier and driven the nail through the roof of the man's mouth, just missing an eye, and going an inch and a half into the brain. The nail was surgically removed and the man was expected to recover.)[28]

Although toothpicks are not intended to be ingested, a lot of things with which they are associated are. Hors d' oeuvres are commonly served with toothpicks holding them together or offering a way to pick them up—or with a supply in a small holder nearby. Many sandwiches, especially the club, would likely suffer spontaneous disassembly without a toothpick. Filets wrapped in bacon, roulades, and other rolled meats often conceal a structurally essential toothpick in their folds. And, of course, the toothpick is an iconic part of the martini. Though neither food nor drink typically comes with a warning label advising the consumer to remove the toothpick, the thoughtful host or hostess will often provide a caveat.

Toothpicks can also be less dangerous than deceitful when included in other things that people put in their mouths. In Michigan, some state senators once complained that five-cent cigars were "deleterious to the health of smokers." When a bunch of different cigars were inspected, they were found to contain—in addition to tobacco—old hat material, horseradish leaves, strands of wire and rope, and wooden toothpicks. There was no agreement among physicians, however, on whether the foreign matter was deleterious or beneficial to health. In defense of the potential benefits was offered the anecdotal evidence of a patient who chewed up a clothesline being cured of consumption. Back in the nineteenth century there was also speculation that asthma could be cured by the smoke of an old felt hat, and that the liver could benefit from a cigar's scrap-iron content.[29]

Still, some people are quite concerned about the purity of what they consume. Martini sippers and drinkers can be an especially finicky lot, quick to debate over proportions of ingredients and whether the drink should be shaken or stirred. According to one precious recipe, for the "driest Martini cocktail in the world," after straight gin has been poured into the glass, it should be stirred with a toothpick that had been but dipped in dry vermouth. Advice for making a "perfect" vodka martini directs filling the glass one-third full of vermouth, to which is added "a splash of olive brine, a squeeze of lime and three dashes of bitters" along with potato vodka—all stirred with "a battery driven beverage mixer." An olive may be added, "but a toothpick is *verboten*, as it tends to contaminate the liquid."[30]

A toothpick was not off-limits to Daniel Malamud, a biochemist, who fixed himself a Gibson martini one evening. After eating the onions off the toothpick, he placed it back in the glass so as not to soil the furniture. As he was draining the martini glass, the toothpick floated into his mouth, getting lodged somewhere in his throat. He ran to the bathroom and tried to make himself throw it up, but this only sent the toothpick into his nasal passage. He then went to the emergency room, where the triage blackboard advertised his condition to all: "Malamud. Toothpick up nose." After Dr. Mary Harlan Murphy removed the object, she and the patient agreed to write about it.[31] They did, and their communication appeared in the *New England Journal of Medicine* as a letter to the editor describing the incident and cautioning drinkers of martinis.[32]

Perhaps the most well-known of tragic toothpick accidents involved Sherwood Anderson, the author of *Winesburg, Ohio* and other writings about the Midwest. In late February 1941, Anderson boarded the SS *Santa Lucia,* which was heading for the Caribbean. The next day he skipped lunch, complaining of feeling "stuffy" and having abdominal cramps. After five days of alternating between feeling better and worse, he was taken off the ship at Colón, at the Atlantic entrance to the Panama Canal, and put in the hospital there. Within days, his condition deteriorated: His heart raced, he lapsed into delirium and then into a coma, and soon he died. The cause was attributed to "an acute intestinal obstruction," but its cause and location remained unspecified. The body was sent to the morgue in Gorgas, where an autopsy was performed. What at first looked like a burst appendix was soon identified as the result of a toothpick projecting through a weak section of the wall of the colon, where a sac had formed. Anderson had died of peritonitis.[33]

The doctor doing the autopsy was baffled as to how a toothpick could get into the large intestine, but another doctor was sure it was the result of drinking a martini. Since Anderson was known to have had "a history of indiscretion in eating and drinking at farewell parties," the doctor assumed that he got tipsy at the bon voyage party and surmised that after four or five martinis Anderson could not "distinguish olive from toothpick," and swallowed both. The hypothesis was tested by asking the widow if he had drunk at the going-away party, and she confirmed that he did indeed, having had five or six martinis, which he liked "very dry—and with an olive." Decades later, the doctor lamented not having publicized the incident, thinking that he might have saved some lives. Death by toothpick was little known at the time, and one national magazine reported that Anderson had died of a brain tumor.[34]

President Harding was known to have "insisted on picking his teeth after every meal, even though the White House butler complained, 'That's being too much a man of the people.'" When Harding died suddenly, after only two years in office, speculation on the cause of death ranged from his doctor's talk of an "acute gastrointestinal attack" to suspicion that his wife had poisoned him. A more recent speculation is that he died from ingesting a toothpick.[35]

The not insignificant number of fatalities attributable to such ingestion would seem to call for the invention of a safety toothpick, and Miami inventor Terry Lane seems to have done just that, albeit unintentionally. He noted in his patent for a "floatable toothpick assembly" that people used "spoons, forks, or indeed their fingers in order to reach into a drink and retrieve the piece of garnish" that was submerged. To make that task easier, he devised a buoy-like toothpick that enabled the prize to be gotten without getting the fingers wet. Although Lane did not seem to think of his bobbing toothpick as having the potential to reduce the number of toothpick ingestions, it might have that beneficial unintended consequence. It would seem that a one-martini drinker would be no more likely to miss the floating assembly than a sober boater would a channel marker. Even a multi-martini drinker might be saved by gagging on the bulbous toothpick and coughing it up before it was too late.[36]

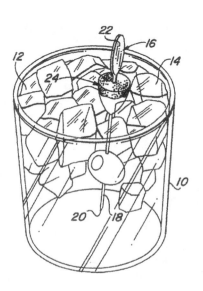

A floating toothpick made to resemble a buoy in appearance and function not only obviates the need to get one's fingers wet in retrieving a garnish from a cocktail but also makes it less likely that the tooth-pick will be accidentally swallowed.

Charles Dickens was a toothpick user, but apparently he never swallowed his, which presumably was not wooden. Since the item did survive him, it was offered for sale in London for about $275.[37] He surely recognized the consequences of ingesting the thing, how-ever, if what he wrote in *A Christmas Carol* is any indication. In that tale, Ebeneezer Scrooge denied believing that he was seeing the ghost of his partner, Jacob Marley. When asked by Marley why he did not believe his senses, Scrooge replied, "Because a little thing affects them. A slight disorder of the stomach makes them cheats. You may be an undigested bit of beef, a blot of mustard, a crumb of cheese, a fragment of an underdone potato. There's more of gravy than of grave about you, whatever you are." Scrooge was actually whistling in the dark, trying to "distract his own attention" from the terror he felt. To divert the specter's gaze from himself, Scrooge took out a toothpick and asked the ghost if he saw it. When Marley acknowledged that he did, Scrooge told him, "I have but to swallow this, and be for the rest of my days persecuted by a legion of goblins, all of my own creation. Humbug, I tell you! Humbug!"[38]

CHAPTER TWENTY-EIGHT

Improving on Perfection

B Y THE BEGINNING of the twentieth century, it had long been established that "every log has 'millions in it' in the shape of toothpicks." A good-sized tree could yield from two to four million of them, with a forty-foot tall one said to give up as many as eleven million flat or three million round toothpicks.[1] In 1932 Forster's Strong plant alone was believed to be producing fifty million picks per day, which would have required as many as two dozen trees daily. Half a century later, when good-sized ones were harder to find, an "average tree" might yield only about four hundred thousand flat toothpicks. But at the end of the century production and demand had also dropped, and in Strong a "mere eight people" were "turning out almost all the toothpicks used in this country, or close to 30 million picks a day," down from an estimated seventy-five million just a decade earlier.[2]

As difficult as it was to get regular and reliable production and sales figures from manufacturers such as Forster, especially when they were trying to market a new and unfamiliar product, it had long been even more problematic to ascertain how many toothpicks that were made and sold were actually used. In the early twentieth century, it was reported that

> there is one article of manufacture that is used so extensively in the United States that no one has an idea of the annual quantity consumed, namely, wooden toothpicks. According to an expert, the number is simply incalculable. Millions upon millions of the tiny wooden slivers are turned out every year from American factories alone, and on top of this tremendous output come importations from Portugal and Japan and other countries nearly as large as the domestic product.[3]

In 1993, toothpicks were said to be found in 96 percent of U.S. homes, with Americans using them "for dental care, for holding hors d'oeuvres together, for cleaning in tight crevices and for just looking cool, as James Dean did with that dangling toothpick in the movie *Giant*."[4] An incalculable number of unused toothpicks just sat in boxes in pantries and kitchen cabinets.

Of course, most wooden toothpicks were used just once and then discarded, a fact that was not lost on their makers. According to the manager of a Forster plant that itself would eventually be "thrown away," most toothpicks "end up in the back of a drawer, lost to memory, adrift in space, randomly replaced the next time someone happens to think of it." As he saw it, "we assume the consumers are losing or throwing away 75 percent of what we do for a living."[5] He was not complaining. In the late twentieth century, when a box containing hundreds was selling for as little as fifty or sixty cents, the toothpick was called "the last great bargain in the United States." What was true in the 1990s, when companies of all kinds were being sold and bought and used and thrown away like disposable items, had been true a century earlier and was a truth that Charles Forster realized when he made his first wooden toothpick in Boston. Others realized it too: "A bunch of wooden toothpicks, which sells in the city for a fraction of a cent must net to the manufacturers a ridiculously small sum, but when they are made and sold by the millions and billions there is money in them."[6] This kind of reasoning led to variety in production, not only for competitive reasons but also to give consumers a choice. If they were given a choice they might buy even more millions.

Not all choices were new. Toothpicks flavored with cinnamon or wintergreen were being produced as early as 1885. The Estate of Charles Forster also made an "aromatic or flavored toothpick," which was essentially a round-ended Velvet "soaked in cinnamon bark until it acquires some of the agreeable and aromatic flavor of it." A box of three hundred, which sold for fifteen cents, was labeled "AROMATIC ANTISEPTIC QUALITY TOOTH PICKS." In smaller print, within the angles of two crossed toothpicks, were the words "smooth polished / superior to / orange wood."[7] A careless reading could lead one to think that the picks themselves were made of orangewood, but in fact the box was only claiming its contents to be superior to the best. At about the same time, the Cutter-Tower Co. was also advertising "Tower's Patent Round End Antiseptic Toothpicks" that were both "aromatic" and "highly flavored." They were also professed to be the "best made" of their type. These toothpicks were stated to "preserve the teeth, sweeten the breath and increase the flow of saliva." In addition to cinnammon, they also came in peppermint and sassafras. During the 1952 U.S. presidential race, flavored toothpicks (peppermint) were provided on campaign airliners for the first time.[8]

As early as the 1940s, school-age boys and girls had learned to make their own flavored toothpicks. Half a century later, the writer Sue Hubbell recalled that when she was in sixth grade, "a bad boy named Hughie Mehaffie regularly soaked toothpicks overnight in liquid cinnamon and brought them to school for everyone to share." Their teacher did not approve of the practice, which of course made Hughie and his classmates crave the picks even more. According to Hubbell, "we would slouch down behind our desktops and savor, especially, that first mouth-burning taste of cinnamon."[9]

Wooden toothpicks so heavily flavored with cinnamon that they have been characterized as "spicy" have been sold under the brand name Fire Picks and have been used by smokers to help them quit their habit, or at least replace cigarettes with a toothpick habit. Hotlix Cinnamon Toothpicks are "hand dipped then dried to lip searing perfection," and their "natural antiseptic quality" is said to be "an effective aid to help cure smoker's breath."[10] Toothpicks made in China flavored with "chile sauce, red pepper, and garlic" are sold as a practical joke ("a *hot* joke for that special friend"); according to the package, the more they are chewed, "the *hotter* these toothpicks become." There was a limit, however, and a Nebraska firm was once advised to take it easy on the cinnamon after "five schoolchildren suffered adverse reactions."[11]

Not every manufacturer believed the hotter the better. The Minto Tooth Pick & Specialty Co., of Saginaw, Michigan, said of its flavored Minto picks that they had a "delightful flavor" and were "more than a tooth-pick, they are a confection." In spite of that, they were "endorsed by the best authorities of the dental profession," who nonetheless remained nameless. The company also claimed that each toothpick was "treated by a secret sanitary process" and that its picks were "sold in Germ-Proof packages only," which were labeled "absolutely sanitary." As if stating it in three different ways in the limited space available on the outside of a small toothpick box were not enough to assure potential customers that they were getting a hygienic product, the toothpicks inside were separated into three large bunches, each wrapped in a wax-paper envelope.[12] In addition to herbs and spices, liquor has also been employed to flavor toothpicks—and possibly to sanitize them at the same time. Scotch and bourbon have been used to produce the "pick with a kick" and add "spirit" to dinner parties. Those with different tastes in liquor could try Teeny-Weeny Dry Martini flavored toothpicks.[13]

More genuinely medicinal ends motivated Edward Petrus of Austin, Texas, to patent a toothpick impregnated with one of a variety of zinc compounds. Since zinc is known to be antibacterial, antifungal, and antiviral, its use in the mouth was expected to help keep gums healthy and thereby prevent periodontal disease, which is often a harbinger of infection elsewhere in the body. In his patent, Petrus pointed out that floss too could be treated with zinc compounds. He further noted that floss and toothpicks could be impregnated or coated with other "therapeutic agents," including antibiotics, bleaching agents, hormones, and vitamins.[14] Of course, such enhancements would boost product use, which would make manufacturers happy.

In the mid-1950s, domestic wooden toothpick production was estimated to be as high as seventy-five billion annually, representing a wholesale value of about five million dollars. This was more than enough to prompt toothpick manufacturers and independent inventors alike to look for new ways to accomplish old ends. In 1960, for example, Carl Kucher of Richmond Hill, New York, patented a combined toothpick and gum massager, which looks like a tool that golfers might use on the putting green.[15] In that same year, Thomas Jackson of Odessa, Texas, patented a combined condiment holder and toothpick to solve the "continual problem" of food service establishments that experience a "substantial loss of condiment holders, that is, salt and pepper shakers, due to theft." His solution, which looks somewhat like an awl with a perforated handle, was intended to be kept "upon the table or counter, so that the customer may use the holder and may thereafter use it as a toothpick and take it with him when he leaves the establishment," which was desirable

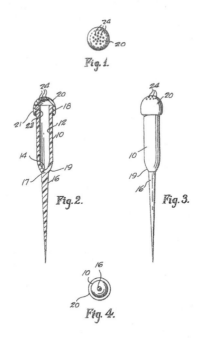

Many combinations of toothpicks and other implements have been patented, including this one that incorporates a toothpick in the base of a saltshaker. The inventor proposed that such a device could also be imprinted with a restaurant's name, but did not say what might happen to the salt when the toothpick was put to use.

because the presumably inexpensive device could be imprinted with the restaurant's name and address. It is unlikely that many patrons would see the genius in the device, however, since using the toothpick would cause the condiment to spill out.[16]

In the mid-1980s, when toothpicks were in 97 percent of American homes and with Americans consuming from fifty million to seventy-five million every day, inventors continued to be motivated. According to Ben Haug, president of Forster at the time, he was then "marketing a new pick as the 'ideal': It has a square middle, so it won't roll when dropped, and rounded edges to satisfy round-pick duties." Certainly a round-bodied toothpick did have a tendency to roll off the table and along the floor, after which it would be considered unsuitable for use. Forster had made round toothpicks until 1984, when it "switched to what toothpick experts call square-centered round, and the rest of us call square." These came to be sold as "square/round tip" toothpicks. They had the clear advantage for the manufacturer that, rather than having to be rounded all along their length, only the ends had to be rounded. Unfortunately, the abbreviated process also produced a toothpick that was not of what might be called classic form. In fact, it is of a shape without grace. Nevertheless, such picks did have their functional advantages, and Forster was able to promote them also as "toothpicks with a square center shank that are easy to grasp."[17] They are that; rather than tending to roll between the fingers when used, they allow for a softer and less tiring grip. They also have the benefit that things skewered with them have less of a tendency to rotate on the pick, an annoying habit of cocktail shrimp. But just as Coca-Cola reverted to making Classic Coke after a disappointing launch of New Coke, so did Forster find that many consumers preferred fully round toothpicks.

The idea of a no-roll toothpick was reminiscent of another wood novelty item often made by the same companies that produced the toothpicks. This was the no-roll clothespin introduced in the 1920s. In the following decades, three distinct types of "pin," each favored by a different part of the country, were marketed. The traditional round clothespin was popular outside the big cities in the Northeast and the Midwest, where clotheslines tended to be in backyards. The spring clothespin, which had been introduced in the late nineteenth century, was sold in the South and West, where clotheslines tended to be made of wire, because cotton ones did not last long in the excessive heat. (Using round clothespins on unyielding metal lines led to splitting, which is why the necks of some round pins were reinforced by being wound with wire.) The square pin was the style of choice in large cities, where people lived in multistory apartment buildings and had clotheslines that stretched from window to window over yards and alleys. Square clothespins did not roll off the windowsill.[18] Perhaps this popular Forster product was the inspiration for the square/round tip toothpick. But there is no telling where inspiration will come from. As Richard J. Corbin remarked during his tenure as Forster chairman and chief executive, "One of the things foremost on my mind every minute I'm awake—I'm thinking about the manufacture of toothpicks."[19]

But who else cares whether the small utilitarian sticks are flat or round or square or are made in America or Japan or China or elsewhere? One reporter likely spoke for most of his readers when he wrote of toothpicks, "Like the mousetrap, they've already been perfected. Like salt and paper clips, who cares what brand name they carry?"[20] But styles and brands and models do matter when it comes to marketing, and that is why Forster and other companies were so protective of their trademarks—and so calculating in their slogans. A box of Diamond Brand "double pointed–tapered–flat polished white birch" toothpicks from about the 1960s claimed to contain "the most perfect toothpick made." The toothpick's quality—but curiously not the quantity contained in the box—was touted on every panel, including the bottom, which bragged that the product was "made in U.S.A. by American workers of American materials." The bottom of the box said also of the Diamond trademark on the top that "its use on a package of wood products insures Diamond quality of both materials and workmanship—Every One Perfect."[21]

Trading on a brand name was nothing new. In the early twentieth century, Perfection brand toothpicks were made and sold by the Perfection Mfg. Co. of Clayton, Michigan. On the box they were claimed to be the "only round-turned double pointed hardwood picks made," by a process carried out on machinery patented by Edward Lamb and Emmor Bales of Clayton. The claim could be made because the round double-pointed picks produced by Forster and protected by Freeman's patent were made by a distinct process employing "compression and polishing." In the Lamb and Bales machine, "a strand or splint of wood is driven forward and subjected to cutters which turn or round off the strand of wood in a nodular form—that is, that portion of the strand intended for a single toothpick is cut large or swelling at its middle part and is reduced almost to a point at each end." However, an early application of the machine produced toothpicks that were far from perfect. The process of severing the individual picks from the continuously formed splint left the ends "more or less frayed and bent," something that was eventually corrected by an improved way of cutting off the picks from the noduled rod.[22]

Today, one can scarcely find a claim of toothpick perfection—explicit or implied—in drugstores or supermarkets. Boxes of toothpicks, especially those containing low-priced imports, say barely more than what it is that they hold. Brands seem less important, since there is little, if any, choice among the boxes of ordinary wooden toothpicks found on the shelf. They do indeed appear to have become a commodity, and the shopper is lucky to find even a small variety of styles. But the mouth- and tooth-care section itself has grown in size and in the abundance of products that it displays. Among the variety of newer toothpick-competitors offered are what have come to be known as brush picks, which are usually made out of plastic with a point integrally tipped with minute bristles. They are advertised to work between the teeth the way a toothbrush does on their outside surfaces.[23] TePe offers a set of interdental brushes that resemble little bottle brushes. A package contains eight sizes, ranging in diameter from 0.4 to 1.3 mm, from which the user can choose the one that best fits the interdental space needing cleaning. The brushes can be bent for easier access to the inside spaces between back teeth.

As we have seen, the hope of improving on both the toothpick and the toothbrush has attracted inventors for some time, and in the late nineteenth century the notion of combining the two devices was especially popular. In 1901, Emma J. Thurston, an inventor from Deming, Washington, was awarded a patent for a toothpick that was handily incorporated into the end of a toothbrush handle. Her "toothpick" was in fact both a pick and a scraper. When one of these was inserted into the handle, the other was exposed to be used as desired. The scraper was thought to be especially helpful in cleaning tartar off false teeth. Another inventor, Semon Eisenberg of Savannah, Georgia, shaped the end of the toothbrush handle into a hook that essentially made it a long-handled toothpick.[24]

The proliferation of toothbrush designs, including those having ergonomic handles and electric-powered heads, caused vast amounts of display space to be taken up by a bewildering selection of styles and colors that can hardly be expected to go with any restrained bathroom décor. The lowly toothpick was pushed aside by these splashy displays, and toothpicks were frequently relegated to the cake-mix aisle, where they can be bought not as toothpicks but as doneness testers. One woman who was looking for toothpicks in a store reported her quest: "Figured they would be near the baking cups, or possibly the dental floss. No luck. Finally stumbled across them in the paper plate section!"[25] Yet the toothbrush and the toothpick should not be so separated. After all, what is the head of a toothbrush but a crowd of little toothpicks (called bristles) that work their way into the nooks and crannies in and between our teeth to dislodge debris and, with the aid of paste, scratch and scrape off plaque? Conversely, the toothpick with a splayed end, such as can be achieved by chewing on it, as if it were a chew stick, can serve as a toothbrush. This old idea was patented in 1905 by James Smith of Princeton, Indiana. His invention flattened one end of a wooden toothpick, thereby separating the fibers there and producing an "outwardly-broadening brush."[26]

The toothpick preceded the toothbrush and may succeed it one day, as the pendulum of fad and fashion swings. Conventional toothbrushes do not have nearly the portability of the toothpick, and the expectation that toothpaste or powder be applied to them makes them all the more inconvenient. Their use requires care to keep the frothy saliva solution from dripping down the front of one's clothes or from splattering on the mirror above the washbasin. The debris-laden waste that we spit into our bathroom sinks at home and flush down the drain with running water or a glass of mouthwash is generally looked upon askance in public washrooms. Though some fastidious workers and diners have been known to use a toothbrush in a public restroom, the practice is certainly no more welcome by many onlookers than is using a toothpick at the table. It is not so easy to make a transparent toothbrush as it is a transparent toothpick, which one inventor believed could be "used without being seen."[27]

It has been a constant goal of inveterate inventors and competing manufacturers to improve not only the way toothpicks are produced but also the nature of the toothpick itself. More ingenious and faster machines were the means to achieve the first objective, but achieving the second has been more elusive. The best of all possible worlds for the manufacturer would be to devise more efficient machinery that made a new kind of toothpick that customers universally thought to be an improvement. But the persistence of flat, round, and variations on these shapes and styles is evidence that no single toothpick design is everyone's ideal. When asked once if a better toothpick could be "built," Fred Beauregard, a Forster product manager in the late twentieth century, said that he spent a lot of time thinking about toothpicks, and he suspected that "we'll find a way to improve on perfection."[28] Such confidence in endless improvement has been and continues to be the story of toothpicks and everything else that is manufactured.

The Butler Did It

T HE PERIOD DURING which the Forster heirs and trustee Oscar Hersey were entangled in legal proceedings over the Estate of Charles Forster had also heralded turbulent times for the entire toothpick industry in America. Recall that a glut in the market caused the bottom to drop out of it within the course of a few weeks early in 1911. The wholesale price of a case of wooden toothpicks had fallen by 80 percent to below the manufacturing cost of the goods. The industry was hit especially hard in Maine, where mills within thirty miles of one another accounted for three-fourths of the world's wooden toothpick production. The plant at Strong was idle for several weeks that winter. The International Manufacturing Co. in Phillips, Maine, which had recently built a new concrete factory building, had orders for soft-wood toothpicks that it could not afford to fill, and instead it planned to use its stock of poplar to make matches. New trustee Frank Butler, working to protect the interests of the Forster estate, attended a conference in Indiana where the industry-wide situation was being discussed.[1]

In 1913, the firm operating as the Estate of Charles Forster was identified as the "largest that is engaged in the exclusive manufacture" of toothpicks, which it did in its mills in Dixfield and Strong. The next largest was likely Maurice's Forster Manufacturing Co., of Dixfield, whose relationship to the Estate was "very close altho not direct or connected in anyway."[2] However, the assimilation after World War I of Maurice's company into the Berst-Forster-Dixfield Co. (and in the 1940s the total suppression of the historically significant Forster name into the monogrammatic designation B.F.D. Co.) presented an opportunity for the awkwardly designated corporation known as "The Estate of Charles Forster" to be confused with Forster Manufacturing, whose name it would eventually adopt.[3]

The ironic conflation of the father's legacy and the son's rebelliousness in the single corporate name no doubt contributed to the unbounded confusion that has prevailed about the origins of and distinctions between the two companies. Historical sketches in newspapers and magazines came to be replete with errors and contradictions. Among the most egregious and stubborn of misrepresentations is the claim that appeared on countless boxes of Forster toothpicks, even as late as the early twenty-first century:

A TRADITION OF QUALITY

Since 1887, when Charles Forster began the first wooden toothpick factory in the United States, the Forster name has meant quality products for American homes.[4]

The myth of 1887 was propagated, if not memorialized for all time, when a celebration of the "centennial" of the Forster firm was introduced by Senator George Mitchell, on behalf of himself and Maine's other U.S. senator, William Cohen, into the *Congressional Record* for 1987:

Forster's beginning was, like the toothpick, humble indeed. But the story behind that indispensable item is one of typical American ingenuity, quite representative of the enterprising spirit for which our Nation is famous.

Although the disposable wooden toothpick can now be found in nearly every home and eating establishment across these United States, 100 years ago in 1887 it was not so. . . .

In 1869, Charles Forster, an uncommonly inventive and ambitious young man, fabricated the first toothpick workshop in the cellar of his Boston area home.[5]

This encomium on the occasion of Forster's self-proclaimed centennial was a model of political rhetoric, full of overarching untruths and underlying contradictions. The toothpick did, of course, exist long before Charles Forster first drew breath, and Benjamin Sturtevant's machines and patents predated the basement workshop.

This was not the only time that toothpicks were to be the topic of discussion in the *Congressional Record*. Later that year, Senator

David Pryor of Arkansas expressed the thought that the Senate had been turned into a toothpick factory. He likened the institution to a "huge giant lumber mill, with all the high technology and the biggest saws in the world, which is making toothpicks." Though Senator Cohen agreed with the implication that the institution did not work very efficiently, he took exception to the metaphor and let it be known that Mainers did "not take kindly to the suggestion that producing toothpicks is something to be scorned."[6]

Whether toothpicks are something to be praised or scorned, it is the odd business that does not want to claim its roots go as deep as possible. Yet, as the real story of the original Forster firm shows, the business had an odd history. The year 1887 has been associated with the establishment by Forster of a company-owned toothpick factory at Strong, but, as we have seen, that was not his first there, and certainly not his first in Maine. The Forster firm could honestly claim—as do educational institutions that transform themselves from provincial colleges into world-class universities over the course of time—that it had been established in some form earlier than it arrived at its eventual incarnation. Given Charles Forster's efforts in Boston and environs in the 1860s, Forster's box-end history could extend itself by another couple of decades of tradition, at least, and as the end of the twentieth century approached, the company could have begun looking ahead to celebrating 125 years of the wooden toothpick industry in America—and perhaps even beyond, to a sesquicentennial.

But that was not to be. In the early 1930s, after the last of the Forster heirs had passed away, Frank Butler was involved in the toothpick industry in a prominent way, being one of three members of the Executive Committee of the Toothpick Institute, the industry's trade association that comprised "practically 100% of the total number of Toothpick Manufacturers in the United States." The Depression-era National Industrial Recovery Act required the industry to develop a code of fair competition that among other things would "equalize production with demand." The signatories to the code included the Berst-Forster-Dixfield Co.; Hardwood Products Co., Inc.; General Woodenware Corp.; and the Estate of Charles Forster, represented by its then president, identified as "Charles W. Butler," which appears to have been a strange typographical error conflating the otherwise unrelated

Charles Forster and Frank W. Butler families.[7] (There were fewer than half a dozen toothpick companies in America at the time, and the number seemed constantly to be changing. At the beginning of the twentieth century there had been about a dozen; at the end of World War II the number would be about twenty; by the 1990s only three major toothpick makers would remain—Forster, "the country's first and largest"; Strong Wood Products; and Diamond Brands of Minneapolis, which also had a Maine presence with a plant in Dixfield.)[8]

The Forster business had become a Butler family affair, with Frank Butler eventually going into partnership in the operation with his son, Benjamin. In 1943, almost ten years after his father had died, Ben Butler and his wife, Natalie, sold their interest in the firm to Theodore Roosevelt Hodgkins, who had worked for the company since 1934, and his wife, Frances, who was Frank Butler's daughter. T.R., as Hodgkins was known, had risen through the ranks quickly, having become in 1936 Forster's president, a position he would hold until 1970, after which he became chairman of the board.[9]

The Estate of Charles Forster would continue its dominant role in the industry through the 1930s and would still on occasion be referred to as operating under that name even in the 1940s.[10] However, the corporate name was legally changed to the Forster Manufacturing Company, Inc., in 1936, a mouthful that was spelled out on the company's letterhead and in its advertisements. In 1950, the mouthful would be officially simplified to "Forster Mfg. Co., Inc." The B.F.D. Co. continued to be a strong competitor, and in the early 1940s it had plants in Plattsburgh, New York; Cloquet, Minnesota; and Oakland, Phillips, and Peru, Maine.[11]

During World War II, most of Forster's production was channeled into military applications, including the manufacture of "tongue blades and applicators" used to treat servicemen. The postwar expansion of the economy gave Forster itself the opportunity to grow, and plants were acquired in Phillips and North Anson, Maine, where candy sticks, skewers, rolling pins, and other wood products were to be manufactured. In 1946, the general offices were moved to Farmington, and the following year a plant was rented in South Portland. In October of that year, the North Anson plant burned down, and two months later the Strong plant was destroyed by a fire initiated when the fracture of a piece of machinery set off a spark igniting wood dust and causing an explosion that "sent flames throughout the entire plant."[12]

All this happened on Hodgkins's watch, and it was under his leadership that the business decided to expand in two distinct directions. First, shortly after the Strong fire, which "could easily have wiped out the company," he resolved "never again to be confined to one site" for making a particular product, and bought plants in East Wilton (an old broom factory) and Mattawamkeag. The former made "ice cream spoons and sticks, cocktail forks and mustard paddles"; the latter, "clothespins, skewers and applicators." Second, Hodgkins further expanded and diversified the product line, branching out from toothpicks and other wooden articles to plastics. Still, Forster was selling "about a freight car of toothpicks every week."[13]

Within a year of the fires, Ideal flat toothpicks were being made on completely new machinery; within another eight months, Worlds Fair round toothpicks were being produced in a newly constructed plant in Strong. At midcentury, with its headquarters located in Farmington, Forster also had plants in Phillips and Portland, though of course not all of them manufactured toothpicks. The growth of the business had come, at least in part, through expansion into other lines of woodenware. Among the products that Forster was producing at this time were shaped soda spoons and forks; round candy sticks; tongue depressors; rolling pins; potato mashers; darning eggs; paper roll plugs; and dowels.[14]

Soon the accounting, purchasing, and sales departments had quadrupled their personnel, so to provide room the engineering department was relocated to the new Strong plant. Continued growth led to a desire for "more modern office facilities," and in 1960 Forster bought a former woolen company plant in Wilton, which made it "finally possible to have, under one roof, the general offices, wood turning operations, machine shop, folding carton department, spring clothespin assembly, wrapping, plastic cutlery packing, engineering, and utilities."[15]

There had come to be an increasing awareness of the value of the Forster name. (Maurice Forster had certainly recognized this when shortly after his father's death he had begun the competing Forster Manufacturing Co. in Dixfield.) Over the years, packages of

Forster toothpicks had changed their appearance—sometimes rad-
ically and sometimes almost imperceptibly. Some of this change
was driven by an attempt to distinguish the Forster product and
packaging from its competitors; some of it seems to have been
driven by the always-changing styles of fashionable typography and
the other graphic arts. In its earliest packaging, when the Forster
name did not even appear on the box, the manufacturer's ego
seemed to be deferring to the brand names and other trademarks
under which its toothpicks were sold by its wholesalers. Often,
early packaging even played down the brand name, emphasizing
instead the physical attributes of the contents. Thus, in the late
nineteenth century, the brand name World's Fair—then spelled
with an apostrophe—was set in much smaller type than "WOOD
TOOTH-PICKS," and was less prominently displayed that the
qualities of their being "COMPRESSED · POLISHED · ROUNDED ·
POINTED." Even the wholesaler's name appeared only in the abbreviated form, "C. T. Co.," inconspicuously tucked into the angles
between the trademarked pair of crossed toothpicks.[16]

Eventually, packages of Worlds Fair toothpicks made by the
Estate of Charles Forster began to be identified as being manufactured by it in Strong, but only by a small seal printed on the box
top's flap, which was of course not visible until the box was opened.
After the corporate name change, boxes of Worlds Fair picks identical in design in almost every other way carried seals identifying
them as being "Manufactured by Foster Mfg. Co., Inc. Factory at
Strong, Maine, U.S.A.," the abbreviation of the firm's name being
made necessary in part by the limited space in which it was printed.
Soon, the design of the flap of these boxes was changed so that the
seal was printed on a thumbnail cutout that could be glued down to
the outside of the box, thus also securing its contents. In this way,
the seal was in clear view, in very much the way the gold medal was
on boxes of Gold Medal toothpicks.[17]

Around the middle of the twentieth century, with an increasing
awareness of the value of branding, the Forster name began to be
given top billing on its packages. Thus, Worlds Fair toothpicks were
packaged as Forster Worlds Fair toothpicks. It was no doubt to distinguish its own packages from those of so many others that the
Forster firm announced (perhaps around the late 1950s, judging by
the typography and the fact that the corporate address was given as

For some time, the manufacturer's name was not visible on the outside of a closed box of Worlds Fair brand round toothpicks, but when opened a seal printed on the flap identified them as having been made at the Forster plant in Strong, Maine. Later, the seal was printed on a thumbnail section of the flap that could be exposed even when the box was closed and sealed.

Farmington), "All Forster packages are BLUE and GOLD now. These colors are your guide to quality woodenware." The Forster colors, brand names, and logos also appeared on some unlikely products, including quart-sized tins of pure maple syrup, "From the Sugar Maple Timberlands Which Produce Your . . . WOODEN-WARE." On the sides of the can, beneath the Forster Ideal and Worlds Fair brand logos, were lists of the company's products grouped according to industrial, institutional, surgical, and retail markets. Flat and round toothpicks headed that last list.[18]

Around 1970, the color scheme on Forster boxes changed to red, white, and blue, which seemed to be the palette that consumers expected on toothpick packages. In fact, boxes containing competing toothpicks were beginning to resemble each other. Contemporary packages of Forster Worlds Fair Brand; Birch Pix (not necessarily birch toothpicks), made by Merrill Woodenware Co., of Merrill, Wisconsin; and National cure-dents, made by Megantic Manufacturing Co., of Lac Megantic, Quebec, all shared the same red, white, and blue color schemes.[19] The packaging of Forster's Ideal and Worlds Fair lines shared other design elements besides colors, but in time the distinctions between the lines came to be dropped, and the only thing that distinguished a box of flat (for-

This ink blotter, dating from about 1960, advertised that Forster products—ranging from toothpicks and eating utensils to spring-action clothespins marketed as paper clips—were uniformly packaged in blue-and-gold colors and sold under the Ideal and Worlds Fair brand names.

merly Ideal) from a box of round (formerly Worlds Fair) Forster toothpicks were the words "flat" and "round."

The descriptive term "flat" associated with toothpicks made in the latter part of the twentieth century bore little connection with what it connoted in earlier times. The original toothpicks that Benjamin Sturtevant and Charles Forster manufactured were made from strips of veneer beveled at both edges. When the best of toothpicks were stamped out of such a strip, the two ends formed wafer-thin points that could be worked into some of the tightest interdental spaces to attack plaque on a considerable amount of tooth surface. Their "flatness" was an artifact of the manufacturing process; it was not their flatness that was remarkable, but their double-pointedness, the quality early packaging emphasized. When round (and also double-pointed) toothpicks were introduced, the term "flat" came to be used to distinguish the different geometries. The nonround toothpicks were sometimes designated "double-pointed tapered flat," even though they were not completely flat. Into the 1960s, Forster was still making its Ideal flat

The growing diversification of products sold under the Forster name included pure maple syrup packaged in tin cans imprinted in the contemporary corporate colors of blue and gold. The sides of the can listed the company's woodenware products.

toothpicks "tapered thin." However, over time the bevels appear to have grown progressively less pronounced, and hence the "points" less pointed. Today's flat toothpicks are truly that: bearing no evidence that the edges of the veneer strip whence they came were beveled or chamfered at all before being fed into the cutting machine. The resulting toothpick is typically one of uniform thickness (that of the veneer) with squared-off blunt tips that cannot be gotten between most teeth. The flat toothpick is, unfortunately, an example of technological degeneration. Evidently toothpick users today are not as discriminating as those of a century ago; otherwise they would demand a more efficacious product for getting into tight interdental spaces. Of course, the flat toothpicks are fine for skewering hors d'oeuvres and testing cakes.

After T. R. Hodgkins's death in 1977, his son, David L. Hodgkins, became chief executive officer and a vice president of the company.[20] Rumors soon began to circulate that he and his sister, Joan Hodgkins Gould, who were the major stockholders, would sell the company, but these rumors were denied. In fact, it was allowed, the company was being held in trust for the benefit of the Hodgkins grandchildren, and no change was expected before it passed on to them. After David Hodgkins died, in 1985, Joan Gould, who remembered a trip to Europe with her father that involved a great deal of "looking over sawmills," became chairman of the board of Forster.[21]

In the meantime, competition from abroad was mounting on all fronts. In 1978, wood-products makers from Maine appealed to the International Trade Commission for relief from "an import flood of low-priced spring-operated clothespins from China, Poland and Romania." The spring type had largely displaced the slotted clothespin, which had a tendency to jam on the clothesline. Spring pins from China, a gross of which could be made and shipped to the United States for $1.20, compared to the $2.04 domestic production cost, were found to be disruptive to the American market, but the Carter administration was reluctant to impose quotas on a country with which it had just signed a trade agreement. Eventually import quotas were imposed on all countries involved, but the terms were such that little protection was guaranteed.[22]

When T. R. Hodgkins died, Forster had gone outside the (Butler/Hodgkins) family to find a new president, and Ben Haug served in that capacity for eight years. Upon his retirement, in 1986, Richard J. Corbin, who specialized in the strategic development and restructuring of manufacturing companies, became president of the firm. Like a true specialist, Corbin immersed himself in the business, ready to tell a reporter that it was "socially acceptable to have a toothpick in your mouth" in the headquarters town of Wilton. He carried toothpicks in his pocket and insisted, "It's our heritage."[23] However, Corbin had assumed the helm during rough times.

Like so many other once-proud American enterprises, the toothpick and wood-utilization industry was not far from moving offshore. On the occasion of celebrating its "100th birthday," in 1987, Forster proudly advertised its grocery and housewares product line, which included "Colored Picks, Flat Toothpicks, Round 'Square-Center' Toothpicks, Rolling Pins, Sandwich/Cocktail Picks, Shish Kabob Skewers, Slotted Clothespins, Spring Clothespins, Plastic Spring Clothespins, Packaged Plastic Cutlery," all of which could be made more cheaply outside Maine and outside the United States. The firm also manufactured items in the categories of recreational equipment (croquet mallets, horseshoes, and bocce balls) and hotel and restaurant supplies (paper chop holders, wrapped toothpicks, sword picks, parasol picks, etc.). In addition, Forster made ice cream sticks and spoons; surgical applicator sticks and tongue depressors; hobby/craft and toy supplies, such as clothespin halves and baby clothespins; and hardware items, such as wooden dowels. At the time, there were plants in East Wilton (one for plastic and one for wood products), Mattawamkeag, Strong, and Wilton, the last of which was by then also the location of the corporate headquarters.[24]

But even the diversification of its product line was not enough to stave off trouble. Whereas in the early 1990s it could still be claimed that "the vast majority of wood products, such as clothespins, toothpicks and craft products, were made in the United States," that was no longer to be the case in the new century. For example, in 1996 there were over 500 wholesale customers for wooden clothespins, but by 2003 the number would have fallen to below 150—a decline of more than 70 percent. The accompanying erosion in business was attributable to imports.[25]

In 1985, the owners of Forster had let it be known that they did wish to sell the business, but even as it was on the market the company sought a permit to expand its East Wilton plant by fifty thousand square feet.[26] When the putative centennial of the Forster business was being celebrated, in 1987, the "stewardship" of the Butler family and its descendants was reported to have dated from around the turn of the century, or about the time that Charles Forster died.[27] As late as 1990, it was stated that "third-generation heirs still head the Forster company," which may have been literally true, but Frank Butler himself was neither a Forster, nor an heir, nor a founder of the original company that he incorporated in 1924 as the Estate of Charles Forster.[28] At the earliest, his involvement with the company began with his appointment as trustee of the estate in 1911. In any case, the three-generation Butler/Hodgkins family connection itself was broken in 1992, when Forster was bought "by a group of investors led by Advent International of Boston." The new owner seemed to care little for history or heritage and certainly not for sentimentality, however, and president Richard Corbin, who had been handpicked by Joan Gould, announced his retirement shortly after the "transition in ownership."[29]

In 1995, Advent's Forster was acquired by Diamond Brands Inc., of which it was to become an independent subsidiary. The Associated Press story on the acquisition repeated the inaccuracy of Forster's being founded in 1887 and propagated a new myth: that "Forster was owned by descendants of the founder until 1992."[30] Thus, the history of the company was bracketed in error.

No matter who owned the business, the toothpick and wood-novelty mills were essential to the small and remote Maine communities in which they were located, remaining among their largest employers. As the twentieth century was drawing to a close, there remained optimism in the Maine woods that the factories and their jobs would continue to exist. One reporter penned a paean to the industry worthy of publication in a literary journal. This found poetry appeared under the headline "Maine's Mills Endure." It carried the dateline Strong and began, "Clattering, hammering din. Lathes and saws laboring. Sweet smell of wet wood, with an unlikely hint of mint."[31]

But poetry on the features pages could not long keep at bay the message embodied in prose and numbers in the newspaper's business section. In 1992, Forster had employed seven hundred people making as many as twenty billion toothpicks and other products annually, with sales estimated to be sixty million dollars. In that same year the corporation had changed its name to the Forster Acquisition Co., and shortly thereafter it became the Wilton Liquidating Co., portending the demise of the business as it had been known.[32]

It had become increasingly difficult to tell the players in the business apart, even with a scorecard. In 2001, the Diamond Brands Operating Corp. went bankrupt, and the following year its assets were bought at auction by Jarden Corp., which had formerly been known as Alltrista Corp. According to the company's chairman and chief executive officer, Martin Franklin, a British-born "dealmaker" whom *Forbes* magazine would describe as "master of the mundane," the acquisition fit into the "idea of building a portfolio of domestic consumable brands with a high market share in niche markets," including such brand names as Ball and Kerr, so familiar to home canners.[33] Among the Diamond plants acquired was a large one in East Wilton, where plastic cutlery and clothespins were injection-molded. Two other plants, one in Cloquet, Minnesota, and the Forster factory in Strong, made wooden toothpicks, matches, and ice-cream and corn-dog sticks. Just before the deal was finalized, Franklin said of Forster that it was "a well-run business and we see no reason to make changes at this point." But within weeks, Jarden announced that it would close the Strong facility, which had been producing annually over seven billion toothpicks, once in more than fourteen varieties. Almost one hundred people would be put out of jobs, and the already struggling town would lose a substantial part of its economic base.[34]

In 2006, the Forster brand name still survived because of its business value, but it was not surviving with dignity and it may not endure for long. I recently bought a box of Forster round toothpicks that tells the tale of the decline of quality toothpicks in a condensed form. The box is printed in the same newer colors that Forster had been using most recently (blue, orange, yellow, and white), though in different combinations, to distinguish its packages on the supermarket shelf. The name "Forster" appears on four sides of the box, and it is printed in the same typeface that had become familiar to so many late-twentieth-century toothpick consumers. However, on one side panel, where I had become accustomed to reading the capsule, if incorrect, story of Charles Forster and the first wooden toothpick factory, there is in small print a record of acquisition: "Forster® is a registered trademark of Alltrista Consumer Products Company," which is identified as a division of Jarden Corporation, whose New York Stock Exchange symbol (JAH) is given. The box also carries the notice "© 2003 Alltrista Consumer Products Company," whose address is given as Muncie, Indiana, but which also could be contacted through its Web site: www.diamondbrands.com. As if this weren't bad enough, on one of the end panels of the box, beneath the Universal Product

Code stripes, in a light sans-serif typeface was the now all-too-familiar legend "Made in China."[35]

This was a shock. The box of 250 round toothpicks resembles in almost all respects a box of 750 flat Forster toothpicks I purchased perhaps a year earlier. But, on closer comparison, I saw that the older box was imprinted in three places with "Made in USA," and the only contact information read, in full, "Forster, P.O. Box 657, Wilton, Maine 04294," on the panel opposite the one containing the familiar text of "A Tradition of Quality." Another recently bought box of Forster "square/round tip" toothpicks was also marked "Made in USA," and identified the brand name Forster and the word in combination with the distinctive nicked line above it as registered trademarks of Alltrista Consumer Products Co., a division of Jarden Corp.

But there was one subtle difference between the boxes of American and Chinese toothpicks, and that was the placement of the nicked rule above the name Forster. I had always admired the thoughtful way that the nick lined up with the notch in the first *r* in Forster, as if the rule were the profile of the blade of a woodworking tool and the letter's notch had been produced by it. On the front of the newer box, the nicked line is displaced a bit to the right, so that the nick is over the space between the *r* and the *s*, evidently having no functional connection to the name that had come to be synonymous with American toothpick making. It is as if to signal that the toothpicks in the box were fashioned with a misaligned tool, which would explain their rough points and gouged shafts (and general lack of uniformity in color and texture). At least the new company had the (perhaps inadvertent) decency to omit the testament encapsulated in "A Tradition of Quality." Can the retirement of the Forster name itself be far behind?

At about the same time, I bought a package of Diamond toothpicks, which came in a bright yellow-and-blue box emblazoned in two places with the legend "Made in the U.S.A." This bordered the image of an American flag, complete with white stars, a blue field, and red stripes. Unfortunately, the square/round toothpicks inside might as well have been made in China. Their color was not uniform, their shafts were not smooth, and their points varied greatly in sharpness and in wholeness. At least one of the 250 picks in the newly opened box had a full quarter of its length broken off.

In the late twentieth century, Forster packaging featured a nicked rule aligned over the notch in the r *of "Forster." The nick did not line up over the notch in the* r *on a package of toothpicks made in China.*

One of the side panels of the package carried the corporate boiler-plate, but with "Diamond" in place of "Forster."[36]

The situation was a sad ending for a once-proud company. In addition to toothpicks, Forster had manufactured quality ice-cream sticks, tongue depressors, clothespins, and just about anything that could be made from veneered or turned hardwood, including yo-yos (called "spinners") and other recreational products.[37] By the late 1980s, Forster was the only American company making croquet sets for the backyard, and it also produced expensive tournament-quality sets sold under the brand name Skowhegan.[38] The company's custom-turning division made "thousands of items, ranging from salt and pepper shakers to wine racks and napkin rings." In its later years, continuing a tradition of welcoming ever-new products to which to apply its woodworking knowledge, Forster had even produced a small quantity of a novelty item known as a Twidd, an "inch-and-a-half-thick slice of a two-and-a-quarter-inch-diameter wooden dowel with holes bored into the opposite faces to facilitate thumb twiddling." The device was invented by Horace Knowles, a retired government speech writer who would seem to have had time on his hands. The "thumb twiddling toy" filled the need for something to assist in the pastime of thumb twiddling, which, when stated in patentese, "consists of the rotation of two thumbs about each other, i.e., orbitally about an imaginary axis between the two thumbs."[39]

Knowles had been turned down by more than a hundred manufacturers before he found Forster. The Twidd did receive some attention when it was introduced in the early 1980s, when people were looking for a "successor to the Pet Rock and other such useless but eye-catching novelties." He—and Forster and the entire town of Strong—may have hoped beyond hope that the Twidd might be it. Stories in the *Wall Street Journal* and the *New York Times* and elsewhere brought Knowles, his Twidd, and, by association, Forster—and Strong—a good deal of publicity, but whatever success it did have was short-lived.[40]

The situation was a far cry from conditions earlier in the century, when the town's economy thrived on the wood utilization industry, of which toothpick making was the most visible component. In 1922, on a menu and program card for the first annual banquet of the Strong Alumni Association, one attendee had printed "TOOTHPICKS" after the dessert and coffee, perhaps thinking that they should not be forgotten in a place they made famous. But by the end of the century the town whose fire trucks had once borne the slogan "Toothpick Capital of the World" had to begin looking for a new image. A "Welcome to Strong" sign that had carried the slogan "The Largest Toothpick Center in the World" was taken down years before the last plant closed in 2003.[41]

The rise and fall of the toothpick industry in Strong and elsewhere in Maine is more than a local story. The toothpick could be the paper clip or the pencil or the zipper—or the business of clothing or tools or toys. In 1946, a manufacturer of zippers took out a large advertisement in the *New York Times* in which he pleaded to labor to stop its strikes, to management to pay higher wages, and to government to trust the industry rather than regulate it. The zipper maker could not get sufficient supplies of tape, metal, labor, or "anything except customers." He insisted that if he and his competitors could get enough of what they needed to make zippers, "the competition would be so keen—that there would be no possibility of increased prices and no danger of inflation, at least so far as zippers are concerned." And he stated that what was true of his business was also "true of Buttons and Dresses and Fabrics and Steel and Autos and Locomotives and Finger Bowls and Tooth Picks . . ."[42] Thus, the story of one is the story of all, and in the story of the toothpick we may read a cautionary tale of anything that comes of invention, enterprise, and manufacturing.

Epilogue

THE CONFUSING TANGLE of fact and fiction about the history of the toothpick and its makers—as manifested in the contradictory information found on the World Wide Web and elsewhere—has proven difficult to unravel. Significant records of the business and the people involved in it seem to have been discarded like used toothpicks. The reliability of contemporary first- and third-person accounts, as contained in newspaper and magazine articles, has to be considered suspect, considering the inconsistencies between various accounts—and given the willingness of Charles Forster, among others, to use deception in the promotion of his products. Virtually all reports have had to be assumed tainted to some degree by the self-serving hype of their sources: the inventors, manufacturers, marketers, and managers.

Words were cheap in the nineteenth-century press, especially during the formative years of the toothpick industry in America. Acquisition and republication of another publication's story, with or without attribution, was common. Thus, repetition is no guarantee of accuracy, and confirmation by a second source can be illusory. Nevertheless, reading the record with a critical eye—and reading between the lines—has provided the basis for piecing together a story.

The most reliable words written about the toothpick as thing and its manufacture as process appear to be in the patent and trademark literature and files—in attested documents whose preparation usually involved attorneys, patent agents, and witnesses who had firsthand knowledge of processes and products. The toothpick, especially in its machine-made wooden form, has proven to be an object of intense interest among inventors. No detail of its design appears to have escaped notice, and no imperfection seems to have been considered too small to study and correct.

The sheer number of patents on toothpicks, machinery to make them, and ancillary equipment to hold, dispense, and facilitate the use of them virtually overwhelms in precision and detail information from all other written sources.[1] Granted, most of the text of a patent is dry reading, reciting as it does the numbered and lettered details of the technical figures, but some inventors embedded in their specifications considerable information about and criticism of the prior art. They accurately and fairly revealed their perceptions of its flaws and shortcomings, and consequently told it like it was. Patents thus record the incremental ratcheting up of a technology from its rough beginnings to its present, which almost always seems to appear to be more sophisticated than it will later prove to be. Though inventors do tend to tout their own inventions, they cannot easily misrepresent what they intend to supersede.

Where the written record has provided no unambiguously sound basis for providing continuity to the story, the artifact itself and its packaging have sometimes served to bridge gaps. Artifacts can be read not only in a cultural sense but also in a technical one. The scars of manufacturing and the claims of packaging speak volumes about how something was made and how it was presented by its maker. The quality of a well-made wooden toothpick—even one over a century old—does not diminish with age, even though the technology of making and marketing it may (or may not) have been surpassed long ago. Like a masterpiece of art, the accomplishment and presence of a well-manufactured product is something to behold. A thing of beauty is indeed a joy forever, even if it is a lowly toothpick.

Just as contradictory quotations gleaned from the World Wide Web provided a cautionary introduction to the history of the toothpick, so advice sought from it provides a warning that the great differences of opinion over toothpick use so prevalent in the later part of the nineteenth century have not yet been resolved. Indeed, the content of the Web in the early twenty-first century provides a veritable recapitulation of the ebb and flow of advice and assertion on health care and etiquette that has from earliest times been associated with the simplest machine.

Dentists seem always to have been ambivalent toward toothpicks. The American Dental Association makes no mention of them in its extensive historical timeline that begins with a Sumerian text from

5000 B.C., in which tooth decay is attributed to "tooth worms," and ends in 1997, when the U.S. Food and Drug Administration approved a new laser treatment for dental caries.[2] It is as if the oldest dental tool, which was used to attack and drive worms out of cavities, was too common to be included. This is odd, indeed, for a profession whose practitioners continue to use heavy-duty specialized double-pointed metal toothpicks to scratch and scrape off plaque and probe for cavities during checkups.

Not all contemporary dentists eschew mention of the toothpick. The members of the New York City–based practice Park Avenue Periodontal Associates "are big advocates of this old-fashioned method of cleaning teeth." On their Web site, the associates go on to explain that "toothpicks get to those last nooks and hiding places that both floss and toothbrushing miss" and cite studies that demonstrate that plaque is removed most effectively when all three tools of dental hygiene are used. At the same time, the practitioners recognize that as old as the toothpick is, it and its use still leave room for improvement. That is why, of course, its use should be supplemented.[3]

The Park Avenue Associates also answer the seemingly naive but perhaps leading question, "Do I hold the toothpick with my fingers?" Their obvious answer is that holding the toothpick in the fingers is "minimally effective" because "it's hard to maneuver the toothpick toward the back of your mouth, and near-impossible to clean the inside (tongue side) of your teeth." They are certainly describing what we all know, and the difficulty of using a straight toothpick in those areas is precisely why toothpicks with curved points have existed since ancient times. Countless inventors have devised descendants of such toothpicks and come up with holders that enable us to get straight-pointed toothpicks into areas otherwise impossible to reach. The holder that the Park Avenue Associates recommend is known as a Perio-Aid.[4]

When I asked my own dentist[5]—who practices not on Park Avenue in New York but on Shannon Road in Durham, North Carolina—about toothpicks, he directed his assistant to get me one of the Perio-Aid holders. It took her a while to find where they were kept. This was excused as being due to the office having just moved into new quarters, but it was my impression that she had not distributed a Perio-Aid in some time. This "recommended toothpick holder" is a five-inch-long plastic device that is bent and offset at each end in a manner suggestive of an auto mechanic's box wrench—or a dentist's professional pick. Each end of the Perio-Aid has a transverse hole into which a toothpick can be inserted and locked into place at the proper angle by means of a collar that can be screwed firmly against the pick. Practically speaking, inserting a full toothpick makes the tool unwieldy and potentially dangerous, and so only half a broken one is expected be used. The other half can be inserted into the other end of the tool to provide a fresh point. The half dozen toothpicks provided with my Perio-Aid have uncharacteristically blunt points, no doubt made deliberately so in order that they not puncture the gums or the roof of the mouth.

The Academy of General Dentistry in its response to the question "Is there any danger in using toothpicks?" reflects the profession's cautious ambivalence about them: "The occasional use of toothpicks to remove food particles is fine. Long term, vigorous use of toothpicks will cause abrasion of the teeth and gingival recession. Habitually leaving a toothpick in the mouth for long periods of time can cause excessive wearing of the teeth. . . . Also, be careful not to break off the tip in your gums." The answer confirms that prolonged toothpick use can leave marks on teeth that may be future fossils. A Chicago dentist seconds the observation when she states, "I can tell when I have a habitual toothpick user in my dental chair. There are the tell-tale signs of toothpick marks."[6] In spite of the caveats, today's dentists generally do not discourage toothpick use, but neither do they promote it as proactively as do the Park Avenue Periodontal Associates.

The general lack of enthusiasm of American dentists for toothpicks may just reflect the fact that their use is currently out of fashion. A typical contemporary "Etiquette Survival Guide" aimed at college students advises them that while dining they should "not use a toothpick, or apply makeup at the table." When a gentle reader asked Miss Manners, "When and where is it acceptable to use a toothpick?" she replied in her inimitable style, "When: When there is something stuck in your teeth. Where: In spaces between your teeth. Oh, and in the bathroom."[7]

There is, however, the recognition that toothpick use is, as it has always been, a matter of culture. The business traveler to China is advised that "toothpicks are usually offered between courses and at the conclusion of a meal. When using a toothpick, cover your mouth with your free hand for concealment." Similar advice is given to the traveler to Vietnam.[8]

In modern Europe, there is an old application of the toothpick that has captured the imagination of young tourist bar-hoppers. In Spain, the custom of eating *tapas,* which are served in the form of many plates of small portions of food, is said to have evolved from the southern "practice of laying a slice of bread or meat across the top of a sherry glass." The northern version of *tapas* is known as *pinchos* (or *pintxos* in Basque), which is said to derive from the Spanish word *pinchar,* meaning "to prick," after the manner in which wooden toothpicks skewered the food. "Pinchos are a distinct set of little dishes that are never served at meals. And you drink wine or beer with them, not sherry." When eating *pinchos* in a bar, the bill is tallied by counting up the used toothpicks.[9]

In America, it is hors d'oeuvres that are eaten off toothpicks, and this use of the implement has remained perfectly respectable. Indeed, it is possible to find rather fancy (if not gaudy and surprisingly expensive) toothpicks designed and sold exclusively for this purpose in fashionable wine, gourmet food, and kitchen gadget stores. The more lowly party picks, which are usually just ordinary wooden toothpicks disguised both in name and appearance to distance them from their humble origins, are readily available in supermarkets. The unadorned and unrenamed wooden toothpick is usually less conspicuously available.

But a toothpick is still a toothpick proudly displayed in one type of large retail outlet in America, and that is in discount stores like Kmart and Wal-Mart. Here, boxes of toothpicks—and usually a selection of styles—are shelved with kitchen-table accessories like napkin holders and salt and pepper shakers. Indeed, this is where one can also buy a toothpick shaker or dispenser for the kitchen table, where many inveterate toothpick users feel the things rightly belong. They also believe that toothpicks belong with the sugar and artificial sweeteners on restaurant tables, as they were in a dining room where I once ate in Carlsbad, New Mexico.

It would appear that in America the use of toothpicks has become largely a matter of class. Unlike in the late nineteenth century, when the urbane crutch-and-toothpick brigade proudly chewed its toothpicks on the steps of fine hotels and restaurants, now it is more the rural and less educated who openly chew theirs in the parking lots, if not at the counter itself, of big-box stores and fast-food establishments. In the longer view, toothpick use appears to have evolved from a hygienic practice of the privileged and fastidious to a habitual one of the working class. I have been waited on at our local state liquor store by a woman with a toothpick in her mouth, whose speech seemed not at all to be impaired by its presence. She simply talked around it. I have also watched a burly laborer in western Maine chewing a toothpick while carrying a meal of fried chicken to the table of a restaurant, but I never saw him remove the toothpick before beginning to eat.

On the other hand, I serve on a national board that includes a distinguished scientist who enjoys picking his teeth after meals. I had never seen any of the other board members join him in the practice until a delegation of us traveled to Sweden and Finland to be briefed on some of their technology. On that trip it was not uncommon for us to find ourselves eating at a table on which toothpicks were set in a holder beside the salt and pepper shakers, and on several occasions I observed several of my colleagues avail themselves of the readily available picks. One of our group habitually broke the toothpick in two before proceeding to use one of the halves in a most inconspicuous manner. Most who used the picks used them openly and with an air of familiarity.

I have also seen university professors who cannot pass the counter of a restaurant without grabbing a few toothpicks—one to be put into service immediately and the rest to be placed in the pocket for later use. In my university's medical center, there is a convenient little eating place that is frequented by medical doctors, faculty members, and administrators. On a recent visit, I saw beside the cash register a plastic cup full of cellophane-topped party picks. My guess is that some patrons had asked for toothpicks and the manager improvised from supplies on hand for the many evening wine-and-cheese receptions held in the room. Though they are now frequently kept out of sight, it is the rare restaurant that, when asked, cannot produce a holder full of (usually individually wrapped) toothpicks for a patron. On a recent stay in a historic and famous resort and spa in Asheville, North Carolina, I noticed that wrapped toothpicks were available in all its restaurants. But whereas they were piled high and deep in a large bowl on the counter of the breakfast room, a smaller and more discreet supply of them was held in a sherry glass in the inn's best dinner restaurant.

The World Wide Web may have the last word in confirming that the toothpick is perceived at the same time as being embraced and eschewed. The Internet domain name toothpick.com has been offered for sale. According to the offerer, "This domain name has an incredible *remember-it* value. Use it to promote a dentistry business, dentistry search and referral business or anything of dental or tooth/teeth significance. Or, if you [*sic*] selling toothpicks, then its [*sic*] obvious why anything else would be a mistake." As of early 2007 the asking price was twenty-five thousand dollars, and as summer approached the domain name remained unclaimed.[10]

NOTES

CHAPTER ONE: THE OLDEST HABIT

1. Hugh de Selincourt, *The Cricket Match* (Oxford: Oxford University Press, 1924), quoted in M. E. Richardson, "In Sickness and in Health," *British Dental Journal* 197, no. 9 (2004), 583.

2. James Joyce, *A Portrait of the Artist as a Young Man* (New York: Viking Press, 1956), 215, 229–35, 247.

3. Mary Joe Clendenin, "People and Toothpicks Make History," http://www.our-town.com/editorials/Edge-Clendenin/edgepicks .htm (Apr. 5, 2006).

4. William A. Agger, Timothy L. McAndrews, and John A. Hlaudy, "On Toothpicking in Early Hominids," *Current Anthropology* 45, no. 3 (2004), 403–4.

5. See, e.g., Christy G. Turner II, "Interproximal Grooving of Teeth: Additional Evidence and Interpretation," *Current Anthropology* 29, no. 4 (1988), 664–65.

6. John Noble Wilford, "But Did They Floss?" *New York Times*, Mar. 7, 1989, C1.

7. A. Siffre, "Note sur une usure spéciale des molaires du squelette de la Quina," *Bulletin de la Société Préhistorique Française* 8 (1911), 741–43.

8. See Christy G. Turner II and Erin Cacciatore, "Interproximal Tooth Grooves in Pacific Basin, East Asian, and New World Populations," *Anthropological Science* 106, supp. (1998), 85–94; Douglas H. Ubelaker, T. W. Phenice, and William M. Bass, "Artificial Interproximal Grooving of the Teeth in American Indians," *American Journal of Physical Anthropology* 30 (1969), 145–50; K. M. Reese, "Toothpicks May Have Been Among Humans' Early Tools," *Chemical and Engineering News*, Mar. 27, 1989, 48.

9. "First Pick," *New Scientist*, Apr. 22, 2000, 19.

10. See, e.g., Paul G. Bahn, "Early Teething Troubles," *Nature* 337 (Feb. 23, 1989), 693.

11. "Ancient Toothpicks Point to Evidence of Early Meat-Eating Hominids," *Journal of Dental Technology* 17, no. 5 (2000), 27; Irwin D. Mandel, "Why Pick on Teeth?" *Journal of the American Dental Association* 121 (Jul. 1990), 129.

12. Peter Ungar, quoted in "Ancient Toothpicks," 27.

13. Ubelaker, Phenice, and Bass, "Artificial Interproximal Grooving," 145.

14. Ibid., 147. See also Vincenzo Fromicola, "Interproximal Grooving of

Teeth: Additional Evidence and Interpretation," *Current Anthropology* 29, no. 4 (Aug.–Oct. 1988), 663–64.

15. See Bahn, "Early Teething Troubles."

16. Robert B. Eckhardt and Andrea L. Piermarini, "Interproximal Grooving of Teeth: Additional Evidence and Interpretation," *Current Anthropology* 29, no. 4 (Aug.–Oct. 1988), 668.

17. Leslea J. Hlusko, "The Oldest Hominid Habit? Experimental Evidence for Toothpicking with Grass Stalks," *Current Anthropology* 44, no. 5 (Dec. 2003), 738.

18. Eckhardt and Piermarini, "Interproximal Grooving," 670; J. A. Wallace, "Approximal Grooving of Teeth," *American Journal of Physical Anthropology* 40 (1974), 385–90.

19. Hlusko, "Oldest Hominid Habit," 738.

20. See Eckhardt and Piermarini, "Interproximal Grooving," 670; P. D. Schulz, "Task Activity and Anterior Tooth Grooving in Prehistoric California Indians," *American Journal of Physical Anthropology* 46 (1977), 87–92.

21. "Ancient Tooth Grooves: Take Your Pick," *Science News* 134, no. 15 (Oct. 8, 1988), 237.

22. Nell Boyce, "Regular Flossing Ancient Humans," *U.S. News & World Report* 135, no. 16 (Nov. 10, 2003), 69.

23. "Fossil Teeth Have Grooves," *News for You*, Apr. 7, 2004, 3.

24. "Toothpicks Popular Through Eons," *New Orleans Times-Picayune*, Dec. 15, 2003, 6; Hlusko, "Oldest Hominid Habit," 739–40.

25. "Grass Stalks Fit Bill for Earliest Toothpicks," *New Scientist Online News*, Nov. 6, 2003; see also Charles Choi, "Oldest Human Custom?" *New Scientist*, Nov. 8, 2003, 10.

26. Malvin E. Ring, "Our Daily Thread," *American Heritage of Invention & Technology* 21, no. 3 (Winter 2006), 32–36.

27. Henry J. Pratt, "The Axe Spares Toothpicks," *Mountain States Collector*, July 1990, http://mountainstatescollector.com/articles.asp?ArticleID=196 (Aug. 18, 2005).

28. "Alaskan Mummies," *New York Times*, Jan. 18, 1875, 5.

29. James Lowe, "A Very, Very Interesting Story About a . . . Toothpick?" *Journal of the Oklahoma State Dental Association* 59, no. 4 (Apr. 1969), 6.

30. Theodore R. Van Dellen, "How to Keep Well," *Washington Post, Times Herald*, Dec. 6, 1966, D11.

31. W. C. McGrew and Caroline E. G. Tutin, "Chimpanzee Tool Use in Dental Grooming," *Nature* 241 (Feb. 16, 1973), 477–78.

32. *Saturday Evening Post* 53, no. 14 (Nov. 1, 1873), 3.

33. L.B.U., "Monkeys," *Merry's Museum for Boys and Girls* 58, no. 3 (Sep. 1870), 139.

34. Jocko, "The Ourang Outang—*Simia satyres*," *Advocate of Science* 1, no. 15 (Jul. 4, 1833), 51.

35. See, e.g., Anjan Sundaram, "Scientists Study Gorilla Who Uses Tools," Associated Press, Oct. 18, 2005.

CHAPTER TWO: ARTIFACTS AND TEXTS

1. Curt Proskauer, "Oral Hygiene in the Ancient and Medieval Orient," *Ciba Symposia* 8, no. 8 (Nov. 1946), 439.
2. J. Menzies Campbell, "Tooth-Picks and Tooth-Brushes," *Dental Magazine and Oral Topics* 81, no. 1 (1964), 24.
3. Vincenzo Guerini, *A History of Dentistry: From the Most Ancient Times Until the End of the Eighteenth Century* (Pound Ridge, N.Y.: Milford House, 1969), 99.
4. Curt Proskauer, "The Care of the Mouth in Greece and Rome," *Ciba Symposia* 8, no. 8 (Nov. 1946), 455–56.
5. Elisabeth Bennion, *Antique Dental Instruments* (London: Sotheby's Publications, 1986), 137.
6. Proskauer, "Care of the Mouth," 455, 456.
7. Sydney Garfield, *Teeth, Teeth, Teeth: A Treatise on Teeth and Related Parts of Man, Land and Water Animals from Earth's Beginning to the Future of Time* (New York: Simon & Schuster, 1969), 28.
8. Campbell, "Tooth-Picks," 24.
9. Ibid.; James Wynbrandt, *The Excruciating History of Dentistry: Toothsome Tales & Oral Oddities from Babylon to Braces* (New York: St. Martin's Press, 1998), 210.
10. Quoted in Guerini, *History of Dentistry*, 94.
11. Ibid., 98.
12. William J. Carter, Bernard B. Butterworth, and Joseph G. Carter, *Ethnodentistry and Dental Folklore* (Overland Park, Kan.: Dental Folklore Books of Kansas City, 1987), 37.
13. Proskauer, "Oral Hygiene," 443.
14. Carter, Butterworth, and Carter, *Ethnodentistry*, 37.
15. Wynbrandt, *Excruciating History*, 210.
16. *Oliver Optic's Magazine: Our Boys and Girls* 5, no. 107 (Jan. 16, 1869), 44.
17. Arden G. Christen and Joan A. Christen, "A Historical Glimpse of Toothpick Use: Etiquette, Oral and Medical Conditions," *Journal of the History of Dentistry* 51, no. 2 (Jul. 2003), 63. Another source states that Muhammad, whose "teeth were well spaced," used a toothpick made of green palm wood. See "Muslim and Non-Muslim Views of Prophet Muhammad," http://www.najaco.com/literature/articles/august2000.htm (May 6, 2005).
18. Carter, Butterworth, and Carter, *Ethnodentistry*, 37. The last words of the French playwright and absurdist Alfred Jarry are reported to have been, "I am dying, please . . . bring me a toothpick." http://www.answers.com/topic/famous-last-words?hltoothpick (Apr. 5, 2006).
19. Garfield, *Teeth, Teeth, Teeth*, 28.
20. *The Works of Ben Jonson*, vol. II, Elibron Classico, www.elibron.com.
21. Act 1, Scene 1.
22. Christen and Christen, "Historical Glimpse," 62.
23. Bennion, *Antique Dental Instruments*, 139.

24. Theodore R. Van Dellen, "How to Keep Well," *Washington Post, Times Herald*, Dec. 6, 1966, D11.

25. Carter, Butterworth, and Carter, *Ethnodentistry*, 37.

26. Bennion, *Antique Dental Instruments*, 139.

27. Malvin E. Ring, *Dentistry: An Illustrated History* (New York: Abrams, 1985), 141. See also Laurence A. Garfin, "Toothpicks and Toothbrushes," *Dental Survey* 40, no. 4 (1964), 102.

28. Quoted in W. G. Probert, "Some Variations of Etiquette," *The Eclectic Magazine of Foreign Literature* 60, no. 2 (Aug. 1894), 212.

29. Ring, *Dentistry*, 141, 117.

30. Hans Sachs, *Der Zahnstocher und seine Geschichte: Eine kultur-geschichtlich-kunstgewerbliche Studie* (Hildesheim, Germany: Georg Olms, 1967).

31. "History of Domestic Things," *New-York Mirror* 10, no. 40 (Apr. 6, 1833), 320.

32. "Cha. Rex. Remem-Obiit-Ber," *Boston Post Boy*, issue 277 (Dec. 6, 1762), [2].

33. *Boston Post Boy*, issue 517 (Jul. 13, 1767), [4].

34. Ring, *Dentistry*, 50.

35. George H. Westley, "The True Story of Hamlet," *Peterson Magazine* 6, no. 9 (Sep. 1896), 909.

36. Christen and Christen, "Historical Glimpse," 63.

CHAPTER THREE: SUCKSACKS AND WHISKERS

1. Arden G. Christen and Joan A. Christen, "A Historical Glimpse of Toothpick Use: Etiquette, Oral and Medical Conditions," *Journal of the History of Dentistry* 51, no. 2 (Jul. 2003), 62.

2. "Botanical Notes," *Scientific American* 74, no. 18 (May 2, 1896), 278.

3. W. K. Higley, "The Cactus," *Current Literature* 25, no. 2 (Feb. 1899), 166.

4. "Under the Dome," *Washington Post*, Jan. 18, 1903, 19.

5. Arthur J. Burdick, "The Plant of Mystery," *American Magazine* 62, no. 2 (Jun. 1906), 204.

6. See, e.g., Baltimore Comb Factory advertisement in *Baltimore Patriot* 32, issue 104 (Oct. 29, 1828), [1].

7. "Will's Wonder-Book," *Merry's Museum for Boys and Girls* 1, no. 6 (Jun. 1868), 224; Alice, "Little Annie," *Saturday Evening Post* 57, no. 45 (Jun. 1, 1878), 7; Erasmus, quoted in Margaret Visser, *The Rituals of Dinner: The Origins, Evolution, Eccentricities, and Meaning of Table Manners* (New York: Grove Weidenfeld, 1991), 324; E.L.S., "Why Bowles Remained a Batchelor," *Puck* 3, no. 69 (Jul. 3, 1878), 11; Lord Playfair, "The Conquest of Waste," *Current Literature* 14, no. 1 (Sep. 1893), 79; Minnesota Historical Society, Item 1981.119.8, described in e-mail from Marcia Anderson to author, Mar. 6, 2006; "Political Miscellany," *Baltimore Patriot & Evening Advertiser* 1, issue 155 (Jul. 1, 1813), [2]; Claire Baer and Paul N. Baer, "A Story of the Toothpick," *Journal of Periodontology* 37 (Mar.–Apr. 1966), 159; "Raccoon Penis Bone 10 pcs Mountain Man

Toothpick E-1064," offered on eBay, Feb. 14, 2006; "Coyote Baculum Penis Bone Oosik Toothpick Brush Wolf," offered on eBay, Mar. 1, 2006; "4 Red Fox Baculum Penis Bone Oosik Toothpick," offered on eBay, Mar. 4, 2006.

8. Christen and Christen, "Historical Glimpse," 62; C. F. Holder, "Animal Traps and Trappers," *St. Nicholas* 12, no. 7 (May 1885), 525.

9. "Great West India Spider," *New-York Mirror,* May 8, 1841, 150.

10. *Scientific American* 2, no. 2 (Oct. 3, 1846), 10.

11. Dan McCowan, "Unusual Phases of Animal Behavior," *Forest and Stream* 96, no. 1 (Jan. 1926), 10.

12. "A Most Monstrous Monster," *Workingman's Advocate* 5, no. 33 (Mar. 29, 1834), 1.

13. Quoted in Malvin E. Ring, "Oddments in Dental History: The Toothpick in Literature," *Bulletin of the History of Dentistry* 30, no. 1 (Apr. 1982), 25.

14. J. Henry Hoffman, "Toothpicks," *Cal* 45, no. 9 (Mar. 1982), 7; Aristotle, "On Marvelous Things Heard," *Minor Works,* Loeb Classical Library, Sect. 7 (Cambridge, Mass.: Harvard University Press, 1980).

15. "Animal Wonders," *Scientific American* 32, no. 15 (Apr. 10, 1875), 230; "La Boudoir," *Monthly Magazine of Belles,* Sep. 1834, 133.

16. Quoted in "Reviews," *The Critic and Good Literature,* May 17, 1884, 230.

17. Elspeth Huxley, *Out in the Midday Sun: My Kenya* (London: Chatto & Windus, 1985), quoted in M. E. Richardson, "In Sickness and in Health," *British Dental Journal* 197, no. 9 (2004), 581–84.

18. Hoffmann, "Toothpicks," 7; *Harper's New Monthly Magazine,* Dec. 1854, 40; "Animal Wonders"; "Book IV," *Littell's Living Age* 29, issue 360 (Apr. 12, 1851), 65; "La Boudoir"; "Bric-a-Brac," *Saturday Evening Post* 61, no. 3 (Aug. 6, 1881), 3.

19. Theodore R. Van Dellen, "How to Keep Well," *Washington Post, Times Herald,* Dec. 6, 1966, D11.

20. "France," *New-York Daily Times,* Nov. 9, 1853, 2.

21. "Humorous," *Saturday Evening Post* 61, no. 6 (Aug. 27, 1881), 14.

22. Alaskan Treasures, "Eskimo Toothpick," author's collection.

23. "Natural History Notes," *Scientific American* 74, no. 16 (Apr. 18, 1896), 247; cf. *British Journal of Dental Science* 39, issue 682 (Jun. 15, 1896), 536.

24. Robert Dunn, "On the Chase for Volcanoes. II—The Hunt for Sea-Lion and for the Great Okmok Crater," *Outing Magazine* 51, no. 5 (Feb. 1908), 540.

25. "British Get Oddities from Many Places," *Washington Post,* Mar. 5, 1925, 20.

26. See "Wien Air Alaska," http://www.timetableimages.com/ttimages/wc1.htm (Aug. 18, 2005).

27. Card containing walrus whisker toothpick, Wien Consolidated Airlines, undated [1969–1973], author's collection.

28. Alaskan Treasures, "Eskimo Toothpicks."

CHAPTER FOUR: POOR GOOSE!

1. William Wolf, "A History of Personal Oral Hygiene—Customs, Methods and Instruments—Yesterday, Today, Tomorrow," *Bulletin of the History of Dentistry*, Oct. 1966, 56.

2. Curt Proskauer, "The Care of the Mouth in Greece and Rome," *Ciba Symposia* 8, no. 8 (1946), 456–57; Vincenzo Guerini, *A History of Dentistry: From the Most Ancient Times Until the End of the Eighteenth Century* (Pound Ridge, N.Y.: Milford House, 1969), 94.

3. Joyce M. Pluta, "History of the Toothpick and Toothbrush," typed manuscript, 1971, 2, Tennessee Dental Association Records, 1935–1980, Box 3, Folder 14 (Microfilm Accession No. 1727), Tennessee State Library and Archives; Guerini, *History of Dentistry*, 94.

4. Henry J. Pratt, "The Axe Spares Toothpicks," *Mountain States Collector*, Jul. 1990, http://mountainstatescollector.com/articles.asp?ArticleID=196. (Unless otherwise noted, Web sites were accessed between spring 2004 and winter 2006.)

5. "Sketch of Myself," *Boston Gazette* 13, issue 49 (Feb. 17, 1803), [1].

6. See, e.g., illustration in Hans Sachs, *Der Zahnstocher und seine Geschichte* (Hildsheim, Germany: Georg Olms, 1967), xi; Arden G. Christen and Joan A. Christen, "A Historical Glimpse of Toothpick Use: Etiquette, Oral and Medical Conditions," *Journal of the History of Dentistry* 51, no. 2 (July 2003), 62; Charles Dickens, *American Notes for General Circulation*, vol. I (London: Chapman and Hall, 1842), 127.

7. "Contingencies," *Vermont Gazette* 18, issue 44 (Oct. 16, 1827), [3].

8. John Beckmann, *A History of Inventions, Discoveries, and Origins*, 4th ed., trans. William Johnston, vol. I (London: Henry G. Bohn, 1846), 409, 407–8, 412.

9. "Quill Pens by the Million," *Boston Daily Globe*, Jun. 28, 1896, 29.

10. "The Inventor of Steel Pens," *Farmer's Cabinet* 68, issue 36 (Mar. 24, 1870), [1]; cf. www.penroom.co.uk.

11. "Utilization of the Goose Quill," *Scientific American* 38, no. 4 (Jan. 26, 1878), 50; "The Season and Crops," *New England Farmer* 14, no. 2 (Feb. 1862), 82.

12. See Ruth Clifford Engs, *Clean Living Movements: American Cycles of Health Reform* (Westport, Conn.: Praeger, 2000); *Home Journal*, Nov. 16, 1850, 2.

13. Dr. Lewis, "The Hair and Teeth," *Water-Cure Journal* 21, no. 6 (Jun. 1856), 129.

14. Box of quill toothpicks, no manufacturer, no date, author's collection; Tower Manufacturing and Novelty Co., 1903 catalog, 3.

15. See, e.g., http://www.flick.com/~liralen/quills/quills.html (Dec. 15, 2005).

16. Box of quill toothpicks.

17. "Preservation of the Teeth," *Saturday Evening Post*, Nov. 12, 1870, 8.

18. "Bailey's Plight," *New York Times*, Mar. 23, 1870, 2.

19. "The Poems of the Cradle," *Punchinello* 1, no. 22 (Aug. 27, 1870), 341.

20. "Miscellaneous Inventions," *Scientific American* 83, no. 11 (Sep. 15,

1900), 172; George W. Schellenbach, "Toothpick," U.S. Patent No. 656,479 (Aug. 21, 1900).

21. "The Evening Lamp," *Christian Union* 31, no. 2 (Jan. 8, 1885), 19.
22. James Wynbrandt, *The Excruciating History of Dentistry: Toothsome Tales & Oral Oddities from Babylon to Braces* (New York: St. Martin's Press, 1998), 212.
23. Sue Hubbell, "Let Us Now Praise the Romantic, Artful, Versatile Toothpick," *Smithsonian* 27, no. 10 (Jan. 1997), 77–78.
24. "Utilization of the Goose Quill," 50.
25. "The Polite Toothpick," *Washington Post*, Jan. 18, 1908, 6.
26. "Utilization of the Goose Quill."
27. Ibid.
28. Ibid.
29. Box of 250 Hygeia quill toothpicks, author's collection; "Mercurochrome Bottle, Old/Quill Toothpicks Orig Box," offered on eBay, May 22, 2006.
30. *Scientific American* 63, no. 4 (Jul. 26, 1890), 52.
31. "The Toothpick," *Boston Globe*, Jul. 26, 1883, 5.
32. http://www.sanex.com/better.html (Apr. 30, 2004); http://www.sanex.com/lore.html (Apr. 15, 2004).
33. "Gossip of Society," *Washington Post*, Aug. 10, 1908, 5; "Society Outside the Capital," *Washington Post*, Jul. 28, 1909, 7; see also "Kings Go into Trade," *Boston Daily Globe*, Jun. 29, 1913, 51.
34. "Fakirs' Funny Frauds," *Boston Globe*, Aug. 27, 1893, 26.
35. Scribble-Scrabble, "For the Cumberland Gazette," *Cumberland* (Mass.) *Gazette*, Mar. 16, 1787, [4].
36. L. B. France, "The Sportsman Tourist," *Forest and Stream* 37, no. 7 (Sep. 3, 1891), 122.
37. "The Monarch of the Pool," *Forest and Stream* 36, no. 15 (Apr. 30, 1891), 292.
38. W. M. Brown, "Mending a Broken Tip," *Forest and Stream* 48, no. 21 (May 22, 1897), 411.
39. "Expulsion by 'Quilling,'" *Medical and Surgical Reporter* 72, no. 4 (Jan. 26, 1895), 146.

CHAPTER FIVE: OUT OF THE WOODS

1. Chris Lefteri, *Wood: Materials for Inspirational Design* (Mies, Switzerland: RotoVision, 2003), 116.
2. Quoted in Elisabeth Bennion, *Antique Dental Instruments* (London: Sotheby's Publications, 1986), 139.
3. Arden G. Christen and Joan A. Christen, "A Historical Glimpse of Toothpick Use: Etiquette, Oral and Medical Conditions," *Journal of the History of Dentistry* 51, no. 2 (Jul. 2003), 62.
4. Bennion, *Antique Dental Instruments*, 139; Mary McNeill Scott, "The Useful Bamboo," *Current Literature* 16, no. 3 (Sep. 1894), 257; letter, Bo Krasse to author, Mar. 21, 2005; Theodore R. Van Dellen, "How to Keep Well," *Washington Post, Times Herald*, Dec. 6, 1966, D11; "The Way Wooden Toothpicks Are Made," *Scientific American* 13, no. 1

(Jul. 1, 1865), 7; A. J. Panshin, E. S. Harrar, J. S. Bethel, and W. J. Baker, *Forest Products: Their Sources, Production, and Utilization,* 2nd ed. (New York: McGraw-Hill, 1962), 304; http://www.dooyoo.co.uk/health-products/toothpicks/404840 (May 6, 2005); "Story of the Origin and Growth of the Toothpick Business Was Outstanding Feature at Dedication Wednesday," *Franklin* (Me.) *Journal,* Nov. 13, 1931, in "Maine Wood Util. Industry," clipping book #124, Reference Office, Maine State Library.

5. J. Henry Hoffman, "Toothpicks," *Cal* 45, no. 9 (Mar. 1982), 6.

6. S. Phillips Day, "Thoughts on Toothpicks," *Missouri Dental Journal* 6, no. 1 (1847), 25.

7. M. D. K. Bremner, "Mouth Care Through the Ages," *Dental Items of Interest* 70 (Jun. 1948), 606.

8. George L. Parmele, "Dental Curios," *Dental Practitioner and Advertiser* 27, no. 2 (Apr. 1896), 63.

9. Curt Proskauer, "The Care of the Mouth in Greece and Rome," *Ciba Symposia* 8, no. 8 (1946), 457.

10. Ibid., 455–56; see also William Wolf, "A History of Personal Oral Hygiene—Customs, Methods and Instruments—Yesterday, Today, Tomorrow," *Bulletin of the History of Dentistry,* Oct. 1966, 55–56.

11. Vincenzo Guerini, *A History of Dentistry: From the Most Ancient Times Until the End of the Eighteenth Century* (Pound Ridge, N.Y.: Milford House, 1969), 98.

12. J. Menzies Campbell, *Dentistry Then and Now* (Glasgow: privately printed, 1963), 291.

13. M. J. Kimery and R. E. Stallard, "The Evolutionary Development and Contemporary Utilization of Various Oral Hygiene Procedures," *Periodontal Abstracts* 16, no. 3 (Sep. 1968), 91.

14. Irwin D. Mandel, "Why Pick on Teeth?" *Journal of the American Dental Association* 121 (Jul. 1990), 129.

15. Kimery and Stallard, "Evolutionary Development," 91–92.

16. Charles Panati, *Panati's Extraordinary Origins of Everyday Things* (New York: Harper & Row, 1987), 208.

17. Wolf, "History of Personal Oral Hygiene," 54–64.

18. Curt Proskauer, "Oral Hygiene in the Ancient and Medieval Orient," *Ciba Symposia* 8, no. 8 (Nov. 1946), 443–44.

19. Leo Kanner, *Folklore of the Teeth* (New York: Macmillan, 1933), 80–81; Campbell, *Dentistry Then and Now,* 290.

20. Kanner, *Folklore of the Teeth,* 81–82.

21. Proskauer, "Oral Hygiene," 444.

22. Sen Nakahara and Kuninori Homma, "Tufted Toothpicks and Teeth Blackening Customs in *Ukiyoe,*" *Bulletin of the History of Dentistry* 34, no. 2 (Oct. 1986), 88.

23. Kanner, *Folklore of the Teeth,* 86.

24. Nakahara and Homma, "Tufted Toothpicks," 89; Proskauer, "Oral Hygiene," 446; see also http://www.nyu.edu/greyart/exhibits/shiseido/images/meiji/mei23.jpg.

25. Quoted in Stuart L. Fischman, "The History of Oral Hygiene

Products: How Far Have We Come in 6000 Years?" *Periodontology 2000* 15 (1997), 8–9; see also Christen and Christen, "Historical Glimpse," 63.

26. "The Use and Abuse of Toothpicks," *Medical and Surgical Reporter* 72, no. 19 (May 11, 1895), 681.

27. W. Irving Thayer, "Saving the Teeth," *Ladies' Home Journal* 11, no. 6 (May 1894), 19.

28. Gustav Kline, "Silver Toothpick with Safety Action," *Dental Abstracts* 2, no. 6 (1957), 372.

CHAPTER SIX: MADE IN PORTUGAL

1. "Extracts from an Address of the American Society of New York for Promoting Domestic Manufacturers, to the People of the United States," *Analectic Magazine* 9 (Jun. 1817), 1, 3.

2. Ibid., 4.

3. F. E. R. de Maar, "The Toothpick Carvers of Lorvão," *Bulletin of the History of Dentistry* 28, no. 1 (Apr. 1980), 11–15.

4. Leo Kanner, *Folklore of the Teeth* (New York: Macmillan, 1933), 94; "Toothpicks and the Like," *National Era* 13, issue 661 (Sep. 1, 1859), 137.

5. "The Toothpick Business," *Washington Post*, Jul. 21, 1900, 9.

6. "Where Toothpicks Are Made," *Boston Globe*, Sep. 16, 1904, 6. A contemporary reference gives the pay at "but a fraction more than 2 cents a day": "Large Output of Toothpicks," *New York Times*, Mar. 15, 1903, 33. Another reference gives the duty on the picks as "only 36 percent" ("The Toothpick Business").

7. See, e.g., "Odd Items from Everywhere," *Boston Globe*, Dec. 23, 1898, 12; J. B. Wilkinson, Jr., "Kindling, Toothpicks, and Brime," *New York Times*, Mar. 18, 1900, 25.

8. "Good Manners Bad Blow to Toothpick Industry," *Washington Post*, Apr. 11, 1950, 10.

9. De Maar, "Toothpick Carvers," 13.

10. Ibid., 13–14.

11. Ibid., 14.

12. Hans Sachs, *Der Zahnstocher und seine Geschichte* (Hildesheim, Germany: Georg Olms, 1967), 48.

13. De Maar, "Toothpick Carvers," 14; paper label on package of ten toothpicks: see http://luceweb.nyhistory.org/luceweb/itemdetail.htm?qmkey=501984 (May 10, 2005).

14. "Large Output of Toothpicks."

15. "Toothpicks and the Like"; de Maar, "Toothpick Carvers," 13.

16. "Toothpicks and the Like."

17. "Portuguese Toothpicks," *Atlanta Constitution*, Sep. 12, 1897, 5.

18. Author's collection; the toothpicks and booklet are contained in a plastic case bearing the notice "Copyright 1957 Tip'n Twinkle, Roanoke, Va."

19. *Saturday Evening Post*, Mar. 29, 1862, 6.

20. "One of the World-Builders," *Californian* 4, no. 23 (Nov. 1881), 387.

CHAPTER SEVEN: CHARLES FORSTER IN PERNAMBUCO

1. Quoted in William J. Carter, Bernard B. Butterworth, and Joseph G. Carter, *Ethnodentistry and Dental Folklore* (Overland Park, Kan.: Dental Folklore Books of Kansas City, 1987), 37. Cf. Malvin E. Ring, "Oddments in Dental History: The Toothpick in Literature," *Bulletin of the History of Dentistry* 30, no. 1 (Apr. 1982), 24–26, who quotes George L. Parmele extensively, including a passage that adds "pots for the German."

2. James Lowe, "A Very, Very Interesting Story About a . . . Toothpick?" *Journal of the Oklahoma State Dental Association* 59, no. 4 (Apr. 1969), 5.

3. J. Menzies Campbell, *Dentistry Then and Now* (Glasgow: privately printed, 1963), 288.

4. "The Grievance of Being Overestimated," *Littell's Living Age*, 150, issue 1944 (Sep. 17, 1881), 756.

5. "He Has His Tooth Pick, and Your Worship's Mess," *Baltimore Patriot* 43, no. 11 (Jul. 15, 1834), [2].

6. "Prehistoric Science en Fete," *Littell's Living Age* 148, issue 1992 (Feb. 5, 1881), 370.

7. Sydney Garfield, *Teeth, Teeth, Teeth: A Treatise on Teeth and Related Parts of Man, Land and Water Animals from Earth's Beginning to the Future of Time* (New York: Simon & Schuster, 1969), 91.

8. William S. Walsh, "In the Azores," *New Peterson Magazine* 3, no. 3 (Mar. 1894), 210.

9. "Brazil and Brazilian Society," trans. Asher Hall, *American Monthly Knickerbocker* 64, no. 2 (Aug. 1864), 133.

10. Mable and Lula Martin, "Two Kankakee Teachers Tell of S. America," *Chicago Tribune*, Jan. 7, 1951, G10.

11. Quoted in "Toothpicks and the Like," *National Era* 13, issue 661 (Sep. 1, 1859), 137.

12. "Charles Forster," Mar. 10, 1901, obituary file, Maine State Historical Society; Thomas Bellows Wyman, *The Genealogies and Estates of Charlestown, in the County of Middlesex and Commonwealth of Massachusetts, 1629–1818* (Boston: David Clapp and Son, 1879), 353. In fact, the Forster name would often be misspelled in what some considered the inferior way of Foster, complicating research into the family history.

13. Edward Jacob Forster, *The Pedigree and Descendants of Jacob Forster, Sen., of Charlestown, Mass.* (Charlestown, Mass.: privately printed, 1870), 8–9.

14. Wyman, *Genealogies and Estates*, 353–54.

15. "For Whom the School Is Named," *Boston Globe*, Jun. 25, 1886, 2.

16. Ibid.

17. Forster, *Pedigree and Descendants*.

18. An image of Charles Forster's oldest child, Charlotte, did appear in the *Los Angeles Times* for Aug. 26, 1908. It would make sense that a portrait of his son, Maurice, would have been placed in a memorial

building dedicated to him in Strong, Maine, in 1931, but a photographic study of his wife only was hanging there in the summer of 2006. However, what appear to be prints of a companion portrait of Maurice do survive in nearby towns (see chs. 19, 21).

19. Forster, *Pedigree and Descendants*, 7–9, 11–12, 19.
20. "Disasters, &c.," *Boston Globe*, Aug. 5, 1872, 7; "South America," *Boston Globe*, Mar. 30, 1872, 2.
21. "South America."
22. Ibid.
23. "Foreign Gossip," *Boston Globe*, Jun. 26, 1873, 3.
24. George B. Stanley, "Charles Forster Fathered the Toothpick Industry and Built First Toothpick Mill," *Rumford Falls* (Me.) *Times*, Aug. 26, 1943, in "Maine Wood Util. Industry," clipping book #124, Reference Office, Maine State Library (hereafter Maine State Library clipping book #124). In the same sentence, Stanley's recollection mistakenly identifies Charles Forster's birthplace as Buckfield, Maine, the town in which his son was born.
25. "Charles Forster," obituary, Maine Historical Society
26. Ibid., and see also obituary, *Daily Eastern* (Me.) *Argus*, Mar. 11, 1901; Forster, *Pedigree and Descendants*, 19; "Story of the Origin and Growth of the Toothpick Business Was Outstanding Feature at Dedication Wednesday," *Franklin* (Me.) *Journal*, November 13, 1931, in Maine State Library clipping book #124; Stanley, "Charles Forster Fathered."
27. "Charles Forster," obituary.
28. Joyce M. Pluta, "History of the Toothpick and Toothbrush," typed manuscript, 1971, 3, Tennessee Dental Association Records, 1935–1980, Box 3, Folder 14 (Microfilm Accession No. 1727), Tennessee State Library and Archives.
29. "A Day in a Past Century," *Boston Globe*, Jun. 30, 1893, 12.
30. "Saint Jonathan," *United States Review*, Feb. 1855, 100–101. A similar claim was made for Costa Rica, where "every peasant carries in his sash" a machete, which was "capable of felling a tree and making a tooth-pick": "Costa Rica and Its Railroad," *Overland Monthly and Out West Magazine* 10, no. 2 (Feb. 1873), 160.
31. Carolyn C. Cooper, "Myth, Rumor, and History: The Yankee Whittling Boy as Hero and Villain," *Technology and Culture* 44 (2003), 85–86.
32. F., "Bachelor's Quarters," *New York Times*, Oct. 28, 1874, 4.
33. "Topics of the Month," *Holdens Dollar Magazine of Criticisms, Biographies, Sketches, Essays, Tales, Etc.* 1, no. 2 (Feb. 1848), 122.
34. *Boston Liberator*, Jun. 22, 1849, 99; cf. *Pittsfield* (Mass.) *Sun*, Jun. 21, 1849, [1, 2].
35. Paul Cobden, "The Little Folks," *Christian Union*, Oct. 4, 1876, 280.
36. "Boston," *New York Times*, Sep. 28, 1853, 2.
37. "Toothpicks and Their Importation," *American Medical Times* 1 (Sep. 29, 1860), 233–34; see also "Toothpicks," *Scientific American* 3, no. 23 (Dec. 1, 1860), 353.

38. Lewis Brackley and Charles Lisherness, comps., *Strong, Maine* (Strong, Me.: Strong Historical Society, 1992), 47.

CHAPTER EIGHT: FROM PEGS TO RICHES

1. "B. F. Sturtevant, the Inventor," obituary, *Boston Daily Advertiser,* Apr. 18, 1890, 5.
2. J. D. Van Slyck, *New England Manufacturers and Manufactories,* vol. II (Boston: Van Slyck and Co., 1879), 612–13; see also Waldemar Kaempffert, ed., *A Popular History of American Invention,* vol. II (New York: Scribner's Sons, 1924), 409.
3. Marjorie Valliere Howe, "Brattleboro History Scrapbook: The Old Shoe Peg Factory," http://www.geocities.com/seekingthephoenix/pq/pegfactory.htm (Jul. 15, 2004).
4. "Affidavit of Alfred H. Batcheller," Application, Statement and Testimony, National Archives, College Park, Maryland (hereinafter NACP or understood), Patent Reissue No. 2,286 (original Patent No. 25,149) extension file, RG 241, Box 259, 31.
5. Ross Thomson, *The Path to Mechanized Shoe Production in the United States* (Chapel Hill: University of North Carolina Press, 1989), 41–42. See also Ross Thomson, "Crossover Inventors and Technological Linkages: American Shoemaking and the Broader Economy, 1848–1901," *Technology and Culture* 32, no. 4 (Oct. 1991), 1032.
6. "How Shoe-Pegs Are Made," *Scientific American* 21, no. 23 (Dec. 4, 1869), 362.
7. Thomson, *Path to Mechanized Shoe Production,* 35.
8. Howe, "Brattleboro History Scrapbook."
9. "Deposition of Benjamin F. Sturtevant," Ans. to Int. 22, Testimony for Applicant, Patent No. 26,627 extension file, RG 241, Box 261, 62.
10. "Affidavit of Calvin T. Sampson," Testimony for Applicant, Patent No. 26,627 extension file, RG 241, Box 261, 87.
11. Thomson, *Path to Mechanized Shoe Production,* 52.
12. Kaempffert, *Popular History,* 409; T. Rowell, "Making Shoe Pegs," U.S. Patent No. X4,027 (Feb. 11, 1825); Thomson, *Path to Mechanized Shoe Production,* 41–42.
13. See, e.g., Stephen K. Baldwin, "Improvement in Machines for Cutting Shoe-Pegs," U.S. Patent No. 2,725 (Jul. 16, 1842), reissued as U.S. Patent Reissue No. 409 (Nov. 4, 1856); see also Thomas A. Robertson, "Machine for Making Wooden Pegs," U.S. Patent No. 4,148 (Aug. 16, 1845).
14. "Peg Factory," *Scientific American* 2, no. 30 (Apr. 17, 1847), 237.
15. Lewis Brackley and Charles Lisherness, comps., *Strong, Maine* (Strong, Me.: Strong Historical Society, 1992), 48.
16. Norman T. Winans and Thaddeus Hyatt, "Manufacture of Splints or Sticks for Friction and Other Matches," U.S. Patent No. 1,867 (Nov. 26, 1840).
17. "Affidavit of Calvin T. Sampson," Patent No. 26,627 extension file, 88.
18. "B. F. Sturtevant Dead," *Boston Globe,* Apr. 18, 1890, 5.

19. "Affidavit of Alfred W. Bacon," Application, Statement and Testimony, Patent Reissue No. 2,286 extension file, 38; "Affidavit of John C. Bickford," Patent Reissue No. 2,286 extension file, 42.

20. B. F. Sturtevant, "Machine for Pegging Boots and Shoes," U.S. Patent No. 17,544 (Jun. 9, 1857); "Machine for Pegging Boots and Shoes," No. 17,998 (Aug. 11, 1857); "Method of Preparing Blanks for Shoe-Pegs," No. 19,282 (Feb. 2, 1858); "Machine for Pegging Boots and Shoes," No. 21,593 (Sep. 21, 1858).

21. B. F. Sturtevant, "Method of Preparing Blanks for Shoe-Pegs," U.S. Patent No. 19,282 (Feb. 2, 1858).

22. Kaempffert, *Popular History*, 409; Thomson, *Path to Mechanized Shoe Production*, 121; "Affidavit of Edward Arnold," Application, Statement and Testimony, Patent Reissue No. 2,286 extension file, 21; Van Slyck, *New England Manufacturers*, 613.

23. George J. Wardwell, "Machine for Pegging Boots and Shoes," U.S. Patent No. 11,346 (Jul. 18, 1854). As was common in the mid-nineteenth century, the letters *I* and *J* were often used interchangeably, leaving open the question of Wardell's middle initial. Although the initial appears twice as *I*—once associated with Wardwell as inventor and once as co-assignee—on the title page of the patent, Wardwell's middle initial is given as *J* on the drawing page, in the opening statement of the patent specification, and also in the "signature" at the end of the patent.

24. Thomson, *Path to Mechanized Shoe Production*, 127; J. J. Greenough, "Machine for Sewing or Stitching All Kinds of Straight Seams," U.S. Patent No. 2,466 (Feb. 21, 1842).

25. John Jas. Greenough, "Machine for Pegging Boots and Shoes," U.S. Patent No. 10,427 (Jan. 17, 1854), Reissue No. 269 (Jul. 4, 1854).

26. *Sturtevant v. Greenough*, Case No. 13,579, Circuit Court, District of Columbia, 23 F. Cas. 336, 1860 U.S. App. LEXIS 612 (Jun. 9, 1860); Van Slyck, *New England Manufacturers*, 613.

27. Elmer Townsend, "Pegging-Machine," U.S. Patent No. 27,085 (Feb. 7, 1860); "Machine for Introducing Pegs and Cement into Soles, &c.," No. 44,029 (Aug. 30, 1864).

28. Thomson, *Path to Mechanized Shoe Production*, 127.

29. B. F. Sturtevant, "Blank for Shoe-Pegging Machines," U.S. Patent No. 25,149 (Aug. 16, 1859), reissued as "Improvement in Blanks for Shoe-Pegs," Reissue No. 6,300 (Feb. 16, 1875); "Lathe Attachment for Cutting Veneers," No. 26,627 (Dec. 27, 1859).

30. "Deposition of Benjamin F. Sturtevant," Ans. to Int. 23, Testimony for Applicant, Patent No. 26,627 extension file, 62–63.

31. See Sturtevant, Patent No. 26,627, col. 1, lines 15–44.

32. "B. F. Sturtevant Dead."

33. Kaempffert, *Popular History*, 409.

34. "Deposition of Benjamin F. Sturtevant," Ans. to Int. 4, Testimony for Applicant, Patent No. 26,627 extension file, 50.

35. "Affidavit of Edward Arnold," Application, Statement and Testimony, Patent Reissue No. 2,286 extension file, 21.

36. "Deposition of Benjamin Davis," Ans. to Int. 6, 7, Testimony for Remonstrants, Patent No. 26,627 extension file, 29.

37. "Deposition of Benjamin F. Sturtevant," Ans. to Int. 13, Testimony for Applicant, Patent No. 26,627 extension file, 56–57.

38. "Affidavit of John A. Rowell," Application, Statement and Testimony, Patent Reissue No. 2,286 extension file, 47.

39. "How Shoe-Pegs Are Made."

40. "Deposition of Benjamin F. Sturtevant," Ans. to Int. 13, Testimony for Applicant, Patent No. 26,627 extension file, 57, 58.

41. Benjamin F. Sturtevant, "Improved Preparation of Shoe Pegs," U.S. Patent No. 35,902 (Jul. 15, 1862); [Vincent Tocco], "History of the B. F. Sturtevant and Westinghouse-Sturtevant Companies," http://www.sturtevantfan.com/index.html (Jul. 15, 2004).

42. Benjamin F. Sturtevant, "Improvement in Fuses for Explosive Shells," U.S. Patents No. 36,037, 36,038, 36,039 (Jul. 29, 1862); Benj. F. Sturtevant, "Improvement in Projectiles for Rifled Ordnance," U.S. Patent No. 36,116 (Aug. 5, 1862).

43. No. 25,149 (Aug. 16, 1859); "Deposition of Benjamin F. Sturtevant," Ans. to Cross-Int. 99, 115, Testimony for Applicant, Patent No. 26,627 extension file, 80, 82.

44. Van Slyck, *New England Manufacturers*, 614; "Statement of the Applicant," Testimony for Applicant, Patent No. 26,627 extension file, 7.

45. "Statement of the Applicant," Testimony for Applicant, Patent No. 26,627, extension file, 7, 11.

46. "Deposition of Caleb King," Ans. to Int. 3, Testimony for Remonstrants, Patent No. 26,627 extension file, 9; "Deposition of Caleb King," Ans. to Int. 1, Testimony for Applicant, Patent No. 26,627 extension file, 37.

47. "Deposition of Edward W. Simmons," Ans. to Int. 2, Testimony for Applicant, Patent No. 26,627 extension file, 24; "Story of the Origin and Growth of the Toothpick Business Was Outstanding Feature at Dedication Wednesday," *Franklin* (Me.) *Journal*, Nov. 13, 1931, "Maine Wood Util. Industry," in Maine State Library clipping book #124.

48. "Deposition of Caleb King," Ans. to Int. 20, Testimony for the Remonstrants, Patent No. 26,627 extension file, 10.

49. "Holland Tunnel Project," http://www.sturtevantfan.com/Holland Tunnel.html (Dec. 16, 2005).

50. Van Slyck, *New England Manufacturers*, 615.

51. Benjamin Franklin Sturtevant, "Improvement in Blowers," U.S. Patent No. 86,469 (Feb. 2, 1869); "Improvement in Rotary Blowers," U.S. Patent No. 86,470 (Feb. 2, 1869). See also Benjamin F. Sturtevant, "Improvement in Blower-Case," U.S. Patent No. 87,523 (Mar. 2, 1869); B. F. Sturtevant, "Improvement in Pressure-Blowers," U.S. Patent No. 92,489 (Jul. 13, 1869).

52. "The Exhibition of the American Institute," *Scientific American* 21, no. 18 (Oct. 30, 1869), 281.

53. Rene Bache, "Inventions That Made Fortunes," *Current Literature* 14, no. 1 (Sep. 1893), 136.

54. Advertisement, *Scientific American* 26, no. 4 (Jan. 20, 1872), 62.

55. "B. F. Sturtevant Dead"; Bache, "Inventions That Made Fortunes."

56. See http://www.sturtevantfan.com (May 6, 2005).

57. George L. Roberts, Argument for the Applicant, Patent No. 26,627 extension file, 17.

58. B. F. Sturtevant, Testimony for Applicant, Patent No. 26,627 extension file, 1.

59. "Statement of the Applicant," Testimony for Applicant, Patent No. 26,627 extension file, 10, 13–14.

60. "Deposition of Edward W. Simmons," Ans. to Cross-Int. 20, Patent No. 26,627 extension file, 28.

61. "Additional Statement of the Applicant," Patent No. 26,627 extension file, 16–18; "Statement of the Applicant," Patent No. 26,627 extension file, 5.

62. "Affidavit of Calvin T. Sampson," Patent No. 26,627 extension file, 85–89; "Affidavit of Alfred H. Batcheller," Patent No. 26,627 extension file, 89–92; "Commissioner's Decision," Dec. 24, 1873, Patent No. 26,627 extension file. The extension was granted contingent on Sturtevant disclaiming the patent's second claim, which involved cutters that divided the veneer into specified widths.

63. "B. F. Sturtevant Dead"; "Under the Death Arch," *Boston Globe,* Apr. 21, 1890, 5.

64. "Summary of Book Accounts Relating to Manufacture of Veneers," Testimony for Applicant, Patent No. 26,627 extension file, 18.

CHAPTER NINE: A FAMILY AFFAIR

1. "Toothpicks and Bobsleds," *Maine Central Messenger,* Apr.–May 1972, 10.

2. "Story of the Origin and Growth of the Toothpick Business Was Outstanding Feature at Dedication Wednesday," *Franklin* (Me.) *Journal,* November 13, 1931, in Maine State Library clipping book #124. See also Lewis Brackley and Charles Lisherness, comps., *Strong, Maine* (Strong, Me.: Strong Historical Society, 1992), 47. It is often said that the toothpicks Forster bought in Brazil were packed in a box, but it may be that they were also available tied in a bundle with a piece of string, wrapped in paper, or both, resembling the way handmade Portuguese toothpicks are packaged to this day.

3. Sue Hubbell, "Let Us Now Praise the Romantic, Artful, Versatile Toothpick," *Smithsonian* 27, no. 10 (Jan. 1997), 78; Robert E. Pike, *Tall Trees, Tough Men* (New York: Norton, 1967), 32.

4. "Story of the Origin and Growth."

5. George B. Stanley, "Charles Forster Fathered the Toothpick Industry and Built First Toothpick Mill" *Rumford Falls* (Me.) *Times,* Aug. 26, 1943, in Maine State Library clipping book #124.

6. "Statement of the Applicant," Testimony for Applicant, NACP, Patent No. 26,627 extension file, RG 241, Box 261, 7.

7. U.S. Patent Office, certified copy from digest of assignments, Patent No. 26,627 extension file.

8. "Deposition of Benjamin F. Sturtevant," Ans. to Int. 4, Testimony for Applicant, Patent No. 26,627 extension file, 50–51.

9. U.S. Patent Office to Benj. F. Sturtevant, letter dated Mar. 30, 1863, Patent No. 26,627 extension file.

10. R. H. Eddy to Commissioner of Patents, letter dated May 4, 1863, Patent No. 26,627 extension file.

11. Benjamin F. Sturtevant, "Improvement in Manufacture of Tooth-Picks," U.S. Patent No. 38,768 (Jun. 2, 1863).

12. Charles Tomlinson, ed., *Cyclopaedia of Useful Arts, Mechanical and Chemical, Manufactures, Mining, and Engineering* (London: George Virtue, 1854), s.v. "Nails."

13. Edward H. Knight, *Knight's American Mechanical Dictionary: Being a Description of Tools, Instruments, Machines, Processes, and Engineering; History of Inventions; General Technological Vocabulary; and Digest of Mechanical Appliances in Science and the Arts* (New York: J. B. Ford, 1875), s.v. "Nail-making Machine."

14. "The Richmond Rolling and Slitting Mill, and Nail Manufactory," *Farmer's Repository* 10, issue 502 (Nov. 19, 1817), [2].

15. "Reissue of Defective Patents," 35 USC 251 (see http://www.bitlaw.com/source/35usc/251.html); see also Nathan Reingold, "U.S. Patent Office Records as Sources for the History of Invention and Technological Property," *Technology and Culture* 1, no. 2 (1960), 161.

16. Benjamin F. Sturtevant, "Improvement in Tooth-Picks," U.S. Patent Reissue No. 1,661 (Apr. 26, 1864); cf. Patent No. 38,768.

17. Sturtevant, Patent Reissue No. 1,661.

18. Ibid.

19. J. D. Van Slyck, *New England Manufacturers and Manufactories*, vol. II (Boston: Van Slyck and Co., 1879), 615; *Dictionary of American Biography* (New York: Scribner's, 1928–36).

20. "Deposition of Benjamin F. Sturtevant," Ans. to Int. 18, Testimony for Applicant, Patent No. 26,627 extension file, 60.

21. U.S. Patent Office, certified copy from digests of assignments, Patent No. 26,627 extension file.

22. "Deposition of Benjamin F. Sturtevant," Ans. to Int. 18, Testimony for Applicant, Patent No. 26,627 extension file, 60.

23. "Summary of Book Accounts Relating to Manufacture of Veneers," Patent No. 26,627 extension file, 18. It may be that Sturtevant was persuaded to take over toothpick manufacturing from Charles Forster because he had to return to Brazil for some business reason.

24. "Deposition of Benjamin F. Sturtevant," Ans. to Int. 18, Testimony for Applicant, Patent No. 26,627 extension file, 60.

25. Stanley, "Charles Forster Fathered"; Edward Jacob Forster, *The Pedigree and Descendants of Jacob Forster, Sen., of Charlestown, Mass.* (Charlestown, Mass.: privately printed, 1870), 19.

26. Stanley, "Charles Forster Fathered"; Forster, *Pedigree and Descendants*, 19; "Obituary," *Portland* (Me.) *Press Herald*, Jan. 8, 1924, 2.

CHAPTER TEN: GOING WHERE THE WOOD IS

1. F. E. R. de Maar, "The Toothpick Carvers of Lorvão," *Bulletin of the History of Dentistry* 28, no. 1 (Apr. 1980), 13–14.

2. "Forster Mfg. Co., Inc. Celebrating Our 100th Birthday," broadside, Wilton Historical Society, vertical file, s.v. "Forster Mfg. Co." It is possible that Forster had a partner in this investment, and the likely candidate would have been the stationer Levi Tower, who will be introduced in the next chapter.

3. "Toothpicks and Their Importation," *American Medical Times* 1 (Sep. 29, 1860), 233–34.

4. "Toothpick Town in Maine," *Boston Globe,* Feb 10, 1907, SM12; George B. Stanley, "Charles Forster Fathered the Toothpick Industry and Built First Toothpick Mill," *Rumford Falls* (Me.) *Times,* Aug. 26, 1943, in Maine State Library clipping book #124.

5. "Story of the Origin and Growth of the Toothpick Business Was Outstanding Feature at Dedication Wednesday," *Franklin* (Me.) *Journal,* November 13, 1931, in Maine State Library clipping book #124.

6. "The Toothpick Industry in Maine," *Manufacturer and Builder* 19, no. 12 (Dec. 1887), 265; "Manufacturing, Mining, and Railroad Items," *Scientific American* 21, no. 9 (Aug. 28, 1869), 139; see also "Wooden Toothpicks," *Manufacturer and Builder* 2, no. 1 (Jan. 1870), 25.

7. "Story of the Origin and Growth."

8. J. C. Brown, "Improvement in Machines for Cutting Splints," U.S. Patent No. 43,177 (Jun. 21, 1864); "The Way Wooden Toothpicks Are Made," *Scientific American* 13, no. 1 (Jul. 1, 1865), 7.

9. Benjamin F. Sturtevant, "Improvement in Tooth-Picks," U.S. Patent Reissue No. 1,661 (Apr. 26, 1864).

10. "Wooden Toothpicks," *Manufacturer and Builder* 2, no. 1 (Jan. 1870), 25; U.S. Patent Reissue No. 1,661.

11. "Toothpick Industry in Maine"; "Wooden Toothpicks"; "Manufacturing, Mining"; "Toothpick Town in Maine"; Edward Jacob Forster, *The Pedigree and Descendants of Jacob Forster* (Charlestown, Mass.: privately printed, 1870), 19; cf. "Forster Mfg. Co.," which has Forster making toothpicks in Cambridgeport, Mass., just prior to moving to Maine, and setting up a "small mill" at Sumner, in Oxford County.

12. Robert C. Stewart, "The Noble & Cooley Center for Historic Preservation Dedicated to Celebrating Yankee Ingenuity," *Society for Industrial Archeology Newsletter* 35, no. 4 (2006), 12; Silas Noble and James P. Cooley, "Improvement in Tooth-Pick Machines," U.S. Patent No. 123,790 (Feb. 20, 1872); "Improvement in Machines for Making Tooth-Picks," U.S. Patent No. 127,360 (May 28, 1872).

13. Noble and Cooley, U.S. Patent No. 127,360.

14. "Toothpick Industry in Maine"; "Manufacturing, Mining." Hand production is assumed to yield an output of five hundred toothpicks per hour over a ten-hour working day; see "Toothpicks and Their Importation" for the output of South American natives.

15. Estes' Imported Orange Wood Toothpicks, ca. 2005; "Story of the Origin and Growth."

16. "Story of the Origin and Growth."

17. Lewis Brackley and Charles Lisherness, comps., *Strong, Maine* (Strong, Me.: Strong Historical Society, 1992), 47.

18. "A North Franklin Enterprise," *Farmington* (Me.) *Chronicle,* Jan. 25, 1883 (photocopy of clipping provided by Jerry J. DeVos).

19. Larry Wolters, "Gag Bag," *Chicago Tribune,* May 24, 1964, I34. Though Forster was still living in Massachusetts when the census was taken in July 1870, it is possible that Freeman had set up the factory in Maine before then.

20. Stanley, "Charles Forster Fathered." There has been speculation that Forster's first plant was in Andover or Minot, two other towns in west-central Maine's Oxford County; see Brackley and Lisherness, *Strong,* 48.

21. "Current Notes," *Boston Globe,* Mar. 26, 1872, 1; "New England," *Boston Globe,* Jul. 10, 1872, 6; "Toothpicks," *Dental Cosmos* 14, no. 11 (1872), 625.

22. "Toothpicks"; "New England."

23. "New England," *Boston Globe,* Dec. 8, 1873, 5; "Brevities," *Atlanta Constitution,* Dec. 19, 1873, 2; "Brevities," *Atlanta Constitution,* Jan. 21, 1874, 2.

24. Stanley, "Charles Forster Fathered." Though the birch tree grows in a "swath more than a thousand miles wide (more or less from New York City to Hudson Bay) and reaches westward and northwestward to the Pacific," trees large enough for building bark canoes or providing satisfactory wood for toothpick making were confined to the much smaller area. See John McPhee, *The Survival of the Bark Canoe* (New York: Farrar, Straus and Giroux, 1975), 17.

25. Wm. E. S. Whitman, *Wealth and Industry of Maine* (Augusta, Me.: Sprague, Owen, and Nash, 1873), 131; Samuel W. Matthews, *Report of Industrial and Labor Statistics, Maine, 1897* (Augusta, Me.: Kennebec Journal Print, 1897), 32.

26. Ibid. According to Brackley and Lisherness, *Strong,* 48, Forster's purchase of the starch factory in 1887 is taken as the year of the founding of the Forster Manufacturing Co.; cf. Miriam Lisherness, "A Memory of Years Long Past," *Lewiston–Auburn* (Me.) *Daily Sun,* Jan. 30, 1964 (copy of clipping provided by Joan Gould).

27. Brackley and Lisherness, *Strong,* 48.

28. *Webster's New International Dictionary of the English Language,* 2nd ed., unabridged, s.v. "locofoco." On the Locofoco Party, cf. Brackley and Lisherness, *Strong,* 48.

29. Robert Goff Stubbs, "A Toothpick Industry in Maine, 1913," typewritten report, Maine State Library, [3].

30. Brackley and Lisherness, *Strong,* 48–49; cf. Lisherness, "Memory of Years Long Past," in which Porter is misspelled Potter.

31. For a town history of Dixfield, see *The Early History of Dixfield, Maine* (Dixfield, Me.: Dixfield Bicentennial Committee and Dixfield Histor-

ical Society, 1976), 7, 10. The place's original name of Holmantown honored Jonathan Holman, who in 1789 went to Boston to purchase the land. Upon incorporation in 1803, the name was changed to Dixfield because Elijah Dix "promised to create a town library." When it was not forthcoming, some outraged citizens complained, and Dix sent a box of old books. Today, the town is proudly served by the Ludden Memorial Library.

32. Ibid., 60.
33. "Toothpick Town in Maine."
34. "Toothpick Patent Run Out," *Manufacturer and Builder* 11, no. 12 (Dec. 1879), 280.
35. Alfred Cole and Charles F. Whitman, *History of Buckfield, Oxford County, Maine, from the Earliest Explorations to the Close of the Year 1900* (Buckfield, Me.: privately printed, 1915), 430–31, 345. According to another source, the plant at Buckfield was established in 1880 and later moved to Andover, and then to Dixfield; see Stubbs, "Toothpick Industry," [1]. As for Forster's being born in Buckfield, see Stanley, "Charles Forster Fathered."
36. "Wooden Toothpicks."
37. "Charles Forster Fathered"; "Pine Tree State Furnishes World with Toothpicks," *Portland* (Me.) *Sunday Telegram,* Jun. 19, 1927, in Maine State Library clipping book #124.
38. "A Fortune in Toothpicks," *Scientific American* 35, no. 23 (Dec. 2, 1876), 356. See also "The American Inventor," *Scientific American* 36, no. 20 (May 19, 1877), 307.
39. Baker's Old Hickory Tooth Pick, label, in "Advertisements and Labels for Miscellaneous Products," Library of Congress, Prints & Photographs Division, LOT 10771.
40. Before legislation enacted in 1861, patent protection was granted for fourteen years, with the possibility of a seven-year extension. Patents issued after the legislation were effective for seventeen years, with no right of extension. See, e.g., Nathan Reingold, "U.S. Patent Office Records as Sources for the History of Invention and Technological Property," *Technology and Culture* 1, no. 2 (1960), 163.
41. "Toothpick Patent Run Out."

CHAPTER ELEVEN: AN ARTICLE FOR THE MILLION

1. Robert Goff Stubbs, "A Toothpick Industry in Maine, 1913," typewritten report, Maine State Library, [1].
2. "Manufacturing, Mining, and Railroad Items," *Scientific American* 21, no. 9 (Aug. 28, 1869), 139.
3. "Story of the Origin and Growth of the Toothpick Business Was Outstanding Feature at Dedication Wednesday," *Franklin* (Me.) *Journal,* November 13, 1931, in Maine State Library clipping book #124.
4. "Fact and Comment," *Atlanta Constitution,* Dec. 7, 1876, 2.
5. "Story of the Origin and Growth."
6. Ibid.
7. Ibid.

8. "L. L. Tower, Dean of Stationers, Dead," *New York Times*, Jun. 20, 1912, 11; George B. Stanley, "Charles Forster Fathered the Toothpick Industry and Built First Toothpick Mill," *Rumford Falls* (Me.) *Times*, Aug. 26, 1943, in Maine State Library clipping book #124.

9. Levi L. Tower, "Design for an Inkstand," U.S. Patent No. 4,291 (Aug. 9, 1870); Daniel M. Somers, "Improvement in Suspension Rings," U.S. Patent No. 106,738 (Aug. 23, 1870).

10. "Copartnership Notices," *New York Times*, May 2, 1871, 6.

11. "L. L. Tower."

12. Greene Consolidated Copper Co., advertisment of stock offering, *Los Angeles Times*, Oct. 30, 1900, I14; see also *New York Times*, June 6, 1900, 12; "L. L. Tower"; Sampson & Murdock Co., *Boston Directory*, 1905.

13. "How's Business?" *Boston Globe*, Jul. 5, 1891, 13; "New England Conference," *Zion's Herald*, Sep. 10, 1902, 1181.

14. "Contributions for the Fifty-fourth (Colored) Regiment," *Liberator*, Apr. 10, 1863, 59; "Golden Anniversary Week," *Boston Globe*, Sep. 27, 1908, 16.

15. Stanley, "Charles Forster Fathered."

16. Ruth Robbins Adamo, "Forster Manufacturing Dream Come True for Couple," *Lewiston* (Me.) *Daily Sun*, Aug. 8, 1988, 11.

17. "Story of the Origin and Growth."

18. Business card in the collection of Peter Stowell, West Gardiner, Me.

19. Although the byline on the newspaper story ("Charles Forster Fathered") reads "George B. Stanley," it is likely that the *B* was a typo for *P* and that George B. Stanley was in fact the same George P. Stanley who in 1910 was identified in the census as a superintendent in a toothpick mill in Dixfield—the Forster plant—and who held patents for various inventions relating to machines for making toothpicks and boxing them.

20. Stanley, "Charles Forster Fathered."

21. "The 100th Anniversary of Forster Manufacturing Co.," *Congressional Record*, 100th Cong., 1st sess., 1987, Senate 133, no. 127 (Jul. 30, 1987), S10957.

22. "Toothpick Town in Maine," *Boston Globe*, Feb. 10, 1907, SM12.

23. "100th Anniversary."

24. "History of the Union Oyster House," http://www.unionoysterhouse .com/Pages/history.html (Apr. 30, 2004).

25. Abby Zimet, "Maine's Mills Endure," *Maine Sunday Telegram*, Apr. 19, 1998, 16B.

26. "Importation of Toothpicks," *Pittsfield* (Mass.) *Sun*, Jun. 21, 1849, [1]; "Toothpicks," *Scientific American* 3, no. 23 (Dec. 1, 1860), 353.

27. "Toothpicks and the Like," *National Era* 13, no. 661 (Sep. 1, 1859), 137; "Obituary Notes," *New York Times*, Feb. 19, 1917, 10.

28. "Atlantic Cable: The General Rejoicing," *New York Times*, Aug. 7, 1858, 5.

29. "The Habits of Good Society," *Vanity Fair* 1, 19 (May 5, 1860), 302.

30. "Toothpicks," *Scientific American* 3, no. 23 (Dec. 1, 1860), 353.

31. Stubbs, "Toothpick Industry," [31].

32. "Wooden Toothpicks," *Manufacturer and Builder* 2, no. 1 (Jan. 1870), 25.

33. "Toothpick Patent Run Out," *Manufacturer and Builder* 11, no. 12 (Dec. 1879), 280; "The Funny Things We Do," *Farmers' Cabinet* 77, issue 18 (Nov. 5, 1878), [1].

34. "New York," *Boston Globe*, Mar. 20, 1872, 2.

35. "Wooden Toothpicks," *Farmers' Cabinet* 72, issue 27 (Jan. 14, 1874), [1].

36. Stubbs, "Toothpick Industry," [1].

37. "Story of the Origin and Growth"; cf. Stanley, "Charles Forster Fathered."

38. "1776–1876," *Philadelphia Inquirer*, July 4–5, 1876, various pages.

39. Ibid.

40. "Story of the Origin and Growth."

41. Alphons Krizek, "Improvement in Toothpicks," U.S. Patent No. 90,855 (Jun. 1, 1869).

42. *Scientific American* 21, no. 2 (Jul. 10, 1869), 31.

43. Quoted in William Grimes, "A Car Culture's Not-So-Complex Creator," *New York Times*, Aug. 12, 2005, E33.

44. *Subject-Matter Index of Patents for Inventions Issued by the United States Patent Office from 1790 to 1873, Inclusive*, vol. III (1874; reprint, New York: Arno Press, 1976).

45. John F. Sturdy, "Combined Case for Pen, Pencil, Knife, Toothpick, &c.," U.S. Patent No. 24,502 (Jun. 21, 1859); Richard Cross, "Watch-Key, Toothpick, and Toggle Combined," U.S. Patent No. 52,687 (Feb. 20, 1866); Henry E. Graham and Richard D. Child, "Improvement in Combined Watch-Key and Toothpick," U.S. Patent No. 97,391 (Nov. 30, 1869).

46. Garret S. Boice, "Improvement in Tooth-Picks," U.S. Patent No. 167,733 (Sep. 14, 1875).

47. Howard B. Stephenson and William H. Bennett, "Improvement in Tooth-Pick Cases or Holders," U.S. Patent No. 173,364 (Feb. 8, 1876).

48. William E. Blake, "Improvement in Tooth-Picks," U.S. Patent No. 106,773 (Aug. 30, 1870).

49. George Clark, Jr., "Improvement in Tooth-Picks," U.S. Patent No. 174,619 (Mar. 14, 1876).

50. American Film Institute catalog.

51. "Wooden Toothpicks"; *Farmer's Cabinet*, Jan. 14, 1874; "The Wooden Toothpick," *Christian Union* 31, no. 2 (Jan. 8, 1885), 19.

52. "The Tooth-Pick Business," *Dental Register* 39, no. 4 (1885), 206.

53. See Tower Manufacturing and Novelty Co. catalog, 1899, 1; "Tooth-Pick Business."

54. "News Notes," *Saturday Evening Post*, 64, no. 26 (Jan. 10, 1885), 13; "Many Items of Interest," *Scientific American* 57, no. 4 (Jul. 23, 1887), 49; "The Toothpick Industry in Maine," *Manufacturer and Builder* 19, no. 12 (Dec. 1887), 265; Edward C. Williams, "Effect of Invention upon Labor and Morals," *Chautauquan* 28, no. 6 (Mar. 1899), 559.

CHAPTER TWELVE: ADVICE AND DISSENT

1. J. Menzies Campbell, *Dentistry Then and Now* ([Glasgow]: privately printed, 1981), 288.
2. Ibid., 286, 288.
3. M.D.K. Bremner, "Mouth Care Through the Ages," *Dental Items of Interest* 70 (Jun. 1948), 606.
4. Quoted in Harvey Green, *Wood: Craft, Culture, History* (New York: Viking, 2006), 277, with "meat" taken to be "meal" and "knife of straw" to be "knife or straw."
5. Quoted in Malvin E. Ring, *Dentistry: An Illustrated History* (New York: Harry N. Abrams, 1985), 141.
6. Quoted in W. G. Probert, "Some Variations of Etiquette," *Eclectic Magazine of Foreign Literature* 60, no. 2 (Aug. 1894), 212.
7. Quoted in Margi Warner, "History of Oral Hygiene Apparatus," *Dental Assistant* 32 (Jul.–Aug. 1963), 22.
8. *Much Ado About Nothing*, Act 2, scene 1.
9. *King John*, Act 1, scene 1.
10. *The Winter's Tale*, Act 4, scene 4.
11. George L. Parmele, "Dental Curios," *Dental Practitioner and Advertiser* 27, no. 2 (Apr. 1896), 63.
12. Laurence A. Garfin, "Toothpicks and Toothbrushes," *Dental Survey* 40, no. 4 (1964), 102, 104.
13. Sue Hubbell, "Let Us Now Praise the Romantic, Artful, Versatile Toothpick," *Smithsonian* 27, no. 10 (Jan. 1997), 76.
14. "History of Domestic Things," *New-York Mirror* 10, no. 40 (Apr. 6, 1833), 320.
15. Ann Peteri, "A History of Oral Hygiene Methods," *Dental Assistant* 39 (Jan. 1970), 25; James Cross Giblin, *From Hand to Mouth: Or, How We Invented Knives, Forks, Spoons, and Chopsticks and the Table Manners to Go with Them* (New York: Crowell, 1987), 52.
16. "Rules for Good Behaviour at Table," *Ladies' Literary Cabinet* 4, no. 18 (Sep. 8, 1821), 139.
17. "Miscellany: On Presentiments," *Massachusetts Spy, or Worcester Gazette* 46, issue 34 (Oct. 22, 1817), [4].
18. Quoted in Harley Spiller to author, e-mail message, May 29, 2006, reporting on the exhibit "Feeding Desire," then running at the Cooper-Hewitt National Design Museum.
19. *Oxford English Dictionary*, s.v. "picktooth." The adjective carried the meaning "of a person's manner, conversation, etc.: leisurely, easy, idle, as if resting and picking one's teeth after a meal; (in later use *esp.*) affectedly indolent or languid."
20. "London," *Connecticut Courant*, issue 706 (Aug. 4, 1778), [1].
21. Campbell, *Dentistry Then and Now*, 289.
22. *Newport Mercury*, issue 1976 (Feb. 25, 1800), [3].
23. *American Watchman* 4, issue 305 (Jul. 8, 1812), [3]; cf. *American Watchman*, issue 304 (Jul. 4, 1812), [3].

24. See Noel D. Turner, *American Silver Flatware, 1837–1910* (South Brunswick, N.J.: A. S. Barnes, 1972), 58.

25. R.J., "Communications," *Southern Rose* 5, no. 16 (Apr. 1, 1837), 125.

26. "Off-Hand Sketches—No. 1," *The New-Yorker* 4, no. 24 (Mar. 3, 1838), 790.

27. J.G., "The Disappointed Anglers, An Indian Tale," *The New-Yorker* 4, no. 15 (Dec. 30, 1837), 643.

28. "A Day on Lake Erie," *The New-Yorker* 6, no. 5 (Oct. 20, 1838), 71.

29. "Foula: The Island of the Birds," *Current Literature* 5, no. 3 (Sep. 1890), 226.

30. Sarah P. E. Hawthorne, "Bits of Wood Lore," *New York Observer and Chronicle* 79, no. 37 (Sep. 12, 1901), 349.

31. "The Toilet—No. 2," *Monthly Magazine of Belles-Lettres and the Arts*, Nov. 1834, 225.

32. Quoted in "Picking One's Teeth," *Pittsfield* (Mass.) *Sun* 34, issue 1,766 (Jul. 24, 1834), [1].

33. "A Good Conundrum," *Pittsfield* (Mass.) *Sun* 41, issue 2090 (Oct. 8, 1840), [1].

34. T.T., "The Life and Opinions of Job Doolittle," *The Knickerbocker* 11, no. 3 (Mar. 1838), 228.

35. "The Stranger I Met at My Club," *The New-Yorker* 4, no. 25 (Mar. 10, 1838), 804.

36. "Boz and the Dramatists," *The Corsair* 1, no. 19 (Jul. 20, 1839), 303.

37. J. H. Ingraham, "Glimpses at Gotham.—No. II," *Ladies' Companion*, Feb. 1839, 177.

38. Alfred Bunn, "The Stage, Before and Behind the Curtain," *The New World* 1, 9 (Aug. 1, 1840), 140.

39. "Advice to a Fool," *Vanity Fair* 2, no. 49 (Dec. 1, 1860), 275.

40. Old Nick, "Taking the Bull by the Horns," *Oliver Optic's Magazine. Our Boys and Girls* 1, no. 25 (Jun. 22, 1867), 297.

41. Ippodamos, "Sporting in Yankee-land," *American Turf Register and Sporting Magazine* 15 (Oct. 1844), 593.

42. "Diary of Town Trifles," *New Mirror* 3, 5 (May 4, 1844), 72.

43. Thomas Chandler Haliburton, *The Clockmaker: Series One, Two, and Three*, ed. George L. Parker (Ottawa: Carleton University Press, 1995), xx, 2.

44. S. Phillips Day, "Thoughts on Toothpicks," *Missouri Dental Journal* 6, no. 1 (1874), 26–27.

45. "A Toothpick Tragedy," *New York Times*, Sep. 3, 1884, 4.

46. "Barnard's 'Lincoln'—A Statue of the Spirit of Democracy or a Defamatory Caricature, Which?" *Current Opinion* 63, no. 5 (Nov. 1917), 339.

47. "Art. II.—Does Crime Increase?" *Pennsylvania Journal of Prison Discipline and Philanthropy* 9, no. 2 (Apr. 1854), 74.

48. "Our City Charities—No. IV," *New York Times*, Apr. 7, 1860, 2.

49. "Much in Little," *Flag of Our Union*, Jan. 9, 1869, 31.

50. *Punchinello* 2, no. 29 (Oct. 15, 1870), 48.

51. J. Henry Hoffmann, "Toothpicks," *Cal* 45, no. 9 (Mar. 1982), 6.

52. "The Street-Venders of New York," *Scribner's Monthly* 1, no. 2 (Dec. 1870), 113.

53. "Peddlers and Beggars," *The Galaxy* 8, no. 4 (Oct. 1869), 567.

54. *Saturday Evening Post*, Jul. 18, 1857, 2.

55. "Varieties," *Appletons' Journal of Literature, Science and Art* 3, no. 63 (Jun. 11, 1870), 670.

56. "Trenton," *New York Times*, Mar. 15, 1873, 4.

57. "Pack Pedlers' Goods," *Boston Globe*, Oct. 15, 1893, 31.

58. "New-York City," *New York Times*, Apr. 3, 1855, 1.

59. F. H. Stauffer, "Necessity of Activity," *Home Magazine* 5, no. 3 (Mar. 1855), 189.

CHAPTER THIRTEEN: THE NASTY INSTRUMENT

1. Dio Lewis, "Notes and Maxims About Health," *Scientific American* 23, no. 22 (Nov. 26, 1870), 337; *Washington* (D.C.) *Evening Star*, Nov. 11, 1870, 2.

2. "Hints to Young Teachers," *Connecticut Common School Journal and Annals of Education* 5, no. 11 (Nov. 1858), 321.

3. "Pickpockets at Albany—Heavy Loss Sustained by a Lady," *New York Times*, Aug. 1, 1864, 5.

4. "Gifts for Gentlemen," *Arthur's Home Magazine* 57 (Jan. 1888), 96.

5. See, e.g., *Scientific American* 47, no. 26 (Dec. 23, 1882), 413; Sophie May, "The Little Hatchet," *Merry's Museum, Parley's Magazine, Woodworth's Cabinet, and the Schoolfellow* 12 (Jul. 1, 1861), 161; "Brilliant Wedding at Mont Vernon," *Farmers' Cabinet* 73, no. 51 (Jun. 30, 1875), [2].

6. *Herald of Health* 3, no. 6 (Jun. 1864), 235; "Notice to the Benevolent," *Forest and Stream* 1, no. 20 (Dec. 25, 1873), 317.

7. Burt G. Wilder, "Six Fingers and Toes," *Old and New* 1, no. 2 (Feb. 1870), 158.

8. "Hagerty and Balch," *New York Times*, Sep. 23, 1871, 3.

9. "About the Hair," *Godey's Lady's Book and Magazine* 73 (Oct. 1866), 304; "Happy Hits," *Washington Post*, Oct. 31, 1902, 6.

10. "Table-Talk," *Appletons' Journal of Literature, Science and Art* 5, no. 112 (May 20, 1871), 594.

11. "The Tooth-Pick Remedy," *New York Times*, Jul. 24, 1878, 4.

12. "Minor Matters and Things," *Appletons' Journal of Literature, Science and Art* 8, no. 195 (Dec. 31, 1872), 707.

13. "Portraits from Vanity Fair," *Forest and Stream* 7, no. 6 (Sep. 14, 1876), 89.

14. "Seasoning," *National Police Gazette* 37, no. 167 (Dec. 4, 1880), 2.

15. "Answers to Correspondents," *Arthur's Illustrated Home Magazine* 41, no. 12 (Dec. 1873), 547.

16. Colonel Lynx, "City Characters," *National Police Gazette*, Oct. 11, 1879, 15; "Puck's Exchanges," *Puck* 1, no. 9 (May 1877), 12.

17. Lynx, "City Characters."

18. *Oxford Dictionary of National Biography* (New York: Oxford Univer-

sity Press, 2004); *Universal Dictionary of the English Language* (New York: Peter Fenton Collier, 1898), s.v. "tooth-pick."

19. H. J. Byron, "Crutch and Toothpick," *Songs of the Rice Surprise Party* (Boston: Louis P. Goullaud, [1880]); C. R. Dockstader, "Toothpick and Crutch," *J. W. Pepper's Favorite Publications of Vocal Banjo Music: Comprising the Latest and Best Stage and Popular Comic and Sentimental Favorites* (Philadelphia: J.W. Pepper, 1883).

20. S. Phillips Day, "Thoughts on Toothpicks," *Missouri Dental Journal* 6, no. 1 (1874), 25–26.

21. Ibid., 27.

22. J. E. Cravens, "Thoughts on Toothpicks, or an American Sermon from an Englishman's Text," *Missouri Dental Journal* 6, no. 1 (1874), 28–29.

23. "Notes from California," *New York Times*, Jan. 19, 1874, 2.

24. "A Toothpick Tragedy," *New York Times*, Sep. 3, 1884, 4.

25. Ibid.

26. "Stringing for a Dollar," *National Police Gazette* 44, no. 357 (Jul. 26, 1884), 6.

27. "The Toothpick," *Boston Globe*, Jul. 26, 1883, 5.

28. "The Yankee in the South," *Scientific American* 48, no. 8 (Feb. 24, 1883), 114.

29. "Mr. Dickson's Strange Fancies," *Washington* (D.C.) *Evening Star*, Mar. 5, 1885, 8.

30. "Toothpicks and Toothpicking," *Washington* (D.C.) *Evening Star*, Sep. 21, 1889, 6.

31. Grace Denio Litchfield, "The Knight of the Black Forest," *Century Illustrated Magazine* 29, no. 2 (Dec. 1884), 180.

32. Theodore Child, "Through the States," *Eclectic Magazine of Foreign Literature* 43, no. 4 (Apr. 1886), 456.

33. "Toothpick Output of United States Is Thirty Billions," *Washington Post*, Oct. 10, 1926, F9; "Affairs in West Washington," *Washington* (D.C.) *Evening Star*, Dec. 1, 1886, 1.

34. "Toothpicks and Toothpicking." The spelling *Bazaar* was not adopted until 1929.

35. "American Habits," *Washington* (D.C.) *Evening Star*, Jun. 5, 1888, 5.

36. Laura J. Rittenhouse, "An Afternoon Call," *Godey's Lady's Book* 120, no. 717 (Mar. 1890), 220.

37. Quoted in *Current Literature* 9, no. 2 (Feb. 1892), 305.

38. "Her Defender Knocked Out," *Washington Post*, Nov. 14, 1892, 5.

39. Rose Terry Cooke, "Some 'Don'ts for Girls,'" *Ladies' Home Journal and Practical Housekeeper* 5, no. 8 (Jul. 1888), 17.

40. *Ladies' Home Journal* 7, no. 9 (Aug. 1890), 24; *Ladies' Home Journal* 10, no. 1 (Dec. 1892), 37.

41. Ruth Ashmore, "The Country Girl," *Ladies' Home Journal* 10, no. 8 (Jul. 1893), 15.

42. Maud C. Cooke, *Social Etiquette, or Manners and Customs of Polite Society, etc.* (London, Ont.: McDermid & Logan, 1896), 186.

43. "Around the Dinner Table," *Good Housekeeping* 19 (Sep. 1894),

109–10; "The Evening Lamp," *Christian Union* 31, no. 2 (Jan. 8, 1885), 19.

44. "An Ugly Habit," *Los Angeles Times,* Dec. 5, 1890, 6.

45. "Questions and Answers," *New York Times,* Sep. 3, 1893, 11; "Uncle Peter's Chair," *Christian Union* 43, no. 2 (Jan. 8, 1891), 56.

46. "Condemning the Toothpick," *Los Angeles Times,* Nov. 11, 1905, II5.

47. *Washington Post,* Jun. 27, 1912, 6.

48. "When There Are Guests in the House," *American Farmer* 10, no. 20 (Oct 15, 1891), 237.

49. "Utilization of the Goose Quill," *Scientific American* 38, no. 4 (Jan. 26, 1878), 50.

50. "Care of the Teeth," *Medical and Surgical Reporter* 71, no. 9 (Sep. 1, 1894), 294.

51. "The Tooth-Pick Business," *Dental Register* 39, no. 4 (1885), 206.

CHAPTER FOURTEEN: THE TOOTHPICK MAN

1. "The Wooden Toothpick," *Christian Union,* Jan. 8, 1885, 19.

2. "Pine Tree State Furnishes World with Toothpicks," *Portland* (Me.) *Sunday Telegram,* Jun. 19, 1927, in Maine State Library clipping book #124.

3. Quoted in "Slings and Arrows," *Boston Globe,* Jan. 23, 1882, 4; see also "New England Industries," *Boston Globe,* Jan. 19, 1882, 3.

4. "The Toothpick," *Boston Globe,* Jul. 26, 1883, 5.

5. Ibid.

6. "Two-Horse Load of Tooth-Picks Daily," *New York Times,* Jun. 4, 1883, 5.

7. According to one source, the annual output of the Harbor Springs factory was 1.8 billion picks, which would mean that it operated about 240 days per year; "Odd Items from Everywhere," *Boston Globe,* Sep. 10, 1888, 8.

8. "The Wooden Toothpick Trade," *Washington Post,* Nov. 11, 1888, 10.

9. "The Toothpick Industry in Maine," *Manufacturer and Builder* 19, no. 12 (Dec. 1887), 265–66; "Many Items of Interest," *Scientific American* 57, no. 4 (Jul. 23, 1887), 49.

10. Robert Donald, "Trusts in the United States," *Eclectic Magazine of Foreign Literature* 52, no. 2 (Aug. 1890), 223.

11. "Anything to Oblige," *National Police Gazette* 45, no. 392 (Mar. 18, 1885), 2; "Many Items of Interest."; "Toothpick Industry in Maine."

12. "Combinations in Trade," *New York Times,* Mar. 19, 1888, 1; Ernest A. Harris, "Machine for Cutting Tooth-Picks," U.S. Patent No. 375,812 (Jan. 3, 1888).

13. *Atlanta Constitution,* Jul. 4, 1889, 4.

14. *Washington Post,* Jun. 10, 1891, 4; see also "Slander's Tongue," *Chicago Tribune,* Jun. 21, 1891, 13.

15. "Editorial Points," *Boston Globe,* Sep. 29, 1883, 4.

16. *Chicago Tribune,* Jan. 9, 1894, 12.

17. "The Toothpick Business," *Washington Post,* Jul. 21, 1900, 9; "The

Wood Novelty Business," *Eleventh Annual Report of the Bureau of Industrial and Labor Statistics for the State of Maine 1897* (Augusta, Me.: Kennebec Journal Print, 1898), 30–38.

18. Charles C. Freeman, "Machine for Polishing and Compressing Tooth-Picks," U.S. Patent No. 358,029 (Feb. 22, 1887).

19. "The Toothpick Factory," *Farmington* (Me.) *Chronicle,* May 24, 1888; John C. F. Scammon, "Machine for Making Toothpicks," U.S. Patent No. 539,011 (May 7, 1895).

20. United States Patent Office to Charles C. Freeman, Jan. 21, 1886, NACP, RG 241, Box 11534, Patent No. 448,647 file.

21. Charles Drew to Patent Office, Dec. 13, 1887, Patent No. 448,647 file.

22. Wm. Burke to Charles C. Freeman, Mar. 30, 1888, Patent No. 448,647 file.

23. Chas. Drew to Commissioner of Patents, Oct. 29, 1889, Patent No. 448,647 file.

24. Wm. Burke to Charles C. Freeman, Nov. 14, 1889, Patent No. 448,647 file.

25. Chas. Drew to Commissioner of Patents, Nov. 23, 1889, Patent No. 448,647 file. The substitute paragraph contained a typographical error, citing Freeman's machinery patent as No. 359,029 instead of 358,029. The error would find its way into the toothpick patent as issued.

26. Robt. P. Harris to C. C. Freeman, Aug. 16, 1890, Patent No. 448,647 file.

27. John J. Halsted to Commissioner of Patents, Aug. 1890, Patent No. 448,647 file.

28. Charles C. Freeman, "Tooth-Pick," U.S. Patent No. 448,647 (Mar. 24, 1891). As issued, the patent bore the familiar notation "No Model," suggesting that while the actual toothpicks submitted may have been acceptable as evidence of a physical embodiment of the invention described in the patent specification, they were not considered formal models.

29. Petition dated Sep. 24, 1885, Patent No. 448,647 file.

30. Silas Noble and James P. Cooley, "Improvement in Tooth-Pick Machines," U.S. Patent No. 123,790 (Feb. 20, 1872).

31. *Time,* Nov. 7, 1969, cited in William J. Carter, Bernard B. Butterworth, and Joseph G. Carter, *Ethnodentistry and Dental Folklore* (Overland Park, Kan.: Dental Folklore Books of Kansas City, 1987), 38.

32. *The Early History of Dixfield, Maine* (Dixfield, Me.: Dixfield Bicentennial Committee and Dixfield Historical Society, 1976), 83; *Phillips* (Me.) *Phonograph,* Jul. 31, 1891.

33. Robert Goff Stubbs, "A Toothpick Industry in Maine, 1913," typewritten report, Maine State Library, [31]; Charles Forster, "Trade-Mark for Wooden Toothpicks," U.S. Trade-Mark No. 35,328 (Oct. 30, 1900).

34. Author's collection; "Very Rare Civil War Period Full Box Wooden Toothpicks," offered on eBay, Oct. 30, 2005.

35. Charles Forster, "Trade-Mark for Wooden Toothpicks," U.S. Patent Office Trade-Mark No. 35,457 (Nov. 20, 1900).

36. Tower Manufacturing and Novelty Co. catalog, 1899, 1; see also 1903 catalog, 1.

37. Box offered on eBay, Apr. 18, 2005; author's collection.

38. "Franklin County," *Farmington* (Me.) *Chronicle,* Jan. 29, 1891.

39. Maurice W. Forster, "Device for Removing Waste Products," U.S. Patent No. 644,855 (Mar. 6, 1900).

40. Charles Forster, "Trade-Mark for Wooden Toothpicks," U.S. Trade-Marks No. 35,325 (Oct. 30, 1900); 35,326 (Oct. 30, 1900); 35,327 (Oct. 30, 1900); 35,328 (Oct. 30, 1900); 35,455 (Nov. 20, 1900); 35,456 (Nov. 20, 1900); 35,457 (Nov. 20, 1900); 35,523 (Dec. 4, 1900); 35,524 (Dec. 4, 1900); 35,525 (Dec. 4, 1900).

41. Stubbs, "Toothpick Industry," [33–34].

42. Author's collection.

43. Alfred Cole and Charles F. Whitman, *A History of Buckfield, Oxford County, Maine, from the Earliest Explorations to the Close of the Year 1900* (Buckfield, Me.: privately printed, 1915), 345; E. C. Bowler, comp. and ed., *An Album of the Attorneys of Maine* (Bethel, Me.: News Publishing Co., 1902), 104.

44. *Hawkins v. Hersey,* 86 Me. 394, 30 A. 14 (1894).

45. Cole and Whitman, *History of Buckfield,* 345; "Story of the Origin and Growth of the Toothpick Business Was Outstanding Feature at Dedication Wednesday," *Franklin* (Me.) *Journal,* November 13, 1931, in Maine State Library clipping book #124.

46. "Charles Forster," obituary, Mar. 10, 1901, Maine Historical Society.

47. *Portland* (Me.) *Directory and Reference Book, 1889,* 47, in Portland Room, Portland Public Library.

48. The will was signed on Oct. 1, 1900, a Monday, and the first trade-mark applications were filed on Oct. 5, 1900, that Friday. See Charles Forster, "Trade-Mark for Wooden Toothpicks," U.S. Trade-Mark No. 35,325.

49. "Charlotte M. Forster," obituary, Apr. 1, 1902, Maine Historical Society.

50. "Charles Forster," obituary.

51. "Toothpick Industry in Maine."

52. "San Diego Woman Defendant," *Los Angeles Times,* Jul. 28, 1909, I5; Bill Caldwell, *Enjoying Maine* (Portland, Me.: Guy Gannett Publishing Co., 1977), 94.

53. On obituaries, see Janice Hume, *Obituaries in American Culture* (Jackson: University Press of Mississippi, 2000), esp. 16, 31, 97.

54. Charles Forster, last will and testament dated Oct. 1, 1900, Registry of Probate Office, Cumberland County (Me.) Probate Court, Docket No. 1507.

55. Ibid.

56. "Toothpick Industry in Maine."

57. Forster, last will and testament.

58. *Inhabitants of Peru v. Estate of Charles Forster,* 109 Me. 226, 83 A. 670 (1912).

59. Cole and Whitman, *History of Buckfield,* 345.

CHAPTER FIFTEEN: CORDS OF PAPER BIRCH

1. "Large Output of Toothpicks," *New York Times,* Mar. 15, 1903, 33; Harvey Green, *Wood: Craft, Culture, History* (New York: Viking, 2006), 140, 142; S. T. Dana, *Paper Birch in the Northeast,* U.S. Department of Agriculture Forest Service Circular 163 (Washington, D.C.: U.S. Government Printing Office, 1909), 5.

2. John McPhee, *The Survival of the Bark Canoe* (New York: Farrar, Straus and Giroux, 1975), 21. For decades, the insolubility of birch bark frustrated medical researchers, who believed that the betulonic acid in it would prove useful in developing a drug effective for fighting prostate cancer. See Delthia Ricks, "Bark Yields Cancer Hope," *Newsday,* Jul. 27, 2006.

3. "Paper Birch or White Birch," http://www.domtar.com/arbre/english/pboul.htm (Jun. 13, 2005); McPhee, *Survival of the Bark Canoe,* 47.

4. Robert Goff Stubbs, "A Toothpick Industry in Maine, 1913," typewritten report, Maine State Library, [6].

5. "Big Toothpick Crop," *Boston Globe,* Apr. 29, 1906, 38; "Large Output of Toothpicks"; Claire Baer and Paul N. Baer, "A Story of the Toothpick," *Journal of Periodontology* 37, no. 2 (Mar.–Apr. 1966), 160.

6. George E. Walsh, "Maine's Wood Novelty Mills," *Scientific American* 87, no. 17 (Oct. 25, 1902), 269. The term "wood utilization industry" is used by the Maine State Library Reference Department in its clipping book (#124) of newspaper stories relating to the making of toothpicks, clothespins, ice-cream sticks, etc.

7. Walsh, "Maine's Wood Novelty Mills."

8. "A North Franklin Enterprise," *Farmington* (Me.) *Chronicle,* Jan. 25, 1883 (photocopy provided by Jerry J. DeVos); Edward C. Williams, "Effect of Invention upon Labor and Morals," *Chautauquan* 28, no. 6 (Mar. 1899), 559.

9. "North Franklin Enterprise."

10. Dana, *Paper Birch,* 5.

11. Robert P. Tristram Coffin, *Kennebec: Cradle of Americans* (New York: Farrar & Rinehart, 1937), 156, 237.

12. "Satisfying the Toothpick Appetite," *Los Angeles Times,* Jun. 7, 1925, E14.

13. Box of Trophy Tooth Picks, undated, purchased on eBay, September 2005.

14. Box of toothpicks offered on eBay, Nov. 14, 2005; McPhee, *Survival of the Bark Canoe,* 17.

15. A. Jeff Martin, "What Is a Cord?" University of Wisconsin Extension, Forestry Facts Publication No. 44, Aug. 1989.

16. A. Jeff Martin, "What Is a Board Foot?" University of Wisconsin Extension, Forestry Facts Publication No. 42, Aug. 1989.
17. "Toothpick Output," *Chicago Daily Tribune,* Jul. 14, 1946, F6; Dana, "Paper Birch," 5, 8. According to a report of a government estimate, "30,000 acres of trees go into the manufacture of toothpicks used annually in the United States." See "Satisfying the Toothpick Appetite."
18. "Articles 'Made in Maine,' " *New York Times,* Nov. 27, 1898, 3. The last spool mill in Maine would be that of American Thread, in Milo, which would close in 1970; Carl Osgood to author, letter, Feb. 24, 2006.
19. "Some Funny Business," *National Police Gazette* 39, no. 234 (Mar. 18, 1882), 2.
20. "Articles 'Made in Maine' "; "Little Toothpick Still Fells Trees," *New York Times,* Oct. 29, 1955, 27.
21. Ideal Toothpick, "Tips on Tooth-Picks," advertising folder, ca. early 1900s.
22. Dana, *Paper Birch,* 8.
23. Stubbs, "Toothpick Industry," [11, 13].
24. "Toothpick Output of United States Is Thirty Billions," *Washington Post,* Oct. 10, 1926, F9.
25. "Deposition of Benjamin F. Sturtevant," Ans. to Int. 15, Testimony for Applicant, NACP, RG 241, Box 261, Patent No. 26,627 extension file; see Ans. to Int. 13, Patent No. 26,627 extension file.
26. Coffin, *Kennebec,* 157.
27. "The Toothpick Business," *Washington Post,* Jul. 21, 1900, 9.
28. McPhee, *Survival of the Bark Canoe,* 18, 46.
29. Stubbs, "Toothpick Industry," [13].
30. Rockwell Kent, *The Road Roller,* 1909.
31. *Strong, Maine: Bicentennial 1801–2001* (Strong, Me.: Strong Bicentennial Committee, 2001), 35; "Toothpick Output of United States."
32. "Forster Stockpiles Wood for Summer Use," *The Woodline* 2, no. 2 (Summer/Fall 1961), 7.
33. George B. Stanley, "Charles Forster Fathered the Toothpick Industry and Built First Toothpick Mill," *Rumford Falls* (Me.) *Times,* Aug. 26, 1943, in Maine State Library clipping book #124; Stubbs, "Toothpick Industry," [9].
34. "Manufacturing Interests," typescript, in Rachel Kidder, "Dixfield," loose-leaf binder, Ludden Memorial Library, Dixfield, Me., L-Ref Maine 974.1 D621, 14.A; Stubbs, "Toothpick Industry," [13].
35. Estate of Charles Forster, Dixfield, Maine, letterhead [1911], included in Stubbs, "Toothpick Industry," [14].
36. "Forster's First Home," photocopy provided by Joan Gould.
37. Estate of Charles Forster, letterhead, in Stubbs, "Toothpick Industry," [14].
38. Stubbs, "Toothpick Industry," [13, 15]; "Toothpick Business."
39. "Toothpick Business."
40. Stubbs, "Toothpick Industry," [18].

41. "Dixfield," anonymous typewritten manuscript, in Kidder, "Dixfield," 4, 5.
42. *Boston Globe,* Nov. 23, 1911, 12.
43. "Criminal Treatment of Trees," *New York Evangelist* 65, no. 38 (Sep. 20, 1894), 6.

CHAPTER SIXTEEN: TRADE SECRETS AND CLOSED DOORS

1. James F. Bastl, "Round or Flat? Take Your Pick," *Chicago Tribune,* Jan. 14, 1982, A18.
2. "Village Philosophy," *New York Times,* Jun. 1, 1884, 4.
3. George P. Stanley and Willis W. Tainter, "Process for Making Toothpicks," U.S. Patent No. 874,131 (Dec. 17, 1907).
4. J. C. Brown, "Improvement in Machines for Cutting Splints," U.S. Patent No. 43,177 (Jun. 21, 1864); *Scientific American* 13, no. 1 (Jul. 1, 1865), 7.
5. See, e.g., "Wooden Toothpicks," *Manufacturer and Builder* 2, no. 1 (Jan. 1, 1870), 25.
6. Charles C. Freeman, "Machine for Polishing and Compressing Toothpicks," U.S. Patent No. 358,029 (Feb. 22, 1887).
7. William H. Dyer, "Machine for Polishing and Rounding Toothpicks," U.S. Patent No. 635,756 (Oct. 31, 1899).
8. Joseph M. Hommel, "Toothpick-Machine," U.S. Patent No. 661,642 (Nov. 13, 1900).
9. Harry A. Dorr, "Cutter for Toothpick-Machines," U.S. Patent No. 795,494 (Jul. 25, 1905).
10. "Obituary Notes," *New York Times,* Mar. 20, 1902, 9.
11. Charles F. Scamman, "Toothpick-Machine," U.S. Patent No. 521,734 (Jun. 19, 1894); Sidney Kaplan, "Jan Earnst Matzeliger and the Making of the Shoe," *Journal of Negro History* 40, no. 1 (Jan. 1955), 8.
12. "Business and Personal Wants," *Scientific American* 95, no. 3 (Jul. 21, 1906), 53.
13. H. E. Mitchell and E. K. Woodard, *New Vineyard and Strong Register, 1902,* reprint (North New Portland, Me.: Western Somerset Historical Society, [1975]), 27, 29, 43. The common Stubbs given name is spelled Phillip in the text but Philip in the census compiled expressly for the register. Both spellings occur in various U.S. censuses, but Philip, which is used here, occurs twice as frequently as Phillip.
14. "Robert Goff Stubbs, '13, 1886–1941," obituary, *Yale Forest School News* 29, no. 3 (1941), 50; Robert Goff Stubbs, "A Toothpick Industry in Maine, 1913," typewritten report, Maine State Library, preface.
15. Stubbs, "Toothpick Industry," [13–16].
16. Ibid., [16, 18].
17. Ibid., [31–31 verso].
18. Ibid., [4, 32, 34].
19. Ibid., [18]; Tower Manufacturing and Novelty Co., 1903 catalog, 2.
20. Stubbs, "Toothpick Industry," [18, 19, 34].
21. Ibid., [23].

22. Ibid., [4, 19, 21, 22, 25]

23. Ruth Robbins Adamo, "Forster Mfg. Came into Being 94 Years Ago," *Lewiston* (Me.) *Daily Sun,* Feb. 27, 1982, 13. See also Cleaves & Company, "Renewable Energy," http://www.cleavco.com/energyforster .htm (Jun. 22, 2005).

24. Stubbs, "Toothpick Industry," [32, 33].

25. Charles C. Freeman, "Tooth-Pick," U.S. Patent No. 448,647 (Mar. 24, 1891); box of Improved Worlds Fair Polished Wood Tooth Picks, author's collection.

26. Forster Mfg. Co., Inc., *Wilton's 150 Years, 1803 . . . 1953,* advertisement; see also *The Woodline* 4, no. 2 (Jul. 1972), masthead.

27. Stubbs, "Toothpick Industry," [33–34]; "Flames Among Funeral Flowers," *Washington Post,* Feb. 4, 1891, 6; Stubbs, "Toothpick Industry," [33–34].

28. A. J. Panshin, E. S. Harrar, J. S. Bethel, and W. J. Baker, *Forest Products: Their Sources, Production, and Utilization,* 2nd ed. (New York: McGraw-Hill, 1962), 304–5.

29. Box of 750 Ideal Toothpicks (ca. 1965), author's collection.

30. Stanley and Tainter, "Process for Making Toothpicks."

31. Panshin et al., *Forest Products,* 304.

32. "Toothpick Town in Maine," *Boston Daily Globe,* Feb. 10, 1907, SM12.

33. "Curious and Interesting," *Washington Post,* Nov. 17, 1889, 12.

34. Henry P. Churchill, "Toothpick Machine," U.S. Patent No. 549,488 (Nov. 12, 1895); "Receiving and Delivering Spout for Toothpick-Machines," U.S. Patent No. 570,744 (Nov. 3, 1896).

35. Maurice W. Forster, "Device for Removing Waste Products," U.S. Patent No. 644,855 (Mar. 6, 1900).

36. Willis W. Tainter and George P. Stanley, "Toothpick-Machine," U.S. Patent No. 848,406 (Mar. 26, 1907); George P. Stanley and Willis W. Tainter, "Toothpick-Machine," U.S. Patent No. 854,116 (May 21, 1907); Simon S. Tainter, Willis W. Tainter, and George P. Stanley, "Machine for Boxing Toothpicks," U.S. Patent No. 869,993 (Nov. 5, 1907).

37. Simon S. Tainter, "Machine for Boxing Toothpicks," U.S. Patent No. 886,091 (Apr. 28, 1908); Simon S. Tainter, Willis W. Tainter, and George P. Stanley, "Boxing-Machine," U.S. Patent No. 953,358 (Mar. 29, 1910); Charles C. Freeman, "Machine for Boxing Toothpicks," U.S. Patent No. 882,879 (Mar. 24, 1908).

38. Stubbs, "Toothpick Industry," [26].

39. Sue Hubbell, "Let Us Now Praise the Romantic, Artful, Versatile Toothpick," *Smithsonian* 27, no. 10 (Jan. 1997), 76.

40. David Lamb, "Toothpick Firms Are Working Hard in the Sticks: The Small, Arcane Industry Is Serious Business, Filled with Fierce Rivalry and Closely Guarded Secrets," *Los Angeles Times,* Nov. 15, 1993, 5; Michaela Cavallaro, "Secrets of the Trade," *Mainebiz,* Jan. 7, 2002, 12.

41. Alltrista Consumer Affairs to author, e-mail message, Jan. 27, 2005.

42. "Toothpick Information," attachment to ibid.

43. Ibid.
44. See, e.g., M. R. Montgomery, "The Thing About Clothespins," *New England Magazine*, May 23, 1982.
45. "The Toothpick Mystery," *Down East*, Mar. 1988, 25; Laralyn Sasaki, "End of an Era: Toothpicks Fall Out of Favor," *Wall Street Journal*, Oct. 3, 1985, 1.
46. Cavallaro, "Secrets of the Trade."
47. Lamb, "Toothpick Firms"; Sasaki, "End of an Era."

CHAPTER SEVENTEEN: OTHER TOOTHPICKS

1. "The Nicknames of the States," *New York Times*, Jul. 30, 1871, 1; see also "Bric-a-Brac," *Saturday Evening Post* 63, no. 5 (Aug. 18, 1883), 3.
2. William B. Worthen, "Arkansas and the Toothpick State Image," *Arkansas Historical Quarterly* 53, no. 2 (Summer 1994), 166.
3. "The Bowie Knife," *New York Illustrated Magazine of Literature and Art* 1, no. 1 (Sep. 20, 1845), 58; "Arkansas," http://www.netstate .com/states/intro/arintro.htm (Oct. 3, 2005).
4. Worthen, "Arkansas," 161–90; "Arkansas."
5. John S. Gibbs, "Knives and Swords," *Arthur's Home Magazine* 62 (Dec. 1892), 1126.
6. "Epitome of the Times," *Atkinson's Saturday Evening Post* 15, no. 771 (May 7, 1836), 2.
7. "Legislation," *American Jurist and Law Magazine* 21, no. 41 (Apr. 1839), 229; "Miscellaneous Items," *Advocate of Peace* 5 (Feb. 1843), 16.
8. "The Troubles at Erie," *New-York Daily Times*, Jan. 5, 1854, 1; George T. Ferris, "Mark Twain," *Appletons' Journal of Literature, Science and Art* 12, no. 276 (Jul. 4, 1874), 15.
9. "On Things in General," *United States Review* 2, no. 4 (Sep. 1853), 281; "The American Inventor," *Scientific American* 36, no. 20 (May 19, 1877), 307.
10. See, e.g., "Journal of a Traveller to California," *Home Journal*, Jul. 28, 1849, 3; "Miscellany," *New-York Mirror* 16, no. 39 (Mar. 23, 1839), 312.
11. "Following the Drum," *Graham's American Monthly Magazine of Literature, Art, and Fashion* 53, no. 2 (Aug. 1858), 145; "A Bit of History," *Forest and Stream* 21, no. 11 (Oct. 11, 1883), 201; "A Pen Picture of Oshkosh," *National Police Gazette* 36, no. 136 (May 1, 1880), 4.
12. "A Sunday-School Excursionist Loose," *National Police Gazette* 34, no. 98 (Aug. 9, 1879), 3; "Desultory Selections," *New-York Mirror* 20, no. 52 (Dec. 24, 1842), 414.
13. "Red Neck Toothpick 6½ in Blade," offered on eBay, Apr. 27, 2005; "Hand Forged Railroad Spike AR-CAN-SAW Toothpick Dagger," offered on eBay, Apr. 4, 2006.
14. "1956 Topps Davy Crockett Card No #53A Bowies Toothpick," offered on eBay, Aug. 22, 2005.
15. "Carl's Jr. Heats Things Up with Spicy Burger Addition," PR Newswire Association, Feb. 23, 2005; "Merrick's Texas Toothpicks," offered on eBay, Feb. 12, 2006; "Merrick Texas Toothpicks 6–8"— 100 ct Box," offered on eBay, Sep. 20, 2006.

16. Ken Smith and David Belcher, "Nit-Picking," *Glasgow Herald,* Jan. 6, 2005; "Toothpick Use Praised at Dentists' Conference," *Los Angeles Times,* Jul. 17, 1950, 27.

17. Stephen E. Weil, "The Proper Business of the Museum: Ideas or Things?" *Rethinking the Museum and Other Meditations* (Washington, D.C.: Smithsonian Institution Press, 1990), 43–56.

18. Ibid.; Victor Ginsburgh and François Mairesse, "Defining a Museum: Suggestions for an Alternative Approach," *Museum Management and Curatorship* 16 (1997), 15–33; Siegfried Giedion, *Mechanization Takes Command: A Contribution to Anonymous History* (New York: Norton, 1969), 3.

19. James Franklin Fitts, "Facetiae of the War," *Galaxy* 6, no. 3 (Sep. 1868), 320; J. F. McClure, "French War Slang," *Bookman* 48, no. 6 (Jan. 1919), 566.

20. [James Cook], *The Three Voyages of Captain James Cook Round the World,* vol. IV (London: Longman, Hurst, Rees, Orme, and Brown, 1821), 195; see also "Antarctic Explorations," *Littell's Living Age* 29, no. 359 (Apr. 5, 1851), 1.

21. "Hyperbole," *New Bedford* (Mass.) *Mercury* 3, issue 44 (Jun. 1, 1810), [4]; see also *Polyanthos* 3 (Mar. 1814), 334.

22. "Tell of Growth of Chautauqua," *Chicago Tribune,* Sep. 19, 1916, 19; "Sylvester A. Long," folder, ca. 1910, http://sdrcdata.lib.uiowa.edu/libsdrc/details.jsp?id=/1ongsa/5.

23. Harold C. Schonberg, " 'Rachmaninoff? Well, I Mean . . . ,' " *New York Times,* Apr. 14, 1968, D17.

24. *Eastern Argus* 5, issue 426 (Oct. 31, 1828), [3].

25. Quoted in Henry Litchfield West, "American Politics," *Forum* 35, no. 1 (Jul. 1903), 3; "Our Views," *Riverside* (Calif.) *Press-Enterprise,* editorial, B10.

26. Vincent Starrett, "Superstitions and Sayings," *Chicago Tribune,* Sep. 5, 1965, J8; "Trouble in Prospect," *Boston Globe,* Oct. 8, 1893, 6.

27. "A Knowing Insect," *Saturday Evening Post,* Feb. 3, 1872, 8.

28. Albert B. Chandler, " 'The Toothpick' on Lake Ontario," *Outing* 34, no. 6 (Sep. 1899), 564; "Captain John Barr," *Forest and Stream* 72, no. 4 (Jan. 23, 1909), 144; "Largest Search-Light in the World," *Current Literature* 14, no. 3 (Nov. 1893), 380.

29. "Old Salts," *Current Literature* 28, no. 3 (Jun. 1900), 348.

30. "1955—Paul Bunyan's Toothpicks—Logging Truck Postcard," offered on eBay, Jun. 7, 2006; author's collection; "Oregon Toothpicks from a Northwest Forest Postcard, 30s," offered on eBay, Jun. 19, 2006.

31. "A Washington Toothpick," Library of Congress, Prints and Photographs Division, http://hdl.loc.gov/loc.pnp/cph.3b15146 (Apr. 15, 2005); William Grimes, "Into the Wild Yet Tranquil Blue Yonder," *New York Times,* Sep. 7, 2005, E1, E7.

32. Cam Hutchinson, "Sharpe Words for Colts Defenders," *Montreal Gazette,* Oct. 9, 2004, C5.

33. Jill Lawrence, "Senator Makes Pitch in N.H. Buttressed by Red Sox Brass," *USA Today,* Nov. 1, 2004, http://www.usatoday.com/news/

politicselections/nation/president/2004-11-01-kerry-campaign_x
.htm; "Voice of the People," letter, *Rock Hill* (S.C.) *Herald,* Oct.
20, 2004, 7A; "Looking North Past That Toothpick They Call a Monument," aerial photograph, Library of Congress, G3851.A35 1930.F3.
34. Wm. A. Phelon, "Forecasts and Recollections," *Baseball Magazine* 18,
no. 3 (Jan. 1917), 100; advertisement, *Forest and Stream* 85, no. 11
(Nov. 1915), 681; see *Daily Star* (U.K.), Nov. 18, 2004, 13.
35. Lucy J. Rider, "Dick Hardin Away at School," *St. Nicholas* 5, no. 6
(Apr. 1878), 386; comment on soup quoted by Andrew Pierce, *The
Times* (U.K.), Aug. 24, 2004, 8.
36. "Roosevelt Talks with 200 Editors," *New York Times,* Apr. 16, 1937, 21.
37. *Louisiana Advertiser* 1, no. 22 (Jun. 8, 1820), [2]; "Pennsylvania in
1850," *Saturday Evening Post* 9, no. 449 (Mar. 6, 1830), 1.
38. "Bric-a-Brac," *Saturday Evening Post* 64, no. 14 (Oct. 18, 1884), 3;
"Advertising Duels," *New York Times,* Jul. 6, 1883, 4.

CHAPTER EIGHTEEN: NEW USES FOR OLD

1. "London, December 5," *Providence Gazette and Country Journal* 2,
issue 74 (Mar. 17, 1764), [2].
2. Madame de Lasteyrie, "Madame de Lafayette," *Hours at Home* 11, no.
6 (Oct. 1870), 552; Claire Baer and Paul L. Baer, "A Story of the
Toothpick," *Journal of Periodontology* 37, no. 2 (Mar.–Apr. 1966), 160.
3. "Obituary," *The Annual Register: A Review of Public Events at Home
and Abroad for the Year 1895* (London: Longmans, Green, 1896), 167;
American Monthly Magazine 1, no. 4 (Jun. 1, 1833), 250; Madame de
Lasteyrie, "Madame de Lafayette," 552.
4. "Toothpick Called Nazi Spying Tool," *New York Times,* Feb. 11,
1942, 16.
5. "Fingers and Toes. II. Minimus.—What the Little Finger Said,"
Hours at Home 11, no. 4 (Aug. 1870), 324.
6. "Remarkable Escape from Prison," *American Quarterly Register and
Magazine* 1, no. 2 (Sep. 1848), 484–90.
7. Denise Marchionda, "Lowell, Mass., Beauty Salon Spends 60 Years in
Hair Styling Business," *Lowell* (Mass.) *Sun,* Dec. 20, 2003.
8. See http://www.tuscanrose.com/128%20pin%20curls.jpg; Edith Bee-
son, "Dressing Dolls Found Creative by One Woman," *New York
Times,* Aug. 14, 1957, 18.
9. "What Toothpicks Do," *Washington* (D.C.) *Evening Star,* Oct. 26,
1886, 6.
10. "The Prize Winners," advertisement, *Chicago Tribune,* Jun. 20,
1929, 22.
11. Elverda M. Phillips, "It Has Its Points," *Cal* 30 (Aug. 1967), 15.
12. John Linton, "Three Tricks for the Little Ones," *Ladies' Home Journal* 14, no. 10 (Sep. 1897), 26.
13. "New Products Introduced," *The Woodline* 7, no. 1 (Feb. 1975), 3.
14. "Forster's Bonanza: The Toothpick Christmas Tree," *The Woodline* 1,
no. 3 (Fall 1960), 2; "The 1961 Christmas Toothpick Promotion," *The
Woodline* 2, no. 2 (Summer/Fall 1961), 5.

15. See, e.g., Beverly Beyette, " 'Toothpick King' Picks Eiffel Tower," *Los Angeles Times*, Jun. 19, 1988, 2; Franki V. Ransom, "Toothpick Artist Credits Higher Power," *Los Angeles Times*, Jul. 20, 1991, 1; Harvey Green, *Wood: Craft, Culture, History* (New York: Viking, 2006), 279.

16. Quoted in Green, *Wood*, 281; Charlotte B. Norris, "Pickbuilder, Unlimited," *Los Angeles Times*, Nov. 11, 1951, J11.

17. Christine N. Ziemba, "Obsessions," *Los Angeles Times*, Feb. 13, 2005, E3; Stephen Talasnik to author, e-mail messages, Jun. 1, 2, 2006; "Gator May Hit World Record," *Baton Rouge Advocate*, Mar. 24, 2005, 10-G; Hart Seely, "36 Hours: Syracuse," *New York Times*, May 19, 2006, D3.

18. "Man from Kolar Picking His Way into Guinness Book," *Press Trust of India*, Feb. 23, 2005; Ann Landers, "Ann Landers Tells Greensboro Reader She Should Butt Out," *Greensboro* (N.C.) *News Record*, Oct. 14, 1993, D2.

19. "Hard Times," *Boston Globe*, Sep. 17, 1913, 12.

20. Pam Belluck, "Evolution Foes Dealt a Defeat in Kansas Vote," *New York Times*, Aug. 3, 2000, A1, A30.

21. *Ladies' Home Journal* 19, no. 12 (Nov. 1902), 46; *Ladies' Home Journal* 18, no. 12 (Nov. 1901), 45; "Durango's First Baby," *National Police Gazette* 38, no. 191 (May 21, 1881), 5; Mildred K. Smith, "In Motherland," *Pictorial Review* 7, no. 7 (Apr. 1906), 37.

22. Emelyn Lincoln Coolidge, "The Young Mother's Calendar," *Ladies' Home Journal* 21, no. 9 (Aug. 1904), 30; *Ladies' Home Journal* 19, no. 6 (May 1902), 36; *Ladies' Home Journal* 19, no. 2 (Jan. 1902), 35.

23. Harry A. Dorr, "Cutter for Toothpick-Machines," U.S. Patents No. 795,494 (Jul. 25, 1905), 931,044 (Aug. 17, 1909); Harry A. Dorr, "Toilet Article," U.S. Patent No. 812,189 (Feb. 13, 1906).

24. Wendy Moonan, "The Modest Masterpieces," *New York Times*, July 23, 2004, E33.

25. Elisabeth Robinson Scovil, *The Care of Children* (Philadelphia: Altemus, c. 1894), 181; "VI—the Care of the Hands," *Ladies' Home Journal* 21, no. 4 (Mar. 1904), 27; Emma E. Walker, "Pretty Girl Papers," *Ladies' Home Journal* 22, no. 11 (Oct. 1905), 23.

26. Rad Sallee, "Houston Vandals Use Tactics from Other Cities in Targeting Starbucks," *Houston Chronicle*, Dec. 2, 2003.

27. James Donald Gray, "A One Hundred Dollar Bill," *Frank Leslie's Popular Monthly* 32, no. 4 (Oct. 1891), 14; Cheryl N. Schmidt, "Toothpick Leads to Indictment in 1991 Killing," *Tampa* (Fla.) *Tribune*, Jun. 5, 2004, 1.

28. "Hardcovers in Brief," *Washington Post Book World*, Sep. 15, 1991, X13; cf. Patricia Holt, *The Bug in the Martini Olive: And Other True Cases from the Files of Hal Lipset, Private Eye* (New York: Little, Brown, 1991), 5, 66, 69.

CHAPTER NINETEEN:

MAURICE, OSCAR, AND THE INDUSTRY

1. "Story of the Origin and Growth of the Toothpick Business Was Outstanding Feature at Dedication Wednesday," *Franklin* (Me.) *Journal*,

November 13, 1931, in Maine State Library clipping book #124; George B. Stanley, "Charles Forster Fathered the Toothpick Industry and Built First Toothpick Mill," *Rumford Falls* (Me.) *Times*, Aug. 26, 1943, in Maine State Library clipping book #124; *Franklin County* (Me.) *Directory, 1899,* 188.

2. "Story of the Origin and Growth."

3. Stanley, "Charles Forster Fathered"; *Scientific American* 83, no. 4 (Jul. 28, 1900), 63; "Inquiry No. 8835," *Scientific American* 99, no. 9 (Aug. 29, 1908), 146, and repeated in several subsequent issues.

4. "Obituary Notes," *New York Times*, Nov. 21, 1915, 19.

5. "Securities at Auction," *New York Times*, Mar. 10, 1910, 11; "New Incorporations," *New York Times*, Jul. 11, 1915, S6.

6. Notes on Dixfield attached to e-mail message from Peter R. Stowell to author, Sep. 18, 2006; Stanley, "Charles Forster Fathered."

7. Oscar H. Hersey, "Trade-Mark for Wooden Toothpicks," U.S. Trade-mark No. 52,205 (May 1, 1906); Oscar H. Hersey, "Trade-Mark for Wooden Toothpicks," U.S. Trademark No. 70,370 (Aug. 25, 1908).

8. U.S. Census, 1910; George P. Stanley, "Cutter for Cutting Toothpicks from Veneer Strips," U.S. Patent No. 812,895 (Feb. 20, 1906). Although the word "waver" may appear to be a typographical error for "wafer," it may not have been. According to a definition at the bottom of the page in *Webster's New International Dictionary* (1929), a waver was "a sapling among felled trees; a twig growing from a stump." To those accustomed to working in wood, this might have been an apt noun used as an adjective to describe the toothpick's rounded chamfered point.

9. Harry A. Dorr, "Cutter for Toothick-Machines," U.S. Patent No. 795,494 (Jul. 25, 1905); Harry A. Dorr, "Cutter for Toothpick-Machines," U.S. Patent No. 931,044 (Aug. 17, 1909). The application for the second patent was filed Dec. 1, 1905.

10. Albert Henry Hall, "Splint-Cutting Machine," U.S. Patent No. 963,141 (Jul. 5, 1910).

11. Charles Forster, "Trade-Mark for Wooden Toothpicks," U.S. Trade-mark No. 35,326 (Oct. 30, 1900).

12. William C. King, "Knife for Cutting Toothpicks," U.S. Patent No. 1,037,494 (Sep. 3, 1912); Edward Greenwood, "Toothpick-Machine," U.S. Patent No. 1,220,096 (Mar. 20, 1917); George P. Stanley, "Machine for Making Toothpicks and the Like," U.S. Patent No. 1,613,623 (Jan. 11, 1927).

13. Forster Manufacturing Company, "Trade-Mark for Toothpicks," U.S. Trademark No. 77,991 (May 24, 1910).

14. "Final Account of Oscar H. Hersey, Executor of the Estate of Charles Forster," Cumberland County (Me.) Probate Registry Office, Docket 1507 file.

15. Charlotte M. Forster, obituary, Apr. 1, 1902, Maine Historical Society; abstract of certificate of death, Portland (Me.) City Clerk Department; copy of statement included in "First Account of Oscar H. Hersey, Trustee," Oct. 26, 1904, Cumberland County (Me.) Probate Registry Office, Docket 1507 file.

16. "First Account of Oscar H. Hersey," Cumberland County (Me.) Probate Registry Office, Docket 1507 file. See also in the same docket file a copy of Hersey's final account as executor, which was dated Jun. 30, 1902, but not sworn to or certified until Oct. 26, 1904.

17. George C. Wing to Benjamin Coffin, letter dated Mar. 26, 1910, Cumberland County (Me.) Probate Registry Office, Docket 1507 file.

18. Various annual accounts and statements of expenses, Cumberland County (Me.) Probate Registry Office, Docket 1507 file.

19. "Second Account of Oscar H. Hersey," Cumberland County (Me.) Probate Registry Office, Docket 1507 file.

20. "Third Account of Oscar H. Hersey," Cumberland County (Me.) Probate Registry Office, Docket 1507 file.

21. Ibid.

22. "Fourth Account," dated Aug. 23, 1907; "Fifth Account," dated Dec. 11, 1908; "Sixth Account of Oscar H. Hersey, Trustee," dated Aug. 6, 1909; see, e.g., dockets of first and second trustee accounts, Cumberland County (Me.) Probate Registry Office, Docket 1507 file.

23. "Big Toothpick Crop," *Boston Globe,* Apr. 29, 1906, 38.

24. Statements of expense and commission account, for year ending July 1, 1907, s.v. July 18, 1906; for year ending July 1, 1908, s.v. Sep. 5, Nov. 13, Feb. 19, Apr. 21, Cumberland County (Me.) Probate Registry Office, Docket 1507 file.

25. *Forster Manufacturing v. Cutter-Tower,* 211 Mass. 219; 97 N.E. 749 (1912); *Forster Mfg. v. Cutter-Tower* 215 Mass. 136; 101 N.E. 1083 (1913).

26. Advertisement, *Zion's Herald,* Apr. 28, 1909, 539. In a lawsuit about this time, the Cutter-Tower Co. was identified as "manufacturer of wood toothpicks, stationers, specialties and advertising novelties," as well as "selling agents of the Franklin Typewriter." It was because of its involvement in the latter capacity that the company became engaged in legal dispute with a Georgia man over the rental of a typewriter. The appeals court upheld a ruling for the plaintiff to the effect that a rental contract remained in force until the rented machine was returned to the renting company. *Cutter-Tower v. Clements,* Court of Appeals of Georgia, 5 Ga. App. 291; 63 S.E. 58; 1908 Ga. App. LEXIS 99.

27. "$50 a Week," advertisement, *Baseball Magazine* 2, no. 1 (Nov. 1908).

28. Advertisement, *Saturday Evening Post,* Sep. 10, 1910; reproduced in *The Woodline* 8, no. 1 (Mar. 1977), 5. The rounding of both ends was made possible and protected by the patent assigned by inventor Albert Hall to Forster Manufacturing: Albert Henry Hall, "Splint-Cutting Machine," U.S. Patent No. 963,141 (Jul. 5, 1910).

29. "Toothpicks Take a Drop," *Boston Globe,* Feb. 14, 1911, 2; "Editorial Points," *Boston Globe,* Feb. 15, 1912, 10.

30. "Cartoons and Comments: Toothpick Journalism," *Puck* 74, no. 1922 (Dec. 31, 1913), 2.

CHAPTER TWENTY: THE TRAGIC HEIRESS

1. "San Diego," *Los Angeles Times,* Nov. 9, 1893, 7.
2. See, e.g., "Pasadena," *Los Angeles Times,* Sep. 13, 1896, 29; "Out-of-Town Society," *Los Angeles Times,* Nov. 7, 1897, 12; "Coronado Beach," *Los Angeles Times,* Dec. 28, 1897, 11.
3. William E. Smythe, *History of San Diego, 1542–1908: An Account of the Rise and Progress of the Pioneer Settlement on the Pacific Coast of the United States,* vol. I: *Old Town* (San Diego: History Co., 1908), 112, 174, 274–75; Clarence Alan McGrew, *City of San Diego and San Diego County: The Birthplace of California* (Chicago: American Historical Society, 1922), 378.
4. Smythe, *History of San Diego,* 18.
5. "Little Hope, Say Doctors," *Los Angeles Times,* Aug. 27, 1908, I1; "San Diego Woman Attempts Suicide," *San Diego* (Calif.) *Union,* Aug. 26, 1908, 1.
6. Elsinore display advertisement, *Los Angeles Times,* Jul. 12, 1908, I5.
7. "Events in Society," *Los Angeles Times,* Dec. 18, 1898, B2; Westlake Hotel advertisement, *Los Angeles Times,* May 28, 1908, I3.
8. "Los Angeles Pacific Balloon Route Excursion," http://www.erha .org/balloon.htm (Dec. 8, 2005); "San Diego Woman Attempts"; "Declares Her Incompetent," *Los Angeles Times,* Oct. 7, 1908, II9.
9. "Seeks Death in Mad Leap," *Los Angeles Times,* Aug. 26, 1908, I11.
10. "San Diego Woman Attempts"; "Seeks Death"; "Declares Her Incompetent."
11. Annual reports, Estate of Charles Forster, Cumberland County (Me.) Probate Registry Office, Docket 1507 file; *Jolls v. Forster,* Superior Court, County of San Diego, California (1908), No. 14536.
12. "Declares Her Incompetent"; "Miss Forster Improves," *Los Angeles Times,* Sep. 4, 1908, I11; "Guardian Is Named for Miss C. Forster," *San Diego Union,* Oct. 6, 1908, 5.
13. "San Diego Woman Defendant," *Los Angeles Times,* Jul. 28, 1909, I5.
14. Ibid.; "Final Account of Oscar H. Hersey, Executor of the Estate of Charles Forster," Jun. 30, 1902, Schedule B, Cumberland County (Me.) Probate Registry Office, Docket 1507 file.
15. "Richest Bachelor of City, Nathan Watts, to Wed a Maine Heiress," *San Diego* (Calif.) *Sun,* Oct. 20, 1909, 1.
16. Ibid.
17. Ibid.
18. Ibid.
19. "San Diego County," *Los Angeles Times,* Jan. 27, 1898, 11.
20. "Resents Publication," *Los Angeles Times,* Oct. 21, 1909, II11; "Seeks Death."
21. The census image is difficult to read, but the Norwegian lodger's name appears to be written as "Bertet Holtz," which may well have been how the census taker heard "Bergit Olsen" pronounced with an accent.

22. Last Will and Testament of Charlotte Bowman Forster, San Diego Historical Society.

23. "In the Matter of the Guardianship of Charlotte B. Forster, an Incompetent Person: Order for Notice," Superior Court of the State of California, County of San Diego, Jan. 20, 1917, Docket 7103, Probate Orders and Decrees Book 65, San Diego Historical Society Research Archives; "Order to Deposit Personal Property," Jan. 25, 1917.

24. Certification of Vital Record, County of San Diego; "Nietzsche Chronicle, 1900," http://www.dartmouth.edu/~fnchron/1900.html.

25. "In the Matter of the Guardianship of Charlotte B. Forster, an Incompetent Person, Decree of Distribution," Mar. 31, 1917, Docket 7141, San Diego Historical Society Research Archives.

CHAPTER TWENTY-ONE:
EXEUNT OSCAR AND MAURICE

1. "Total Cash Expenses of H. H. Hersey, Trustee, from July 1, 1908, to July 1, 1909," Cumberland County (Me.) Probate Registry Office, Docket 1507 file, s.v. "Oct. & Nov."

2. "Total Cash Expenses of H. H. Hersey, Trustee, from July 1, 1908, to July 1, 1909," Cumberland County (Me.) Probate Registry Office, Docket 1507 file, s.v. Dec. 2, Dec. 4, Dec. 22, Dec. 23 (two items).

3. Richard P. Mallett, *The Last 100 Years: A Glimpse of the Farmington We Have Known* (Wilton, Me.: Wilton Printed Products, 1991), 125.

4. E. C. Bowler, comp. and ed., *An Album of the Attorneys of Maine* (Bethel, Me.: News Publishing Co., 1902), 136.

5. Frank W. Butler to Benjamin Coffin, note, Sep. 6, 1909; Estate of Charles Forster, fourth, fifth, and sixth trustee's accounts, Cumberland County (Me.) Probate Registry Office, Docket 1507 file.

6. See *Inhabitants of Peru v. Estate of Charles Forster* 109 Me. 226; 83 A. 670 (1912).

7. Statement of Maurice W. Forster, dated March 20, 1909; Petition for Additional Bond, May 6, 1909, Cumberland County (Me.) Probate Registry Office, Docket 1507 file.

8. Declaration of Frank W. Butler and Robert T. Whitehouse, Mar. 25, 1910, Cumberland County (Me.) Probate Registry Office, Docket 1507 file.

9. "Total Cash Expenses of O. H. Hersey, Trustee, From July 1, 1908 to July 1, 1909," Cumberland County (Me.) Probate Registry Office, Docket 1507 file.

10. "Story of the Origin and Growth of the Toothpick Business Was Outstanding Feature at Dedication Wednesday," *Franklin* (Me.) *Journal*, Nov. 13, 1931, in Maine State Library clipping book #124.

11. Oxford, Maine, Supreme Judicial Court, May Term 1912 Oxford SJC, case file 52, Maine State Archives.

12. Letterhead of Estate of Charles Forster [1911], in Robert Goff Stubbs, "A Toothpick Industry in Maine, 1913," typewritten report, Maine State Library, [14].

13. "Story of the Origin and Growth."

14. Forster Manufacturing Company, "Trade-Mark for Toothpicks," U.S. Trademarks No. 87,196 and 87,197 (Jul. 2, 1912).

15. Willis W. Tainter, "Sight-Glass for Toothpick-Packing Machines and the Like," U.S. Patent No. 1,033,128 (Jul. 23, 1912).

16. Albert H. Hall, "Machine for Stacking Toothpicks and the Like," U.S. Patent No. 1,024,773 (Apr. 30, 1912); "Mechanism for Stacking and Conveying Toothpicks," U.S. Patent No. 1,024,774 (Apr. 30, 1912); Maurice W. Forster, "Packing-Machine for Toothpicks and the Like," U.S. Patent No. 1,026,665 (May 21, 1912).

17. George B. Stanley, "Charles Forster Fathered the Toothpick Industry and Built First Toothpick Mill," *Rumford Falls* (Me.) *Times*, Aug. 26, 1943, in Maine State Library clipping book #124.

18. Quoted in "Editorial Points," *Boston Globe*, Aug. 8, 1913, 10.

19. George P. Stanley, "Charles Forster Fathered the Toothpick Industry and Built First Toothpick Mill," typescript by Peter R. Stowell based on original newspaper article; "Berst-Forster-Dixfield Company of Oakland Changes Its Name to B.F.D. Co.; Is Town's Major Industry," *Waterville* (Me.) *Sun*, May 1, 1945, in Maine State Library clipping book #124. People of the Maine county were known as "Oxford Bears," a reference that could be traced back to the early nineteenth century. Then, newspaper editors from coastal cities "satirized the rural yeomen as 'backwoods bears' whose 'huge paws' were suited to the plough—not to the pen." Oxford residents embraced the image and signed letters to the editor with appropriate allusions. In time, local teams and other groups adopted the bear as their mascot and nicknamed themselves accordingly. It was thus natural for a Dixfield toothpick factory to incorporate images of bears in its product packaging. See "Bears, Bears, and More Bears: About the Bicentennial Logos," http://www.bethelhistorical.org/Oxford_County_1805-2005.html.

20. "Berst-Forster-Dixfield Company"; Stanley, "Charles Forster Fathered."

21. Harland L. Durrell, "BFD Is Oakland's Largest Industry," *Waterville* (Me.) *Morning Sentinel*, Feb. 23, 1948. See also "Berst-Forster-Dixfield Company"; Yvette Raymond, "Mill Closing Marks the End of Manufacturing in Peru Area," *Lewiston* (Me.) *Sun Journal*, Nov. 27, 1988, Peru column.

22. Forster Mills Company corporation files, Department of the Secretary of State (Me.), Bureau of Corporations, Elections and Commissions, Archives Box 131.

23. Forster Corporation file, Department of the Secretary of State (Me.), Bureau of Corporations, Elections, and Commissions, Archives Box 131.

24. "Maurice W. Forster," *Portland* (Me.) *Press Herald*, Jan. 8, 1924, 3.

25. Ibid.

26. "Deaths," *Portland* (Me.) *Press Herald*, Jan. 9, 1924, 2; "Maurice W. Forster."

27. Franklin County (Me.) Probate Court, Docket 7683 file; "Dixfield

School Photos," http://www.dixfieldcitizennews.net/DixfieldSchool.html. Maurice W. Forster married Lelia M. Randall on Oct. 12, 1899.

28. Franklin County (Me.) Probate Court, Docket 7683 file.

29. Ibid.

30. Ibid. When the Forster Corporation was organized in 1916, only $300 of its $100,000 capital stock was paid in. At that time, Maurice Forster owned one share of the corporation's common stock, which carried a par value of $100.

31. "Certificate of Organization of a Corporation under the General Law," filed Feb. 4, 1924, Department of the Secretary of State (Me.), File 19240005 D 19100001183204.

32. Ibid.

33. Petition submitted by Frank W. Butler, Feb. 9, 1924, Cumberland County (Me.) Probate Registry Office, Docket 1507 file.

34. Ruth Robbins Adamo, "Forster Manufacturing Dream Come True for Couple," *Lewiston* (Me.) *Sun*, Aug. 8, 1988, 11.

35. Franklin County (Me.) Probate Court, Docket 8534 file.

36. "Story of the Origin and Growth."

CHAPTER TWENTY-TWO: FOREIGN AFFAIRS

1. W. Irwin, "The Banner of the Pork and Soap," *Current Literature* 32, no. 3 (Mar. 1902), 288.

2. Price information from J. Lauchheimer & Co. advertisement, *New York Times*, Apr. 2, 1905, 10.

3. "Foreign Trade Opportunities," *New York Times*, Aug. 26, 1910, 11.

4. "Foreign Trade Opportunities," *New York Times*, Dec. 17, 1913, 17.

5. "Foreign Trade Opportunities," No. 7,842, *New York Times*, Jan. 5, 1912, 16.

6. "The Tooth-Pick Business," *Dental Register* 39, no. 4 (1885), 206; see, e.g., the online encyclopedia *Wikipedia*, s.v. "Toothpick."

7. "Toothpick Factory, Strong," http://www.mainepreservation.com/dayswork/dayswork5.shtml; "The Toothpick Factory," *Farmington* (Me.) *Chronicle*, May 24, 1888; "In the Big Industries," *Los Angeles Times*, Feb. 24, 1907, VI20.

8. "Mexico's Toothpick Famine," *Boston Globe*, Dec. 6, 1914, 52.

9. "Radio Toothpicks Order," *New York Times*, Jul. 17, 1923, 2.

10. "Maine Becoming Toothpick State," *Los Angeles Times*, Mar. 26, 1925, 5.

11. Forster Mfg. Co., Inc., centennial advertisement, Wilton (Me.) Historical Society vertical file, s.v. "Forster Mfg. Co."

12. "Story of the Origin and Growth of the Toothpick Business Was Outstanding Feature at Dedication Wednesday," *Franklin* (Me.) *Journal*, Nov. 13, 1931, in Maine State Library clipping book #124.

13. James Joyce, *A Portrait of the Artist as a Young Man* (New York: Viking Press, 1956), 215, 229–30.

14. "Articles 'Made in Maine,'" *New York Times*, Nov. 27, 1898, 3.

15. Mary Bellis, "Popsicle—the History of the Popsicle," http://inventors.about.com/library/inventors/blpopsicle.htm (Mar. 22, 2005).

16. See, e.g., "BFD Assures Young Fry Plenty of Stick Ice Cream," *Waterville* (Me.) *Morning Sentinel*, Feb. 28, 1950, in Maine State Library clipping book #124.

17. *Wikipedia*, s.v. "Spoon," http://en.wikipedia.org/wiki/Spoon (Nov. 23, 2005); Eva D. Bachelder, "Consume 1,000 Cords of Birch Yearly in Manufacture of Cocktail Sticks," *Waterville* (Me.) *Morning Sentinel*, Feb. 21, 1947, in Maine State Library clipping book #124.

18. "Articles 'Made in Maine' "; Edward H. Carlson, "Making Lollypop Sticks Major Maine Industry," *Portland* (Me.) *Evening Express*, undated clipping, in Maine State Library clipping book #124.

19. "Little Toothpick Still Fells Trees," *New York Times*, Oct. 29, 1955, 27; Hammacher Schlemmer display advertisement, *New York Times*, Jun. 24, 1967, 5.

20. "Le Negri Goose Quill Toothpicks," Caswell-Massey Co. Ltd. Apothecary Catalogue, 1977–1978, 63, Box 52, Toiletries Folder (Ad Ephemeral Boxes), John W. Hartman Center for Sales, Advertising & Marketing History, Special Collections Library, Duke University.

21. "The New Way of Life?" *The Woodline* 4, no. 1 (Apr. 1972), 2.

22. "Forster Advertising Goes National," *The Woodline* 4, no. 3 (Sep. 1972), 1.

23. Alan S. Oser, "Clothespins: China, Trade and a Factory in Maine," *New York Times*, Oct. 14, 1979, F3; Neil Genzlinger, "Forster: 'We're Not Interested in Selling,' " *Waterville* (Me.) *Moring Sentinel*, Mar. 4, 1978; Alan S. Oser, "The Forster Legacy Lives on in Wilton," *Maine Sunday Telegram*, Nov. 4, 1979; Laralyn Sasaki, "End of an Era: Toothpicks Fall Out of Favor," *Wall Street Journal*, Oct. 3, 1985, 1.

24. National Recovery Administration, *Code of Fair Competition for the Toothpick Industry*. Registry No. 329-03 (Washington, D.C.: U.S. Government Printing Office, 1933), 4.

25. Canadian International Trade Tribunal, "Flat Wooden Toothpicks Originating in or Exported from the United States of America and Produced by or on Behalf of Forster Mfg. Co. Inc. of Wilton, Maine, Its Successors and Assigns," Inquiry No. NQ-91-005, 1992; "Toothpick Dumping Investigated," *North American Report on Free Trade* 1, no. 31 (Sep. 2, 1991), 8; "Trade Tribunal Inquires into Toothpick Dumping," *North American Report on Free Trade* 1, 45 (Dec. 9, 1991), 4; "Destinations and Discoveries in Midwestern Ontario," http://www.ddontario.com/reviews/index.cgi?db=Current+Issue&r=293&q= (Jun. 23, 2006).

26. "Toothpick Output of United States Is Thirty Billions," *Washington Post*, Oct. 10, 1926, F9; David Lamb, "Toothpick Firms Are Working Hard in the Sticks," *Los Angeles Times*, Nov. 15, 1993, 5.

27. Kitchen Compliments, box of 800 round toothpicks, purchased summer 2004, Brunswick, Me.

28. Advertisement, *New York Times*, Feb. 13, 1968, 2; "Large Output of Toothpicks," *New York Times*, Mar. 15, 1903, 33.

29. I was given a few packages of these in summer 2005 by Deirdre Barton, proprietor of Weatherbird, a boutique in Damariscotta, Maine.

30. Sen Nakahara and Kuninori Homma, "Tufted Toothpicks and Teeth Blackening Customs in *Ukiyoe*," *Bulletin of the History of Dentistry* 34, no. 2 (Oct. 1986), 88; Claire Baer and Paul N. Baer, "A Story of the Toothpick," *Journal of Periodontology* 37, no. 2 (Mar.–Apr. 1966), 161; see, e.g., the verse by Mary McNeill Scott, "The Bamboo," *St. Nicholas* 20, 5 (Mar, 1893), 352, in which a Japanese woodcutter makes from a single large stalk "a parasol-frame, and an intricate game, / And ribs to a paper fan; / A sole to his shoe, and a toothpick or two" in addition to larger things like a tub and a pail. On Japanese words, Charles Worthen to author, e-mail messages, May 31–Jun. 1, 2007.

31. William J. Carter, Bernard B. Butterworth, and Joseph G. Carter, *Ethnodentistry and Dental Folklore* (Overland Park, Kan.: Dental Folklore Books of Kansas City, 1987), 38.

32. "Toothpick Output"; "Large Output of Toothpicks."

33. "Large Output of Toothpicks"; "The Toothpick Business," *Washington Post*, Jul. 21, 1900, 9; Don C. Seitz, "The Japanese Overload," *North American Review* 197, no. 691 (Jun. 1913), 733.

34. Oscar Hersey, Fifth Trustee Account, Estate of Charles Forster, Cumberland County (Me.) Probate Records Office, Docket 1507 file.

35. M. D. K. Bremner, "Mouth Care Through the Ages," *Dental Items of Interest* 70 (Jun. 1948), 607.

36. "American Toothpicks," http://homepage1.nifty.com/kmisty/etooth pick.htm.

37. John Heskett, *Toothpicks and Logos: Design in Everyday Life* (Oxford: Oxford University Press, 2002), 41–43; Stephen Talasnik to author, e-mail message, May 2, 2005; see also "Wise in the Ways of Japanese Toothpicks," http://www.stare.com/2002/weak34.html (Apr. 3, 2005).

38. Nakahara and Homma, "Tufted Toothpicks," 89–90.

39. Baer and Baer, "Story of the Toothpick," 159.

40. Author's collection; "Fortune Toothpicks in Cool Vintage Wood Box from Japan," offered on eBay, December 11, 2005.

41. Pin-Chin Huang, "Toothpick Assembly," U.S. Patent No. 6,044,848 (Apr. 4, 2000). Boxes of bamboo toothpicks from Thailand carry the notice, "Should be stored in dry place" ("Triple Grooven [Bamboo Toothpick] Square Box," offered on eBay Apr. 25, 2006). Bamboo used in construction will rot if it remains wet without air circulating about it: "Bamboo: A Remarkable Building Material," http://www.sutton.com/resourcecentre/HomeMaintenance/bamboo.html (Apr. 22, 2006).

42. Wadsworth Atheneum Museum of Art, Hartford, Conn. The knotted toothpicks now in author's collection were compliments of the chef.

43. Fox Run Kitchens, 300 Bamboo Food Picks, purchased at Now You're Cooking, Bath, Me., June 2006.

44. "Chinese Dandyism, &c.," *Casket* no. 7 (Jul. 1830), 322; cf. "A Chinese Dandy," *Robert Merry's Museum* 7, no. 5 (May 1844), 149.

45. Sozodont, advertisement, *New York Times*, Oct. 13, 1881, 5; Eric Curtis, *Hand to Mouth: Essays on the Art of Dentistry* (Chicago: Quintessence Publishing, 2002), 124.

46. Bo Krasse to author, letter, Mar. 21, 2005, e-mail messages, Apr. 1, 2005, and Sep. 20, 2006; see also Bo Krasse, "Hur TePe kom Till," typewritten manuscript, undated, in author's files; "TePe: We Care for Healthy Smiles," brochure (Malmö, Sweden: TePe Munhygienpro-dukter AB, n.d.), 2.

47. "Very Old Toothpick Packet (Interdens)," offered on eBay, Apr. 19, 2005.

48. Available at http://shop.store.yahoo.com/evryaustralian/teatretoot .html (Mar. 16, 2005).

49. Package of twenty-five TePe Medium Dental Sticks with fluoride.

CHAPTER TWENTY-THREE:
OLD GOLD AND GOOD WOOD

1. Brooklyn Jewelry Mfg. Co. advertisement, *Harper's Weekly* 11, no. 6 (1886), 723.

2. "Old Gold Sought to Aid Democrats," *New York Times*, May 3, 1936, 26.

3. Ann Sledge, "Restoration Hardware Coming to Greenwich, West-port," *Fairfield County* (Conn.) *Business Journal*, 38, issue 22 (May 31, 1999), 7; Thyra Porter, "No More Wooden Toothpicks: Accessories Going Upscale," *Home Furnishing News*, Oct. 27, 2003, 59.

4. James Huneker, "The Seven Arts," *Puck* 75, no. 1942 (May 23, 1914), 16.

5. Offered on eBay, Apr. 16, 2005, and Apr. 22, 2005.

6. Offered on eBay, Apr. 27, 2005.

7. "Little Toothpick Still Fells Trees," *New York Times*, Oct. 29, 1955, 27.

8. "Toothpick Town in Maine," *Boston Globe*, Feb. 10, 1907, SM12.

9. However, I was pleased to find an individual interdental stick offered with each meal service on an SAS flight from Chicago to Stockholm in summer 2006. The stick came handsomely laminated between two pieces of cellophane, one clear and one black, thus at the same time exposing what the little package contained and framing its contents as something to be admired.

10. "From a London Paper, April 26," *New-Hampshire Sentinel* 21, issue 1057 (Jul. 10, 1819), [4].

11. Arthur O. Corey, "Tooth-Pick," U.S. Patent No. 420,531 (Feb. 4, 1890).

12. James Edwin Hills, "Toothpick," U.S. Patent No. 725,081 (Apr. 14, 1903); Albert E. Lickman, "Toothpick," U.S. Patent No. 817,978 (Apr. 17, 1906); George M. Browning, "Toothpick and Method of Making Same," U.S. Patent No. 1,462,062 (Jul. 17, 1923).

13. Robert Ross Freeman, "Toothpick," U.S. Patent No. 797,423 (Aug. 15, 1905).

14. Albert H. Baird, "Composition Toothpick," U.S. Patent No. 973,842 (Oct. 25, 1910).

15. Francis H. Grant, "Tooth Cleanser," U.S. Patent No. 2,008,206 (Jul. 16, 1935).

16. Russell Edward Lunday, "Toothpick," U.S. Patent No. 1,465,522 (Aug. 21, 1923).

17. Rolf Barman, "Toothpicks," U.S. Patent No. 3,563,253 (Feb. 16, 1971).

18. Edward M. Barnett, "Periodontal and Dental Cleanser and Periodontal Stimulator," U.S. Patent No. 3,775,848 (Dec. 4, 1973).

19. Ibid.

20. George F. Grant, "Golf-Tee," U.S. Patent No. 638,920 (Dec. 12, 1899). See also http://inventors.about.com/library/inventors/blgolf teehtm.htm (May 9, 2005). That a dentist invented the golf tee was brought to my attention by my own dentist, Baxter B. Sapp, Jr.

21. John N. Cole, *Maine Trivia* (Nashville, Tenn.: Rutledge Hill Press), 100.

22. "Vintage 1950's Pick-n-Tee Golf Tees Cocktail Toothpicks," offered on eBay, Oct. 4, 2006.

23. "Matrimonial Failure!" *New York Times*, Dec. 8, 1907, SM8; M. E. Richardson, "In Sickness and in Health," *British Dental Journal* 197 (2004), 583.

24. Charles Dickens, *The Old Curiosity Shop*, chapter 28.

CHAPTER TWENTY-FOUR: BOXED AND UNBOXED

1. Clifford W. Estes Co., "Our History," http://www.estesco.com/history.html (Jun. 23, 2006).

2. Harvey Green, *Wood: Craft, Culture, History* (New York: Viking, 2006), 282.

3. Estes Co., "Our History"; E. B. Estes & Sons, price lists and circulars, Lloyd Library (Cincinnati, Ohio) archives, Box 21, Folder 5.

4. Author's collection; "Antique Estes Wood Barrel Box Advertising Toothpicks," offered on eBay, Jun. 23, 2006.

5. Estes & Sons, price lists.

6. Estes Co., "Our History."

7. Distributed by E. B. Estes Division, Box No. 779, Westbrook, Me., 04098; "Antique Estes Wood Barrel Box."

8. "Tried to Smoke Toothpicks," *Washington Post*, Oct. 14, 1894, 19.

9. Oscar H. Hersey, Sixth Account, Estate of Charles Forster, Aug. 6, 1909, Cumberland County (Me.) Probate Registry Office, Docket 1507 file.

10. Examples pasted into Robert Goff Stubbs, "A Toothpick Industry in Maine, 1913," typewritten report, Maine State Library, [31 verso].

11. Duluth Trading Co., http://www.duluthtrading.com/store/and more/ recreation camping/37111.aspx; "Terminator Toothpicks," http:// www.bookofjoe.com/2005/12/29/index.html (Mar. 21, 2006); "Nancy Kerns," http://www.baltimoreapplique.com/nancykerns.html (Mar. 21, 2006).

12. Author's collection.

13. "Packaging of Toothpicks Has Changed Since Turn of the Century," *The Woodline* 4, no. 3 (Sep. 1972), 4; Ben Z. Swanson, Jr., "Toothpick Boxes from the Holt Brothers Print Shop Archive," typed manuscript, courtesy of Ben Swanson, Baltimore, Md.

14. "A Salute to Knowlton & McLeary on Its 100th Brithday," *The Woodline* 3, no. 2 (Sep. 1971), 3; see also "Toothpicks and Bobsleds," *Maine Central Messenger* 11, 2 (Apr.–May 1972), 10–11; Ruth Robbins Adamo, "Forster Manufacturing Dream Come True for Couple," *Lewiston* (Me.) *Sun*, Aug. 8, 1988, 11.
15. George Buch, "Toothpick-box," U.S. Patent No. 550,247 (Nov. 26, 1895). Rights to this patent were assigned to Fred C. Gallman, of Aristotle, New York.
16. Oscar Weik, "Toothpick-Container," U.S. Patent No. 1,198,980 (Sep. 19, 1916).
17. Graham Greene, *The Human Factor* (New York: Simon & Schuster, 1978), quoted in M. E. Richardson, "In Sickness and in Health," *British Dental Journal* 197, no. 9 (2004), 583.
18. Purchased summer 2005, after my son pointed it out to me.
19. A. Auerbach to author, postcard, Mar. 30, 2006.
20. Sarah Jenkins, "EAPG [Early American Pattern Glass] Toothpick Holders," *Toothpick Bulletin* 30, 2 (Mar. 2006), 1; display ad, *Boston Globe*, Feb. 2, 1908, 9; "An R.S. Germany Toothpick," offered on eBay, May 18, 2006.
21. Dave Olson, "DL [Detroit Lakes] Woman's Hobby Dates Back to 1950s Germany," *Toothpick Bulletin* 30, no. 6 (Jul. 2006), 11, reprinted from *The Forum of Fargo-Moorhead* (N.D.-Minn.).
22. See, e.g., Neila and Tom Bredehoft and Jo and Bob Sanford, *Glass Toothpick Holders* (Paducah, Ky.: Collector Books, 1999).
23. Harry Hamilton, "The Personally-Conducted Tourist," *Puck* 49, no. 1270 (Jul. 3, 1901), 5; "Seven-Sevenths of a Christmas Day," *Ladies' Home Journal* 24, no. 1 (Dec. 1906), 17.
24. National Toothpick Holder Collectors' Society, P.O. Box 852, Archer City, Tex. 76351; see also www.nthcs.org.
25. Bergdorf Goodman, display advertisement, *New York Times*, Dec. 15, 2002, 6.
26. Quoted from "Coding Is the Easy Part . . . ," http://www.zdnet.com/5208-10532-0.html?forumID=1&threadID=8762&messageID=176169&start=−1 (Sep. 29, 2005).
27. Abby Zimet, "Millions of Toothpicks Churned Out," *Maine Sunday Telegram*, Apr. 19, 1998, 16B; Laralyn Sasaki, "End of an Era: Toothpicks Fall Out of Favor," *Wall Street Journal*, Oct. 3, 1985, 1.
28. Theodore R. Hodgkins, "Dispensing Box Construction," U.S. Patent No. 3,251,530 (May 17, 1966).
29. Ibid.
30. Box of 750 Ideal toothpicks (ca. 1965) in author's collection.
31. Box of 250 Forster round toothpicks, purchased at Wal-Mart, Brunswick, Me., Aug. 2005. On the idea of instructions, see Paul Mijksenaar and Piet Westendorp, *Open Here: The Art of Instructional Design* (New York: Joost Elffers Books, 1999).
32. Jonas Sahlin, "Toothpick-Stand," U.S. Patent No. 894,702 (Jul. 28, 1908).
33. Reinaldo Rela Zattoni, "Safety Package of Toothpick Holder," U.S.

Patent No. 4,083,487 (Apr. 11, 1978); Hermogenes Rella, "Safety Package of Toothpick Holder," U.S. Patent No. 4,163,492 (Aug. 7, 1979).

34. Box of 750 Forster Ideal flat toothpicks, offered on eBay, Aug. 3, 2005.

35. Faunce Pendexter, "Strong Wood Turning Corporation One of Maine's Busiest Firms," *Lewiston* (Me.) *Journal*, Jan. 2, 1954, magazine section, 6-A; Sherlock Holmes, "Take Your Pick," *Maine Magazine*, Mar.–Apr. 1995, photocopy of Regional page, provided by Maine Historical Society.

CHAPTER TWENTY-FIVE:
WOODPECKERS AND OTHER DISPENSERS

1. *Atlanta Constitution*, Dec. 9, 1875, 1.

2. Harold W. C. Prommel, "Sanitary Toothpick-Dispenser," U.S. Patent No. 1,146,447 (Jul. 13, 1915).

3. Forest Jackson, "Toothpick-Holder," U.S. Patent No. 1,212,357 (Jan. 16, 1917).

4. Frederick C. Lynde, "Automatic Vending Apparatus," U.S. Patent No. 366,490 (Jul. 12, 1887) and U.S. Patent No. 371,146 (Oct. 4, 1887); Horace C. Walker, "Match-Delivering Machine," U.S. Patent No. 1,173,069 (Feb. 22, 1916); Gustav A. Hultenius, "Dispensing Device," U.S. Patent No. 1,627,647 (May 10, 1927); Lawrie L. Witter, "Cigarette Dispenser," U.S. Patent No. 1,986,658 (Jan. 1, 1935); Richard Mayer, "Cigarette Dispenser and Lighter," U.S. Patent No. 2,083,464 (Jun. 8, 1937).

5. "Even Toothpicks Are Going to Be Standardized," *Washington Post*, Nov. 26, 1941, 1. For a parody of the proposed standard, see H. I. Phillips, "The Once Over," *Washington Post*, Nov. 27, 1941, 15.

6. Alfred J. Schupp, "Dispensing Device for Cylindrical Objects," U.S. Patent No. 2,752,065 (Jun. 26, 1956); "Little Toothpick Still Fells Trees," *New York Times*, Oct. 29, 1955, 27.

7. Heinrich Staub, "Receptacle for Toothpicks," U.S. Patent No. 545,546 (Sep. 3, 1895).

8. Louis H. Tangen, "Server for Toothpicks or Articles of Similar Splint-Like Form," U.S. Patent No. 1,004,933 (Oct. 3, 1911).

9. See, e.g., Randal C. Archibold and John M. Broder, "Plane Lands Safely in Los Angeles After Landing Gear Becomes Stuck," *New York Times*, Sep. 22, 2005, A14; Abednego R. Hughes, "Single-Delivery Toothpick-Holder," U.S. Patent No. 1,077,715 (Nov. 4. 1913).

10. Charles Koenig, "Design for a Toothpick Dispenser," U.S. Patent Des. No. 151,773 (Nov. 16, 1948). See also Michael Nicosia, "Design for a Tooth Pick Dispenser," U.S. Patent Des. No. 118,290 (Dec. 26, 1939). A Victorian precursor was offered as a "Cast Iron Woodpecker Toothpick Holder" on eBay, Aug. 10, 2004. For a collector's perspective, see "Mechanical Woodpecker Toothpick or Match Holders," http://home.att.net/~cruller1/woodpecker.story.html.

11. Johan Folke Gustafson and Carl Holger Gustafson, "Cigarette Dis-

penser," U.S. Patent No. 2,562,666 (Jul. 31, 1951); Cecil C. Adams, "Cigarette Dispensing Device," U.S. Patent No. 2,954,143 (Sep. 27, 1960).

12. Hsu C. Pan, "Toothpick Dispenser," U.S. Patent No. 4,998,644 (Mar. 12, 1991); Des. No. 316,313 (Apr. 16, 1991); author's collection.

13. Curtis T. Manz, "Toothpick Dispensing Device," U.S. Patent No. 4,271,979 (Jun. 9, 1981).

14. See, e.g., Alexander Nelson, "Toothpick Dispenser," U.S. Patent No. 4,522,314 (Jun. 11, 1985).

15. "The Polite Toothpick," *Washington Post,* Jan. 18, 1908, 6; "A Japanese Dinner," *Washington Post,* Nov. 25, 1888, 10.

16. National Industrial Recovery Administration, "Hearing on Code of Fair Practices and Competition, Presented by Drinking Straw and Toothpick Industry," Jan. 19, 1934, transcript, 9, 17, 64.

17. See Alexander Herz, "Apparatus for Incasing Toothpicks and the Like," U.S. Patent No. 883,803 (Apr. 7, 1908).

18. *Hygeia Antiseptic Toothpick Co. v. United States,* 1 Ct. Cust. 497; 1911 Ct. Cust. LEXIS 88; T.D. 31529 (1911).

19. Howard E. Barlow, "Toothpick-Machine," U.S. Patent No. 796,394 (Aug. 1, 1905).

20. "Councilman Miller's Reminder," *Boston Globe,* Aug. 28, 1900, 7.

21. "Celluloid Toothpick Carlisle PA Commercial College," offered on eBay, Aug. 2005.

22. Allen C. Harriman, "Web Carried Toothpick Dispenser," U.S. Patent No. 3,709,403 (Jan. 9, 1973); "The Public Must Have . . . ," classified advertisement, *New York Times,* Nov. 4, 1977, 45.

23. Takao Makishima, Shigeru Inaba, and Yasuyoshi Wada, "Portable Toothpick Case," U.S. Patent No. 4,518,099 (May 21, 1985).

24. Gustavus R. Schimmel, "Tooth-Pick Package," U.S. Patent No. 281,731 (Jul. 24, 1883).

25. Joseph A. Doll, "Toothpick," U.S. Patent No. 2,035,425 (Mar. 24, 1936).

26. Robert B. Briggs, "Plastic Toothpicks," U.S. Patent No. 2,760,628 (Aug. 28, 1956).

27. William T. Cameron, "Toothpicks and Method of Making Same," U.S. Patent No. 2,762,501 (Sep. 11, 1956); see also David Pinkas, "Pick Array," U.S. Patent No 3,438,486 (Apr. 15, 1969).

28. Author's collection.

29. Makishima, Inaba, and Wada, "Portable Toothpick Case"; William D. Manciocchi, Jr., "Antiseptic Toothpick," U.S. Patent No. 4,509,541 (Apr. 9, 1985).

30. John H. Feaster, "Combination Toothpick and Stirring or Eating Implement," U.S. Patent No. 2,877,547 (Mar. 17, 1959).

31. George S. Adolfson, "Combination Toothpick and Beverage Stirrer," U.S. Patent No. 4,159,182 (Jun. 26, 1979); Robert P. Julius, "Combined Fork and Separatable Toothpick," U.S. Patent No. Des. 254,229 (Feb. 19, 1980); Mark C. Sanders, "Eating Utensil with

Toothpick Incorporated Therein," U.S. Patent Des. No. 463,221 (Sep. 24, 2002); see http://fatman-seoul.blogspot.com/20041001 fatman-seoularchive.html (Apr. 3, 2005).

32. Lisa E. Bell, "Combined Package for Popcorn and Toothpick," U.S. Patent Des. No. 355,593 (Feb. 21, 1995); "30 Hand Folded Origami Holders & Toothpicks: Tsumayouji," offered on eBay, Aug. 20, 2005; "Personalized Toothpick Packs! Wedding Favors," offered on eBay, Apr. 13, 2005.

CHAPTER TWENTY-SIX:
TALKING ROUND A TOOTHPICK

1. See Colm Tóibín, "Lady Gregory's Toothbrush," *New York Review of Books* 48, no. 13 (Aug. 9, 2001); Lady Gregory is also quoted in Kostya Kennedy, "Clean Sweep," *US Airways*, May 2006, 43.

2. "Democratic Gentility," *Current Literature* 30, no. 4 (Apr. 1901), 390.

3. C. Hanbury Williams, "Fifteen Hundred Miles on Fresh Water," *Eclectic Magazine of Foreign Literature* 136, no. 4 (Apr. 1901), 487; "Where Toothpicks May Be Used," *Ladies' Home Journal* 18, no. 4 (Mar. 1901), 43; *Methodist Review* 18, no. 5 (Sep. 1902), 826.

4. William T. Davis, *Plymouth Memories of an Octogenarian* (Plymouth, Mass.: privately printed, 1906), 519.

5. Emily Wheaton, "The Russells in Chicago," *Ladies' Home Journal* 19, no. 1 (Dec. 1901), 9.

6. Laura A. Smith, "What I Found Out as a Business Girl," *Ladies' Home Journal* 24, no. 11 (Oct. 1907), 28; "The Lady from Philadelphia," *Ladies' Home Journal* 20, no. 6 (May 1903), 33.

7. Emma E. Walker, "Pretty Girl Papers," *Ladies' Home Journal* 22, no. 8 (Jul. 1905), 33.

8. Robert Blight, "The Nuts of Commerce," *Outing* 39, no. 6 (Mar. 1902), 710.

9. See, e.g., "Pot-Pourri," *Potter's American Monthly* 19, no. 128 (Aug. 1882), 215; see also "The Night Car's Freight," *New York Times*, Mar. 11, 1883, 4.

10. "Clang of the Wooden Shoon," *Boston Globe*, Jul. 6, 1884, 2; "Women's Shoes and Hosiery," *New York Times*, Apr. 20, 1902, SM11; display advertisement, *New York Times*, Mar 23, 1903, 14.

11. Charles L. Paige, "Thoreau," *Forest and Stream* 61, no. 16 (Oct. 17, 1903), 294; "The Growth of Business in Mexico," *Overland Monthly and Out West Magazine* 56, no. 1 (Jul. 1910), 44; display advertisement, *New York Times*, Nov. 14, 1926, BR23.

12. See Beatwear website, http://www.beatwear.co.uk/acatalog/Winkle Picker.html (Mar. 16, 2006); see http://en.wiktionary.org/wiki/winkle-picker (Mar. 16, 2006).

13. "Balaklava Outdone," *Puck* 57, no. 1457 (Feb. 1, 1905), 7; "With Timber Scarce," *Puck* 68, no. 1750 (Sep. 14, 1910), 13; "Puck's Views and Reviews," *Puck* 57, no. 1470 (May 3, 1905), 2.

14. "The Confessions of a Literary Drummer," *Bookman* 31, no. 2 (Apr. 1910), 164.

15. "The Confessions of a Literary Press Agent," *Bookman* 24, no. 4 (Dec. 1906), 335; "The Columbus (O.) Gun Club," *Forest and Stream* 70, no. 10 (Mar. 7, 1908), 389.

16. Berton Braley, "Desire Unfulfilled," *Puck* 68, 1762 (Dec. 7, 1910), 15.

17. Viola Burhans, "The Inner Claw," *McBride's Magazine* 97, no. 577 (Jan. 1916), 88.

18. Frederic J. Haskin, "Quick-Lunch Rooms," *Washington Post*, Mar. 29, 1910, 11.

19. E. M. Bishop, "Is the Toothpick Passing?" *New York Times*, May 3, 1915, 10.

20. See Patrick McSherry, "The Sinking of U.S. Navy Collier Merrimac," http://www.spanamwar.com/merrimac.htm (Jan. 25, 2006); "The Rewards of War," *New York Times*, Jun. 24, 1898, 6.

21. Harry Thurston Peck, "Retouches," *Bookman* 9, no. 2 (Apr. 1899), 157.

22. Frank G. Carpenter, "Gossip About the President," *Los Angeles Times*, Dec. 5, 1897, B1; "Is the President Extravagant?" *Outlook* 76, no. 15 (Apr. 9, 1904), 868; H. I. Phillips, "The Once Over," *Washington Post*, Nov. 27, 1941, 15.

23. A. Maurice Low, "Enthusiasm Along Line," *Boston Globe*, Jul. 17, 1896, 3.

24. "Victors' Fun," *Boston Globe*, Nov. 7, 1896, 6.

25. "Election Freaks and Fancies," *Chicago Tribune*, Oct. 25, 1900, 12.

26. "He Wins $3000.," *Boston Globe*, Nov. 8, 1900, 3.

27. "Bar Bryan Toothpick," *New-York Tribune*, Aug. 31, 1906, 3; "The Bryan Toothpick Barred," *Atlanta Constitution*, Sep. 27, 1906, A9.

28. "Bland-Allison Act, 1878," http://www.u-s-history.com/pages/h718.html (Jan. 23, 2006).

29. "Stewart's Free Coinage Bill," *Chicago Tribune*, Jun. 7, 1892, 12; "Bland Will Fight, It Is Said," *New York Times*, Jul. 15, 1893, 8.

30. "Bimetallism," *Wikipedia*, http://en.wikipedia.org/wiki/Bimetallism (Jan. 23, 2006).

31. "Bar Bryan Toothpick."

32. Ibid.; "What Fools These Mortals Be," *Puck* 64, no. 1651 (Oct. 21, 1908), 2.

33. "Editorial Pen Points," *Los Angeles Times*, Jul. 25, 1908, II4; *Webster's New Biographical Dictionary* (Springfield, Mass.: Merriam-Webster, 1988), s.v. "Bryan, William Jennings."

34. W. A. Evans, "How to Keep Well," *Chicago Tribune*, Feb. 23, 1921, 8.

35. "Lost and Found," *New York Times*, Jan. 30, 1919, 22.

36. The Staff, "Pen Points," *Los Angeles Times*, Jun. 5, 1923, II4; display advertisement, *New York Times*, Nov. 28, 1933, 18.

37. Gerard B. Lambert, *All Out of Step* (Garden City, N.Y.: Doubleday, 1956), 170–71.

38. See, e.g., display advertisement, *New York Times*, Nov. 29, 1958, 3; "Lost and Found," *New York Times*, May 14, 1971, 82; display advertisement, *New York Times*, Dec. 7, 1975, 485.

39. "The Journal of Timothy Tubby," *Bookman* 55, no. 2 (Apr. 1922),

139; "Killer Is Doomed in Federal Court," *New York Times*, Dec. 15, 1935, 12.

40. "Judge Whittles as 3 Fight Death," *Chicago Defender*, Sep. 10, 1949, 3.

41. James Lowe, "A Very, Very Interesting Story About a . . . Toothpick?" *Journal of the Oklahoma State Dental Association* 59, no. 4 (Apr. 1969), 6.

42. Amy Vanderbilt, *Amy Vanderbilt's Complete Book of Etiquette* (Garden City, N.Y.: Doubleday, 1952), 252.

43. Amy Vanderbilt, *Amy Vanderbilt's Etiquette* (Garden City, N.Y.: Doubleday, 1972), 334, 365; see also Elizabeth L. Post, *Emily Post's Etiquette*, 14th ed. (New York: Harper & Row, 1984).

44. Letitia Baldrige, *The Amy Vanderbilt Complete Book of Etiquette: A Guide to Contemporary Living*, rev. and exp. ed. (Garden City, N.Y.: Doubleday, 1978), 417, 424.

45. Mary C. Churchill, "Eagleton Institute to 'Change,' " *New York Times*, Jun. 9, 1974, 78.

46. Ann Japenga, "Promoting Social Harmony with Positive TV Images," *Los Angeles Times*, Jan. 9, 1985, 1.

47. Peter Bennett, "Tusk Scrubber a Number One Pick," *Los Angeles Times*, Dec. 14, 1984, L8.

48. Ibid.; "Groom Makes the Best Batters Look Cheap," *Los Angeles Times*, Jul. 28, 1912, 19.

49. "Surveying What's Left When G.M. Leaves Town," *New York Times*, Dec. 20, 1989, C18; Rob Hotakainen, "Seventh Congressional District," *Minneapolis Star Tribune*, Oct. 11, 2004, 1B; Virginia Groark, "Revamped Illinois Tollway Oases to Offer Travelers More Options," *Chicago Tribune*, Jul. 1, 2004.

50. Margaret Visser, *The Rituals of Dinner: The Origins, Evolution, Eccentricities, and Meaning of Table Manners* (New York: Grove Weidenfeld, 1991), 324; Nigel Farndale, "The Wit and Wisdom of James Hewitt," *London Sunday Telegraph*, Mar. 27, 2005, 21.

51. Visser, *Rituals of Dinner*, 325.

52. Charlotte Ford, *21st-Century Etiquette: Charlotte Ford's Guide to Manners for the Modern Age* (Guilford, Conn.: Lyons Press, 2001), 51.

53. Oscar de Mejo, *The Professor of Etiquette: A Guide to the Do's and Don'ts of Civilized Living, with an Additional Section on the Toothpick* (New York: Philomel Books, 1992), [1–3, 21].

54. Ibid., [23–25].

55. Ibid., [26–29].

56. Henry McLemore, "The Lighter Side," *Los Angeles Times*, Oct. 24, 1945, 7; "The Lowly Toothpick," *Washington Post*, Jan. 31, 1927, 6.

57. Ann Landers, "She Shouldn't Wed Cheapskate Widower," *Greensboro* (N.C.) *News Record*, Dec. 8, 1993, D5; Ann Landers, "Mother Tries to Teach Child Honesty," *Greensboro* (N.C.) *News Record*, Sep. 24, 1994, D10.

58. Judith Martin, "Picking: A Fight," *Washington Post*, Nov. 11, 2001, F4.

CHAPTER TWENTY-SEVEN: THE FATAL MARTINI

1. Claire Baer and Paul N. Baer, "A Story of the Toothpick," *Journal of Periodontology* 37 (Mar.–Apr. 1966), 159.
2. "A Toothpick Tragedy," *New York Times*, Sep. 3, 1884, 4.
3. *Salem* (Mass.) *Gazette*, Dec. 22, 1812, [3]; "Dies from Toothpick Wound," *Chicago Tribune*, Mar. 20, 1901, 1; "Nelson as an Angler," *Forest and Stream* 66, no. 16 (Apr. 21, 1906), 637; William H. Holton, "Constitutional Hemorrhage," *American Journal of the Medical Sciences* 134 (Apr. 1874), 414.
4. "The Deadly Toothpick," *Atlanta Constitution*, Aug. 26, 1889, 4.
5. "Chewing Wooden Toothpicks," *Boston Globe*, Nov. 18, 1889, 2; "The Deadly Toothpick," *Washington* (D.C.) *Evening Star*, Aug. 13, 1889, 3.
6. "Deadly Toothpick," *Washington Evening Star*.
7. L. D. Budnick, "Toothpick-Related Injuries in the United States, 1979 through 1982," *Journal of the American Medical Association* 252, no. 6 (Aug. 10, 1984), 796; see also "Toothpick-Related Injury," *American Family Physician* 31, no. 4 (1985), 230.
8. "Doctor Finds Toothpicks Injure 8,000 Yearly," *New York Times*, Aug. 12, 1984, 21.
9. "A Toothpick Maker Dissents," *New York Times*, Aug. 12, 1984, 21.
10. "Death Rather than Law," *National Police Gazette* 51 (Feb. 25, 1888), 3; Clarence J. Blake, "Otology," *American Journal of Medical Sciences* 131, no. 2 (Feb. 1906), 363.
11. Benjamin Collins Brodie, "Extraction of Foreign Bodies," *Medical Examiner, and Record of Medical Science* 7, no. 16 (Aug. 10, 1844), 182; see also B. C. Brodie, "Clinics," *Medical News and Library* 2, no. 19 (Jul. 1844), 54.
12. *Medical and Surgical Reporter* 77, no. 11 (Sep. 11, 1897), 349.
13. See, e.g., "Cheap Tooth Brushes and Tooth Picks," *Scientific American* 70, no. 11 (Mar. 17, 1894), 168; J. E. Cravens, "Thoughts on Toothpicks, or an American Sermon from an Englishman's Text," *Missouri Dental Journal* 6, no. 1 (1874), 27.
14. "Toilet of a Roman Lady," *Godey's Lady's Book and Magazine* 51 (Sep. 1855), 254; "From a London Paper," *Salem* (Mass.) *Gazette* 33, issue 100 (Dec. 14, 1819), [4]; "Danger from Picking the Teeth with Pins," *Graham Journal of Health and Longevity* 2, no. 25 (Dec. 15, 1838), 388.
15. *Medical and Surgical Reporter* 54, no. 5 (Jan. 30, 1886), 160; cf. *Medical and Surgical Reporter* 46, no. 8 (Feb. 25, 1882), 223.
16. A. J. Ochsner, "A Case of Actinomycosis," *Medical News* 58, no. 4 (Jan. 24, 1891), 97.
17. "Humorous," *Saturday Evening Post* 61, no. 2 (Jul. 30, 1881), 14.
18. F. D. T., "Ode to a Toothpick," *Dental Review* 5, no. 9 (1891), 744. The ode is quoted from the *Dental Review* but without mention of an author, even by initials, in Arden G. Christen and Joan A. Christen, "A Historical Glimpse of Toothpick Use: Etiquette, Oral and Medical Conditions," *Journal of the History of Dentistry* 51, no. 2 (Jul. 2003),

61–62. The line and stanza breaks of the former version are followed here, as is its punctuation.

19. "Miseries of Human Life," *Boston Gazette* 34, issue 28 (Apr. 1, 1811), [4].

20. "Death in the Toothpick," *Dental Record* 10 (1890), 86; "Wooden Tooth-Picks," *Dental Register* 36, 6 (1882), 306; Sydney Garfield, *Teeth, Teeth, Teeth: A Treatise on Teeth and Related Parts of Man, Land and Water Animals from Earth's Beginning to the Future of Time* (New York: Simon & Schuster, 1969), 91.

21. "Pugilistic News," *National Police Gazette* 51, no. 525 (Oct. 8, 1887), 10; "Puck's Personal Intelligence," *Puck* 1, no. 12 (May 1877), 4; *National Police Gazette* 39, no. 215 (Nov. 5, 1881), 12.

22. "Fair 'Third Term' Held a Possibility," *New York Times*, Jun. 10, 1940, 20.

23. Franklin R. Cockerill III, Walter R. Wilson, and Robert E. van Scoy, "Traveling Toothpicks," *Mayo Clinic Proceedings* 58, no. 9 (1983), 613; ["Dear Ann Landers,"] *Greensboro* (N.C.) *News Record*, Oct. 7, 1995, D2; "Chicago Surgical Society," *Medical News* 81, no. 22 (Nov. 29, 1902), 1053.

24. K. Hoegh, "Report of a Case of Perforative Peritonitis from a Foreign Body in Colon Ascendens," *Medical News* 68, no. 8 (Feb. 22, 1896), 207, quoted in B. H. Kean, *M.D.: One Doctor's Adventures Among the Famous and Infamous from the Jungles of Panama to a Park Avenue Practice* (New York: Ballantine Books, 1990), 97.

25. Quoted in "Swallows Toothpick, Dies," *New York Times*, Jan. 18, 1948, 36.

26. "Sands Hotel Casino, Red Box of Toothpicks, Mint Cond.," offered on eBay, Apr. 17, 2006.

27. Don Colburn, "Swallowed Toothpick Becomes Deadly Time Bomb," *Washington Post*, Mar. 28, 1989, z.05.

28. Pat Rich, "Don't Swallow Toothpicks," *Toronto Medical Post* 34, no. 4 (Jan. 27, 1998), 17; "Cause of Man's Toothache? A 4-Inch Nail," *New York Times*, national ed., Jan. 17, 2005, A13.

29. "An Analysis of the Five-Cent Cigar," *Washington Post*, Mar. 8, 1881, 3.

30. "Desert Martini Recipe," http://www.drinksmixer.com/drink9050.html (Jan. 17, 2005); "The Creation of a Perfect Martini Is No Great Mystery,"http://www.moderndrunkardmagazine.com/issues/03_03/03-03perfectmartinin.htm (Jan. 17, 2005).

31. "How to Snort a Martini," *Science News* 130 (Nov. 1, 1986), 280.

32. Quoted in Kean, *M.D.*, 96–97.

33. Kean, *M.D.*, 92–96; "Explains Writer's Death," *New York Times*, Apr. 10, 1941, 15.

34. Kean, *M.D.*, 93, 95–97.

35. Eric Curtis, *Hand to Mouth: Essays on the Art of Dentistry* (Chicago: Quintessence, 2002), 125.

36. Terry Lane, "Floatable Toothpick Assembly," U.S. Patent No. 5,337,766 (Aug. 16, 1994).

37. "Toothpick," *Los Angeles Times*, May 19, 1931, A4. At the exchange

rates of the time, the offering price in London would have been about £57. See "The History of the Dollar Exchange Rate," http://www.miketodd.net/encyc/dollhist.htm.

38. Charles Dickens, *A Christmas Carol*, Stave 1.

CHAPTER TWENTY-EIGHT:
IMPROVING ON PERFECTION

1. "Toothpick Patent Run Out," *Manufacturer and Builder* 11, no. 12 (Dec. 1879), 280; David Lamb, "Toothpick Firms Are Working Hard in the Sticks: The Small, Arcane Industry Is Serious Business, Filled with Fierce Rivalry and Closely Guarded Secrets," *Los Angeles Times*, Nov. 15, 1993, 5; Sherlock Holmes, "Take Your Pick," *Maine Magazine*, Mar.–Apr. 1995, copy of article in Maine Historical Society files, s.v. "Forster, Charles"; Henry J. Pratt, "Toothpicks," *Toothpick Bulletin* 30, no. 5 (Jun. 2006), 8–9.

2. "Strong, Toothpick Town, Makes 50,000,000 Daily," *Lewiston* (Me.) *Journal*, Jun. 30, 1932, in Maine State Library clipping book #124; Laralyn Sasaki, "End of an Era: Toothpicks Fall Out of Favor," *Wall Street Journal*, Oct. 3. 1985, 1; Abby Zimet, "Maine's Mills Endure," *Maine Sunday Telegram*, Apr. 19, 1998, 1B; Pratt, "Toothpicks," 8.

3. "Large Output of Toothpicks," *New York Times*, Mar. 15, 1903, 33.

4. Lamb, "Toothpick Firms."

5. Harland L. Durrell, "BFD Is Oakland's Largest Industry," *Waterville* (Me.) *Morning Sentinel*, Feb. 23, 1948, in Maine State Library clipping book #124; Steve Clark, quoted in Zimet, "Maine's Mills Endure," 16B.

6. Lamb, "Toothpick Firms"; George E. Walsh, "Maine's Wood Novelty Mills," *Scientific American* 87, no. 17 (Oct. 25, 1902), 269.

7. Robert Goff Stubbs, "A Toothpick Industry in Maine, 1913," typewritten report, Maine State Library, [35].

8. Advertisement, *Zion's Herald*, Apr. 28, 1909, 539; "Here's a Footnote to Stevenson Tour," *New York Times*, Sep. 6, 1952, 10.

9. Sue Hubbell, "Tales of the Toothpick," *Utne Reader* 81 (May–Jun. 1997), 31.

10. "Tobacco Free and Up to Snuff," *New York Times*, Aug. 6, 1990, C2; "10 Tubes of HotLix Cinnamon Toothpicks," offered on eBay, Apr. 21, 2006.

11. Package of Hot Toothpicks, bought in San Francisco's Chinatown, Dec. 2004; Sue Hubbell, "Let Us Now Praise the Romantic, Artful, Versatile Toothpick," *Smithsonian* 27, no. 10 (Jan. 1997), 76.

12. Author's collection; box of Minto Flavored Tooth-Picks (ca. 1950s or 1960s), purchased on eBay, Jan. 2006.

13. Display advertisement, Mendelsohn's, New Rochelle, N.Y., *New York Times*, Jun. 28, 1964, 34; "Vintage Burbon [*sic*] & Dry Martini Flavored Toothpicks," offered on eBay, Apr. 11, 2006.

14. Edward J. Petrus, "Therapeutic Toothpick for Treating Oral and Systemic Diseases," U.S. Patent No. 5,875,798 (Mar. 2, 1999); Teresa Riordan, "Patents," *New York Times*, Mar. 8, 1999, C11.

15. "Little Toothpick Still Fells Trees," *New York Times*, Oct. 29, 1955, 27; Carl. J. Kucher, "Combined Toothpick and Gum Massager," U.S. Patent No. 2,925,087 (Feb. 16, 1960).

16. Thomas M. Jackson, "Combined Condiment Holder and Toothpick," U.S. Patent No. 2,931,370 (Apr. 5, 1960).

17. Peter Bennett, "Tusk Scrubber a Number One Pick," *Los Angeles Times*, Dec. 14, 1984, L8; Sasaki, "End of an Era"; Zimet, "Maine's Mills Endure"; Lamb, "Toothpick Firms."

18. M. R. Montgomery, "The Thing About Clothespins," *New England Magazine*, May 23, 1982.

19. Quoted in "Toothpick Maker to Have New Owner," *New York Times*, Jun. 19, 1992, D3.

20. Lamb, "Toothpick Firms."

21. Box of Diamond Brand Tooth Picks (ca. 1960s), offered on eBay Jan. 24, 2006.

22. "MI: Calyton [*sic*] Illus Perfection Toothpicks Ad Cover," offered on eBay Apr. 3, 2006; Edward M. Lamb and Emmor Bales, "Machine for Making Toothpicks," U.S. Patent No. 588,645 (Aug. 24, 1897); Charles C. Freeman, "Tooth-pick," U.S. Patent No. 448,647 (Mar. 24, 1891); Edward M. Lamb and Emmor Bales, "Machine for Making Toothpicks," U.S. Patent No. 634,689 (Oct. 10, 1899).

23. Sold under the brand name The Doctor's.

24. Emma J. Thurston, "Toothpick," U.S. Patent No. 682,892 (Sep. 17, 1901); Semon Eisenberg, "Toothbrush," U.S. Patent No. 1,784,986 (Dec. 16, 1930).

25. Mary Krallmann, "A Kitchen Checklist," http://archives.lincolndaily news.com/2001/Mar/30/Features_new/perspectives.shtml (Jul. 7, 2006).

26. James L. Smith, "Toothpick," U.S. Patent No. 792,471 (Jun. 13, 1905).

27. James Edwin Hills, "Toothpick," U.S. Patent No. 714,901 (Dec. 2, 1902).

28. Lamb, "Toothpick Firms."

CHAPTER TWENTY-NINE: THE BUTLER DID IT

1. "Toothpicks Take a Drop," *Boston Globe*, Feb. 14, 1911, 2.

2. Robert Goff Stubbs, "A Toothpick Industry in Maine, 1913," type-written report, Maine State Library, [2].

3. See, e.g., Charles Forster, "Trade-Mark for Wooden Toothpicks," U.S. Trademark No. 35,328 (Oct. 30, 1900). The trademark was renewed to the Estate of Charles Forster, then republished by the Forster Manufacturing Company, Inc. in 1948, and subsequently renewed to that same company in 1950.

4. Quoted from end panel of box of 750 Forster flat toothpicks, purchased 2004.

5. "The 100th Anniversary of Forster Manufacturing Co.," *Congressional Record* 133, no. 127 (Jul. 30, 1987), S10958.

6. William Cohen, "Don't Knock Toothpick Factory," *Daily Kennebec*

(Me.) *Journal*, Jan. 16, 1988; see also "Diamond Brands Buys Forster Toothpick Factory in Wilton," *Bangor* (Me.) *Daily News*, Mar. 8, 1995; Julie Johnson, "Pryor Thinks It's a 'Great Place' but Not for Making Toothpicks," *New York Times*, Dec. 21, 1987, B10.

7. National Recovery Administration, *Code of Fair Competition for the Toothpick Industry*, Registry No. 329–03 (Washington, D.C.: Government Printing Office, 1933), 1, 3, 6;

8. Laralyn Sasaki, "End of an Era: Toothpicks Fall Out of Favor," *Wall Street Journal*, Oct. 3, 1985, 1; see also "Toothpick Maker to Have New Owner," *New York Times*, June 19, 1992, D3; David Lamb, "Toothpick Firms Are Working Hard in the Sticks," *Los Angeles Times*, Nov. 15, 1993, 5; Henry J. Pratt, "Toothpicks," *Toothpick Bulletin* 30, no. 5 (Jun. 2006), 9.

9. Ruth Robbins Adamo, "Forster Manufacturing Dream Come True for Couple," *Lewiston* (Me.) *Sun*, Aug. 8, 1988, 11; "Theodore R. Hodgkins, Retired Forster Co. President, Dies," *Maine Sunday Telegram*, Dec. 18, 1977, 2A.

10. National Recovery Administration, *Code of Fair Competition*, 6; see also George B. Stanley, "Charles Forster Fathered the Toothpick Industry and Built First Toothpick Mill," *Rumford Falls* (Me.) *Times*, Aug. 26, 1943, in Maine State Library clipping book #124.

11. "Change of Name," State of Maine, filed Dec. 1, 1936, The Forster Manufacturing Company, Inc., Change of Name, 103–32, Department of the Secretary of State, Bureau of Corporations, Elections and Commissions, Charter No. 19240005 D, 1910000183202, Aug. 7, 1950; Stanley, "Charles Forster Fathered."

12. Lewis Brackley and Charles Lisherness, comps., *Strong, Maine* (Strong, Me.: Strong Historical Society, 1992), 49, 195.

13. Neil Genzlinger, "Forster: 'We're Not Interested in Selling,' " *Waterville* (Me.) *Morning Sentinel*, Mar. 4, 1978; Brackley and Lisherness, *Strong*, 49; Phil Casey, "Toothpicks, Clothespins . . . and Croquet," *Washington Post, Times Herald*, May 12, 1972, B3.

14. Forster Mfg. Co. Inc. letterhead, Feb. 3, 1950, Wilton Historical Society, vertical file, s.v. "Forster Mfg. Co."

15. Brackley and Lisherness, *Strong*, 49; Ruth Adamo, *A History of the Town of Wilton* (Rumford, Me.: Rumford Publishing Co., 1971), 56.

16. Tower Manufacturing and Novelty Co., 1899 catalog, 1.

17. Various boxes of Worlds Fair and Gold Medal toothpicks in author's collection; Forster Manufacturing Company, "Trade-Mark for Toothpicks," U.S. Trademark No. 83,401 (Sep. 12, 1911).

18. Ink blotter, Forster Mfg. Co., Inc., undated, author's collection; author's collection; "Forster Pure Maple Syrup Tin Farmington Maine Qt.," offered on eBay, May 11, 2006.

19. Group of three boxes of toothpicks offered on eBay, Jan. 2006; now in author's collection.

20. Genzlinger, "Forster." David Hodgkins also served as president of the Wilton printing firm of Knowlton & McLeary, in which his father had acquired an interest in 1945. The Forster firm had bought the

remaining interest in the company in 1950. See "A Salute to Knowlton & McLeary on Its 100th Birthday," *The Woodline* 3, no. 2 (Sep. 1971), 3.

21. Genzlinger, "Forster"; Adamo, "Forster Manufacturing Dream."

22. Alan S. Oser, "The Forster Legacy Lives on in Wilton," *Maine Sunday Telegram,* Nov. 4, 1979.

23. Adamo, "Forster Manufacturing Dream"; "Toothpick Maker to Have New Owner."

24. See "The Croquet Capital of the World," *Yankee* 54, no. 6 (1990), 92–93; "Vintage Steel FORSTER Competitor HORSESHOES Ring Toss," offered on eBay, May 12, 2006; "Forster BOCCE BALLS Competitors Game," offered on eBay, May 12, 2006; "Forster Mfg. Co., Inc. Celebrating Our 100th Birthday," broadside, Wilton Historical Society, vertical file, s.v. "Forster Mfg. Co."

25. "Federal Grant to Aid 45 Workers Laid Off in Strong," *Portland* (Me.) *Press Herald,* Oct. 11, 2003, Saturday Business page.

26. Mary Anne Lagasse, "Forster Manufacturing Up for Sale," *Bangor* (Me.) *Daily News,* Sep. 26, 1985; Ruth R. Adamo, "Forster Manufacturing Permit is Approved for Expansion," newspaper clipping, 1989, Wilton Historical Society, vertical file, s.v. "Forster Mfg. Co."

27. Betty Jespersen, "Forster: It All Began with Toothpicks," 1987 newspaper clipping, photocopy courtesy of Joan Gould.

28. See Pratt, "Toothpicks," 9; "Certificate of Organization of a Corporation Under the General Law," State of Maine, filed Jan. 29, 1924, 19240005 D 1910000183204, Department of the Secretary of State, Bureau of Corporations, Elections and Commissions.

29. Ruth Robbins Adamo, "Forster President Is Resigning," *Lewiston* (Me.) *Sun-Journal,* Mar. 10, 1993, 9; Adamo, "Forster Manufacturing Dream."

30. "Competitor Buys Historic Maine Toothpick Firm," *Portland* (Me.) *Press Herald,* Mar. 8, 1995, p. 6B.

31. Abby Zimet, "Maine's Mills Endure," *Portland* (Me.) *Sunday Telegram,* Apr. 19, 1998, pp. 1B, 16B.

32. "Toothpick Maker to Have New Owner"; see http://icrs.informe.org/nei-sos-icrs/ICRS?CorpSumm=19240005+D (Oct. 4, 2005).

33. Donna Perry, "Forster's Parent Company Files for Chapter 11," *Lewiston* (Me.) *Sun-Journal,* May 24, 2001, B1; Jerry W. Jackson, "Home-Products Firm Explores Buying Pencil-Maker Dixon Ticonderoga," *Orlando* (Fla.) *Sentinel,* Jan. 13, 2004; quoted in Joseph Pryweller, "Jarden Buying Diamond Assets," *Plastics News,* Nov. 25, 2002, p. 3; Jane Brissett, "Jarden Corp. to Cut Hourly Staff at Cloquet, Minn., Alltrista Plant," *Duluth* (Minn.) *News-Tribune,* Jun. 25, 2004.

34. Pryweller, "Jarden Buying Diamond"; Betty Jesperson, "Forster's Owners Plan Few Changes," *Maine Morning Sentinel,* Jan. 20, 2003, B1; Donna M. Perry, "Forster Wood Plant Closing," *Lewiston* (Me.) *Sun-Journal,* Feb. 1, 2003, A1; Betty Jespersen, "Forster Toothpick Plant to Close," *MaineToday.com,* http://business.mainetoday.com/news/030201forsters.shtml (May 30, 2004).

35. Box of 250-count Forster round toothpicks, purchased at Wal-Mart, Brunswick, Maine, summer 2005. As late as 1972, Forster had stated that "the company has not bought and does not intend to buy a single item from abroad to be sold under the Forster label." However, the statement soon had to be retracted, and Forster had to admit that it "did purchase frilled tooth picks from Japan in the past." See "Our Slip Is Showing," *The Woodline* 4, no. 2 (Jul. 1972), 2.

36. Box of 250-count Diamond square/round toothpicks, purchased at Shaw's supermarket, Bath, Maine, summer 2005.

37. "14 Forster Worlds Fair Brand Wooden Yo Yo in Display Bx," offered on eBay, Oct. 29, 2005; Sarah Coffey, "Maine's Last Toothpick Rolls Off Line as Strong Plant Closes," Associated Press State and Local Wire, May 18, 2003.

38. "Croquet OK for Wilton, Forster," newspaper clipping, 1989, Wilton Historical Society, vertical file, s.v. "Forster Mfg. Co."; *Portland* (Me.) *Press Herald,* Apr. 28, 1984, 4.

39. "Twiddle, Twiddle," *Down East,* May 1983, 30–31; Horace A. Knowles, "Thumb Twiddling Toy," U.S. Patent No. 4,227,342 (Oct. 14, 1980).

40. Bill Haynes, "Wilton Wood Firm Serves Twiddlers," *Portland* (Me.) *Press Herald,* April 11, 1983, 1.

41. Brackley and Lisherness, *Strong,* 145, 47; Sarah Coffey, "Gritting Their Teeth," *Portland* (Me.) *Press Herald,* May 19, 2003, 1B. A photo accompanying the story shows the slogan on "the original 1979 fire truck" to read, "Toothpick Mfg. Capital of the World."

42. David Silberman, "A Plea from a 'Bewildered Small Business Man,'" display advertisement, *New York Times,* Jan. 26, 1946, 9.

EPILOGUE

1. The informal and incomplete list of U.S. patents compiled in the course of writing this book contains about five hundred entries.

2. American Dental Association, "History of Dentistry," http://www.ada.org/public/resources/history (Dec. 28, 2005).

3. Park Avenue Periodontal Associates, "FAQs About Flossing and Toothpicks," http://parkaveperio.com/faq/flospick.htm (Dec. 28, 2005).

4. John C. Marquis, "Toothpick Holder," U.S. Patent No. 3,892,040 (Jul. 1, 1975). Made by Marquis Dental Mfg. Co., Aurora, Colorado.

5. Baxter B. Sapp, Jr., DDS.

6. Academy of General Dentistry, "Dental Cleanings/Hygiene Tips," http://www.agd.org/consumer/topics/cleaning/main.asp (Dec. 28, 2005); April Grandinetti, DDS, quoted in "Warn Patients of the Hazards of Toothpicks," http://www.geocities.com/drkhosla1/News/news56.html (Dec. 28, 2005).

7. Florida State University Career Center, "The Etiquette Survival Guide," http://www.career.fsu.edu/ccis/guides/etiquette.html (Dec. 29, 2005); Judith Martin, "Miss Manners," http://lifestyle

.msn.com/Relationships/Article.aspx?cp-documentid=102714 (Dec. 29, 2005).

8. Executive Planet, "Chinese Business Culture," http://www.execu tiveplanet.com/business-culture-in/132272562564.html (Dec. 29, 2005); Kwintessential Language and Culture Specialists, "Vietnam," http://www.kwintessential.co.uk/resources/global-etiquette/viet nam.html (Dec. 29, 2005).

9. Marlene Parrish, "Bilbao Bar-Hopping Reveals Pinchos, the Perfect Party Food," http://www.post-gazette.com/pg/05349/622268.stm (Dec. 29, 2005).

10. Http://www.toothpick.com (Dec. 28, 2005).

ACKNOWLEDGMENTS

A great deal of the research for and writing of this book was done during the academic year 2005–2006, while I was on sabbatical leave from Duke University. For their roles in recommending and approving that leave, I am grateful to Roni Avissar, my chairman at the time; Kristina Johnson, my dean; and Peter Lange, my provost. My sabbatical year was partly supported by a grant from the Alfred P. Sloan Foundation, and I am grateful to Doron Weber of Sloan for his confidence in this project.

My first inkling that the toothpick might provide an interesting case history for exploring various aspects of success and failure in design came while I was preparing a series of three Lewis Clark Vanuxem lectures to be delivered in the fall of 2004 at Princeton University. Indeed, at one time I had thought that one of those lectures would be devoted to the story of the toothpick, as would a chapter of the book to be based on the lectures. As suggested in the preface to this present book, I came to realize that the seemingly small topic was too unexplored for a single lecture then and too complicated for a single chapter later.

Ashbel Green, my editor and friend at Alfred A. Knopf, was enthusiastic about the idea when I proposed it to him, and I am pleased that this is now the ninth book that I have published with this distinguished editor and publishing house. As always, it has been a pleasure to work with the editorial and production staff at Knopf, and especially with Sara Sherbill, Robert Olsson, and Ellen Feldman, for whom no detail was too small to deserve attention. And once again I am grateful to Chip Kidd for a creative jacket design.

Ambitious projects require the help of many individuals and institutions to bring them to fruition. Early supporters of this one included John Staudenmaier, S.J., of the University of Detroit Mercy and The Henry Ford and editor of *Technology and Culture;*

and Seymour Mauskopf, my Duke history colleague. Several individuals associated with the libraries of Duke University provided considerable help during the early stages of my research into the technological history of the toothpick. Foremost among these was Eric J. Smith, of the Perkins Library reference department, who helped me identify and obtain copies of relevant patents and trademarks—and who seems never to tire of my cryptic inquiries of all kinds. Linda Martinez, Duke's engineering librarian, and Marcos Rodriguez, then on the staff of the Vesic Engineering Library, were very cooperative in securing difficult-to-locate materials. A variety of specialist librarians at Duke were also helpful, including Janie Morris, reference librarian in the Rare Book, Manuscript, and Special Collections Library; Lynn Eaton, reference archivist with the John W. Hartman Center for Sales, Advertising and Marketing History, in the Special Collections Library; Suzanne Porter, curator of the History of Medicine Collections in Duke's Medical Center Library; and Linda Purnell of Duke's Interlibrary Loan Department.

Dave Morrison, of the Marriott Library at the University of Utah, helped in providing copies of some relevant trademarks, as did Michele Hayslett, of the D. H. Hill Library at North Carolina State University. Marjorie Ciarlante, of the National Archives II, at College Park, Maryland, helped me understand and identify patent, patent assignment, and patent extension files, which proved to hold unique information on the beginnings of the wooden toothpick industry in America. Without this material, it would not have been possible to tell the full story of how the technology developed and of who produced some of the earliest machine-made toothpicks.

I cast my net far and wide in exploring various technological and cultural aspects of the toothpick. Early in my research, the dental historian Malvin E. Ring was very helpful in providing numerous references to his own work and to that of others on the subject. When I asked my longtime correspondent, Professor Jan Hult, of Chalmers University, about toothpicks in Sweden, he put me in contact with Professor Bo Krasse of the Faculty of Odontology at Malmö University, who was involved with the development of the Swedish-made TePe toothpicks. He in turn encouraged me to visit the TePe factory in Malmö, and I am grateful to Bertil Eklund, head of the company, for personally providing a guided tour of the plant.

For my research in Maine, I am greatly indebted to many people, among whom is Stephanie Philbrick, of the Maine Historical Society Library, who was especially helpful in responding to my early vague and uninformed inquires about Charles Forster and the early toothpick industry in Maine. She not only provided essential biographical information that oriented me to the Forster family but also pointed the way to other resources in the state. Others in Maine who were of considerable help were Cathryn Wilson, reference librarian at the Maine State Library, who brought to my attention her department's "Maine Wood Util. Industry" clipping book, and Emily Schroeder, of the same department, who clarified that the abbreviation stood for "utilization." Many others at the Maine State Library provided prompt responses to my e-mail inquiries and courteous service during my visits. I am also thankful for the help of Pat Burdick, special collections librarian at Colby College; Tom Gaffney, special collections librarian in the Portland Room of the Portland Public Library, as well as Margot McCain, of that same room; Dawn Pratt, Skip Pike, and Frank Stevens, of the Strong Historical Society, who provided special glimpses into the toothpick industry in that town; Margaret Malley, director of the Ludden Memorial Library in Dixfield; Warren Rollins and Pam Brown, of the Wilton Historical Society; Beverly Gilbert, of the Strong Library (formerly the Forster Memorial Library), who called my attention to the photographic portrait of Lelia Forster hanging in the library; and Timothy R. Poulin, of the Bureau of Corporations, Elections and Commissions of the Department of the Secretary of State.

Others who helped in Maine include Stanley and Joan Gould, of Wilton and Baltimore, for providing information on the history of the Forster firm; Maureen Quinlan, of the Law School Library at the University of Maine, Portland; Nicholas Noyes and Jamie Kingman, of the Reference Library of the Maine Historical Society; Everett Tilton, of the Zadoc Long Free Library of Buckfield; Bernadette Boisvert, of the reference department of the Lewiston Public Library; Vaughan Gagne, of the Wilton Free Public Library; Diana Demers, of the Weld Free Public Library, who opened it up on a Saturday morning for a couple of visitors from the coast who wanted to see a portrait of a toothpick manufacturer's wife; Sean Minear, of the Weld Historical Society; Benjamin Proud, of special collections at the University of Maine, and Melvin Johnson, of the Folger Library there; Emily Scribner, librarian of the Medical Library, Franklin Memorial Hospital, Farmington, for bringing to my attention their Forster Room and for making available to me copies of *The Woodline,* the newsletter of the Forster Mfg. Co.; Mrs. Gould, who donated the artifacts and photographs in the hospital's Forster collection, for granting permission to use images of the materials and for encouraging me to stop by an old mill nearby, where other pictures still hung; Jack Lambert, of Jarden Plastic Solutions, who let me look at the pictures hanging in the administration offices of the old Forster plant in East Wilton, thereby enabling me to corroborate that the portrait suspected to be of Maurice Forster was indeed of him; Joey Kinsey, of the Printing Warehouse in Wilton, for sharing his extensive knowledge of Forster plants and procedures; and Peter Stowell, of West Gardiner, who sent me a compact disc containing outstanding images of Dixfield structures and documents and followed up with many helpful resources and clarifications of who was who in Dixfield and environs.

Much of the story of the Forster toothpick business proved to be contained in probate and other court and vital records in Maine. For help in gaining me access to these, I am grateful to Joyce S. Morton, register of probate, Franklin County Probate Court; Mary A. Graffam, of the Registry of Probate Office, Cumberland County Probate Court, who found files that were assumed to have been lost in a fire; Anthony Douin and Anne Small, of the Maine State Archives; and Christine Horne, of the Vital Records Department, Portland City Hall. Other individuals in Maine who contributed to my knowledge of toothpicks and their makers and associates included our neighbor Deirdre Barton, proprietor of Weatherbird, a gourmet boutique in Damariscotta, for graciously telling me about and providing me with packages of orangewood toothpicks imported from Portugal; Paul H. Mills, of Farmington; and Carl Osgood, of Richmond, Virginia, and Surry, Maine.

I was also helped by many people outside Maine, including J. Marc Greuther, curator of industrial collections at The Henry Ford; Mark F. Hall, reference librarian at the Library of Congress;

Carla G. Heister, of the Graves Forestry and Environmental Studies Library at Yale University; Henry F. Scannell, curator of microtext and newspapers at the Boston Public Library; Ted Hansen and Brian Youmans, research volunteers at the Cambridge Historical Society; Jeffrey M. Flannery, of the Manuscript Division, Library of Congress; Jan Grenci, of the Prints and Photographs Division, Library of Congress; Jane Kenealy, archivist at the San Diego Historical Society, who helped me flesh out the story of Charlotte Forster; Jim Roan, of the library in the National Museum of American History of the Smithsonian Institution; Kathlene Ferris, head archivist at the Center for Southwest Research, University of New Mexico; Marcia Anderson, of the Minnesota Historical Society; Dan Lewis, of the Huntington Library; Bernard Finn, of the National Museum of American History, who brought to my attention the concept of the National Toothpick Museum; Robert Post, who put me in touch with the Philadelphia Free Library, where Karen Lightner, curator of the print and picture collection, and Jim Quinn, head of the newspaper and microfilm department, tried in vain to help me find the purported image of Charles Forster throwing toothpicks from a wagon; Angie Kindig, of the University of Notre Dame Archives; Anna K. Heran, archivist at the Lloyd Library and Museum; and Andrea Matlak, archivist at the American Dental Association Library.

Among others who were helpful were Charles R. Siple, an inveterate and endlessly interesting draftsman and correspondent, who addresses his envelopes with more thought and care than many of us take with their contents, and who created the illustration incorporating a toothpick and a pencil in the balancing knife and fork trick; Vincent Tocco, who provided an introduction to the patents of Benjamin Sturtevant; Lois Hirt, a registered dental hygienist who provided insight into the profession's view of the toothpick; Baxter B. Sapp, Jr., my own dentist, who gave me a Perio-Aid toothpick holder and put me on to the story of the golf tee; the artist Stephen Talasnik, who told me of the toothpick model he made as a youngster, who shared with me his experiences with Japanese toothpicks, and who referred me to Charles Worthen, who clarified the terminology used for them; Sandra Wheeler, of Virginia Commonwealth University, who also shared anecdotes about Japanese toothpicks; Steven Lubar, of Brown University, who provided insight into patent archives; railroad historian Jerry DeVos, who shared some of his clipping files on the early toothpick industry in Maine, as well as some of his photos of plants and rolling stock; Ralph Vartabedian, of the *Los Angeles Times;* Ben Z. Swanson, Jr., a dealer in dental antiques, who shared his compilation of toothpick-box records; and Harley Spiller, alias Inspector Collector, who offered to show me his collection of toothpicks.

In addition, I should like to acknowledge the largely anonymous people who conceive and assemble the digital databases that make newspaper and other archives available and searchable on the Internet. Many of the most valuable of these are accessible only through institutional subscription, and I am grateful to the Duke University Libraries and to the Society for American Baseball Research for subscribing. In addition, I should like to thank the pseudonymous sellers of goods on eBay, whose pictures and descriptions of toothpicks and related collectibles enabled me to view at leisure and in comfort more things than I would ever have been able to find, let alone see, visiting Saturday flea markets and crowded antique malls.

As usual, I am also grateful to my family for their help. My son, Stephen, provided me with examples of toothpicks from restaurants and helped me find the golf tee patent. My daughter, Karen, who helped orient me in finding and citing court cases and supplied me with toothpicks made on the Pacific Rim, read and commented on relevant chapters of the manuscript. And, as indicated in the preface, my wife, Catherine, has once again done more than anyone should hope to expect from a spouse or friend. Needless to say, it is I and not any of those who have helped me who am responsible for any errors that might lurk in this book.

—*HP*

ILLUSTRATIONS AND CREDITS

All images digitally rendered and enhanced by Catherine Petroski, unless otherwise noted.

13 Toothpicks incorporated into ancient silver filigree jewelry. *From image in author's collection.*

17 Sixteenth-century gondolier picking his teeth with large pointed stick. *From Alessandro Caravia,* Naspo Bizaro.

24 Walrus-whisker toothpick distributed by an airline. *From author's collection.*

26 Young boy displaying quill toothpicks for sale, around 1800. *From Hans Sachs,* Der Zahnstocher und seine Geschichte.

29 A goose's wing feather sliced up for making different *articles de plume. From* Scientific American, *1878.*

30 Late-nineteenth-century press used to point quill toothpicks. *From* Scientific American.

31 Die enabling both ends of quill toothpick to be cut simultaneously. *From* Scientific American.

32 Quill toothpick showing differently cut ends. *From author's collection; photograph by Catherine Petroski.*

40 Japanese woodblock print of Meiji woman using a chew stick. Ukiyoe *by Yoshitoshi Tsukioka;* image in author's collection.

46 Portuguese toothpicks with elaborately handcarved shafts. *From author's collection; photograph by Catherine Petroski.*

65 Knife blade and presser bar of Sturtevant's lathe attachment. *From U.S. Patent No. 26,627 extension file, National Archives at College Park, Md.*

68 Benjamin Franklin Sturtevant, in his forties. *From J. D. Van Slyck,* New England Manufacturers and Manufactories.

75 Patent drawing of veneer coil blank used for making shoe pegs. *From U.S. Patent No. 25,149.*

78 Pattern of cutting nails from sheet iron. *From Charles Tomlinson, ed.,* Cyclopaedia of Useful Arts.

79 Patent drawing of doubly beveled veneer prepared for use in toothpick-making machinery. *From U.S. Patent No. 38,768.*

86 J. C. Brown's patented machine for making toothpicks and other wooden splints. *From U.S. Patent No. 43,177.*

INDEX

Italicized page numbers refer to illustrations and their captions.

Q-tips, 209
Queen Elizabeth as an Old Woman,
    16
quick-lunch rooms, 295
"quilling," 35
quill pens, 25–6, 27–8, 33–4
quills, 21, 25, 26–7, 29, 30, 109
quill toothpicks, 16, 21, 25–6, *26,*
    27–35, *32,* 117, 124, 126, 260,
    296, 299, 308, 310–11
  chewing of, 28–9, 138, 297, 298,
    299–300
  French, *26,* 29–32, 138, 242, 244,
    246, 310
  imported, 27, 138, 243, 246
  making of, 27–8, 29, 30–1, *30, 31,*
    32–3, 138
  vs. wooden, 29, 32, 38, 41, 105,
    126, 138, 242, 247, 257, 310
  as writing instruments, 202–3

Rabelais, François, 21
radio, first order received by, 244
Rain Man (film), 275–6
Rancho de la Nación, 223
Raynham, Mass., 67
Recife, Brazil, 53
  see also Pernambuco (city)
"redneck toothpick," 194
"Red River toothpick," 194
reed pen, 26
Rella, Hermogenes, 277–8
Renaissance, 16, 19
restaurants, toothpicks in, 101,
    102–3, 103–4, 140, 279–80,
    243, 258, 281, 285, 291, 294–5,
    313, 314, 346–8
  see also hotels, and toothpicks
Restoration Hardware (chain), 257–8
ribbon pegwood, 64–5, *65,* 72, 73,
    74, *75,* 82
  and toothpicks, 74–83, *79*
  see also pegwood
Richardson, Thomas, 108
Richelieu (cardinal), 114
Richleighs, the, 293
Richmond, Va., nail mill at, 79
Rio de Janeiro, Brazil, 56

Ritz (hotel), 285
Roberts, George L., 70
Roger and Me (film), 304
Romania, 32, 339
  Princess of, 32–3
Rome, ancient, 5, 13–14, 19, 37, 38,
    103
Roosevelt, Theodore, 198, 300
rosewood toothpicks, 179–80, 270
Royal Antiseptic Tooth Pick Co.,
    212
rubber band, used as floss, 138
rubber toothpicks, 109, 260, *262,*
    263
Russians, use of fork by, 114
Rutgers University, 303

Sachs, Hans, 18
Saginaw, Mich., 322
Sahlin, Jonas, 277
St. Jonathan (personification), 55
St. Louis Republic, 135
saliva, squirting between teeth, 7
saltshakers, 284, 323–4, *324,* 344,
    351, 352
Sampson, Calvin T., 60, 61–2, 72
Samuel Cupples Wooden Ware Co.,
    151
Sanborn, William A., 239
sand, smalt made from, 267, 269
Sanders, Mark C., 289
San Diego, Calif., 223, 224, 226–8,
    229
San Diego Sun, 226
Sands Hotel and Casino (Las
    Vegas), 314
sandwich picks, *252, 253,* 315
Sandy River and Rangeley Lakes
    Railroad, 167, *168*
Sandy River Railroad, 92, 167, 243
sanitary toothpick, providing a,
    279–80, 281, 284, 285, 286–7,
    322–3
San Juan Capistrano, Calif., 223
San Luis Potosí, Mexico, 50
San Marcos Land Co., 228
Santa Lucia (steamship), 316
Santa Margarita Rancho, 223

A NOTE ABOUT THE AUTHOR

Henry Petroski is the Aleksandar S. Vesic Professor of Civil Engineering and a professor of history at Duke University. The author of thirteen previous books, he lives in Durham, North Carolina, and spends summers in Arrowsic, Maine.

A NOTE ON THE TYPE

The text of this book was set in Ehrhardt, a typeface based on the specimens of "Dutch" types found at the Ehrhardt foundry in Leipzig. The original design of the face was the work of Nicholas Kis, a Hungarian punch cutter known to have worked in Amsterdam from 1680 to 1689. The modern version of Ehrhardt was cut by the Monotype Corporation of London in 1937.

Composed by North Market Street Graphics, Lancaster, Pennsylvania

Printed and bound by R. R. Donnelley, Harrisonburg, Virginia

Book design by Robert C. Olsson